Rotating Disc Cavity Flow and Heat Transfer

Wiley-ASME Press Series

Mixed-flow Pumps: Modeling, Simulation, and Measurements
Wei Li, Leilei Ji, Ramesh Agarwal, Weidong Shi, and Ling Zhou

Thermal Spreading and Contact Resistance: Fundamentals and Applications
Yuri S. Muzychka and M. Michael Yovanovich

Analysis of ASME Boiler, Pressure Vessel, and Nuclear Components in the Creep Range
Maan H. Jawad and Robert I. Jetter

Voltage-Enhanced Processing of Biomass and Biochar
Gerardo Diaz

Pressure Oscillation in Biomedical Diagnostics and Therapy
Ahmed Al-Jumaily and Lulu Wang

Robust Control: Youla Parameterization Method
Farhad Assadian and Kevin R. Mellon

Metrology and Instrumentation: Practical Applications for Engineering and Manufacturing
Samir Mekid

Fabrication of Process Equipment
Owen Greulich and Maan H. Jawad

Engineering Practice with Oilfield and Drilling Applications
Donald W. Dareing

Flow-Induced Vibration Handbook for Nuclear and Process Equipment
Michel J. Pettigrew, Colette E. Taylor, and Nigel J. Fisher

Vibrations of Linear Piezostructures
Andrew J. Kurdila and Pablo A. Tarazaga

Bearing Dynamic Coefficients in Rotordynamics: Computation Methods and Practical Applications
Lukasz Brenkacz

Advanced Multifunctional Lightweight Aerostructures: Design, Development, and Implementation
Kamran Behdinan and Rasool Moradi-Dastjerdi

Vibration Assisted Machining: Theory, Modelling and Applications
Li-Rong Zheng, Dr. Wanqun Chen, and Dehong Huo

Two-Phase Heat Transfer
Mirza Mohammed Shah

Computer Vision for Structural Dynamics and Health Monitoring
Dongming Feng and Maria Q. Feng

Theory of Solid-Propellant Nonsteady Combustion
Vasily B. Novozhilov and Boris V. Novozhilov

Introduction to Plastics Engineering
Vijay K. Stokes

Fundamentals of Heat Engines: Reciprocating and Gas Turbine Internal Combustion Engines
Jamil Ghojel

Offshore Compliant Platforms: Analysis, Design, and Experimental Studies
Srinivasan Chandrasekara and R. Nagavinothini

Computer Aided Design and Manufacturing
Zhuming Bi and Xiaoqin Wang

Pumps and Compressors
Marc Borremans

Corrosion and Materials in Hydrocarbon Production: A Compendium of Operational and Engineering Aspects
Bijan Kermani and Don Harrop

Design and Analysis of Centrifugal Compressors
Rene Van den Braembussche

Case Studies in Fluid Mechanics with Sensitivities to Governing Variables
M. Kemal Atesmen

The Monte Carlo Ray-Trace Method in Radiation Heat Transfer and Applied Optics
J. Robert Mahan

Dynamics of Particles and Rigid Bodies: A Self-Learning Approach
Mohammed F. Daqaq

Primer on Engineering Standards, Expanded Textbook Edition
Maan H. Jawad and Owen R. Greulich

Engineering Optimization: Applications, Methods and Analysis
R. Russell Rhinehart

Compact Heat Exchangers: Analysis, Design and Optimization using FEM and CFD Approach
C. Ranganayakulu and Kankanhalli N. Seetharamu

Robust Adaptive Control for Fractional-Order Systems with Disturbance and Saturation
Mou Chen, Shuyi Shao, and Peng Shi

Robot Manipulator Redundancy Resolution
Yunong Zhang and Long Jin

Stress in ASME Pressure Vessels, Boilers, and Nuclear Components
Maan H. Jawad

Combined Cooling, Heating, and Power Systems: Modeling, Optimization, and Operation
Yang Shi, Mingxi Liu, and Fang

Applications of Mathematical Heat Transfer and Fluid Flow Models in Engineering and Medicine
Abram S. Dorfman

Bioprocessing Piping and Equipment Design: A Companion Guide for the ASME BPE Standard
William M. (Bill) Huitt

Nonlinear Regression Modeling for Engineering Applications: Modeling, Model Validation, and Enabling Design of Experiments
R. Russell Rhinehart

Geothermal Heat Pump and Heat Engine Systems: Theory and Practice
Andrew D. Chiasson

Fundamentals of Mechanical Vibrations
Liang-Wu Cai

Introduction to Dynamics and Control in Mechanical Engineering Systems
Cho W.S. To

Thermal Explosion: Theory and Application
Vasily B. Novozhilov

Rotating Disc Cavity Flow and Heat Transfer
John W. Chew

This Work is a co-publication between John Wiley & Sons, Inc. and ASME Press

Rotating Disc Cavity Flow and Heat Transfer

John W. Chew
Emeritus Professor of Mechanical Engineering
School of Engineering, University of Surrey
Guildford, Surrey, United Kingdom

ASME PRESS
WILEY

This Work is a co-publication between John Wiley & Sons, Inc. and ASME Press.

Library of Congress Cataloging-in-Publication Data Applied for:

Hardback ISBN: 9781394343270

Cover Design: Wiley
Cover Image by John and Rukshana Chew. Features a high pressure turbine disc from the Rolls-Royce/Snecma Olympus 593 turbojet. This engine powered the Concorde supersonic passer aircraft that was in service from 1976 to 2003.

Set in 9.5/12.5pt STIXTwoText by Straive, Chennai, India
SKY10154805_042126

To my wife, Rukshana, and our children, Leroy, Aaron and Carissa.

Contents

Preface

I was first introduced to rotating disc flows when starting my doctorate studies in 1978. I recall a conversation with Professors Fred Bayley and Mike Owen during which Fred suggested that research in turbomachinery rotating disc flows would outlast us all. He was right. An indication of the technical progress made since then is that the computational fluid dynamics (CFD) solutions I obtained during my doctorate studies using national computing facilities could now be generated in a few minutes using widely available CFD software on a standard personal computer. CFD for disc flows at the time was confined to axisymmetric, steady flows, and suitability of the available turbulence models to this application was uncertain. Now, 3D unsteady CFD solutions with turbulence models are routinely used in industry, and direct numerical simulations resolving the turbulent eddies are used in research to further understanding of flow behaviour in rotating disc systems. The combined outcome of CFD, experimental and analytical research has continuously improved understanding of flow mechanisms and predictive capability.

Working at Rolls-Royce from 1982 until 2000, I was able to extend my knowledge of rotating flows, progress research results and understanding to practical engineering calculation methods, and learn much more. On welcoming me to the company, the Director of Engineering advised me that no one in the company understood all that went on in a jet engine, although some people knew a remarkable lot. It was a privilege to work with such people and to see how vital their experience and knowledge were when engine testing or in-service experience threw up new and unexpected problems. Some of the most fascinating projects from my point of view involved investigation of flow phenomena (using analysis and CFD) for which there was no obvious cause. The occurrence and solution of these problems illustrates the essential interaction of design calculation methods, experimental testing and detailed analysis.

At the University of Surrey from 2000 to 2025, I have been able to maintain links with industry while also developing design and analytical tools and investigating some of the remaining mysteries in rotating flows and internal air systems. Increasing awareness and understanding of flow unsteadiness and its importance to some aspects of internal air system performance has been a theme during this period. Worldwide interest and research activity in rotating disc flows has also increased, and the number of research publications on rotating flows has risen dramatically. Reviewing the archived research papers has become a daunting task for new workers in the subject. It now seems timely to produce a book bringing together an account and examples of established results with some of the more recent developments and explanations of the physics for rotating disc flows. I hope this book will help others to apply and advance knowledge.

I consider myself fortunate to have spent my career in the fascinating areas of fluid dynamics and turbomachinery and to have participated in the advancement of knowledge in this area. These are subjects that keep on giving to those of us that enjoy research. The book would not have been possible without support and encouragement along the way from many former and present colleagues and friends. Most recently, I am thankful to Chris Barnes and Tim Scanlon at Rolls-Royce, Diogo Berta Pitz at the Federal University of Paraná, Todd Ebert at Siemens Energy, Feng Gao at Beihang University, and Zixiang Sun and Michael Perkris at the University of Surrey for their comments and suggestions on draft sections of the book. The remaining faults are, of course, my responsibility.

This book draws on many previous publications and permission to re-use and adapt material from Rolls-Royce plc, the American Society of Mechanical Engineers, the American Physical Society, Cambridge University Press, Elsevier, the Japan Society of Mechanical Engineers, the National Aeronautics and Space Administration, the Royal Society, and Sage Publishing is gratefully acknowledged.

Frequently Used Notation

Bold symbols denote a vector quantity.

Stationary (r, ϕ, z) and rotating (r, θ, z) cylindrical coordinate systems are used.

Coordinates (s, n) in planes of constant ϕ or θ are parallel and normal to a surface.

a	disc or rotor inner radius
b	disc or rotor outer radius
Bo	Buoyancy number $= Ro/(\beta\Delta T)^{1/2}$
C_d	discharge coefficient
C_D	drag coefficient
C_M	moment coefficient $= 2M/\rho\Omega^2 b^5$
C_p	specific heat at constant pressure
C_v	specific heat at constant volume
C_W	non-dimensional mass flow $= \dot{m}/\mu b$
$\mathbf{div}$	divergence operator $\stackrel{\text{def}}{=} \left(\frac{\partial(r-)}{r\partial r}, \frac{\partial(-)}{r\partial\phi}, \frac{\partial(-)}{\partial z}\right)$
DR	purge to annulus flow density ratio
e	internal energy
E	total energy
E_{rot}	total energy in a rotating system
Ec	Eckert number $= \Omega^2 b^2/C_p \Delta T_{ref}$
Gr	rotational Grashof number $= \beta\Delta T_{ref} Re_\phi^2$ unless stated otherwise
h	enthalpy
htc	heat transfer coefficient $= q_0/(T_{ref} - T_0)$
I	rothalpy $= C_p T_t - \Omega r v_\phi = C_p T_{t,rel} - \Omega^2 r^2/2$
k	turbulent kinetic energy per unit mass $= (\overline{\acute{u}_r \acute{u}_r} + \overline{\acute{u}_\theta \acute{u}_\theta} + \overline{\acute{u}_z \acute{u}_z})/2$, or constant in ingress model
k_δ	freestream swirl fraction outside boundary layer $= v_{\phi,\delta}/\Omega r$
l_m	mixing length
L	representative length scale or length dimension
K	kinetic energy per unit mass $= \left(v_r^2 + v_\phi^2 + v_z^2\right)/2$
m	mass
$\dot{m}$	mass flow rate
m^*	non-dimensional mass exchange rate $= \dot{m}_{ex}/2\pi s_a a \rho_{\text{ref}} \sqrt{\beta\Delta T^*}\Omega b$
M	moment (torque) or Mach number $= V/\sqrt{\gamma RT}$ or mass dimension

n	wall normal coordinate in planes of constant ϕ or θ, or inverse of exponent in power law velocity profile
n^+	wall normal coordinate in wall units $= n\sqrt{\rho\tau_0}/\mu$
Nu	Nusselt number
p	pressure
p^*	rotary stagnation pressure
Pr	Prandtl number $= \mu_{ref}C_p/\kappa_{ref}$
q	heat flux (per unit area)
Q	heat flow rate
r	radius
R	gas constant $= C_p - C_v$
R_f	recovery factor in adiabatic wall temperature
R_{fac}	recovery factor in energy equation wall function
Ra	rotational Rayleigh number $PrGr$
Re_k	local rotational Reynolds number $= \rho\|1 - k_\delta\|\Omega r^2/\mu$
Re_δ	local disc boundary layer Reynolds number $= \|\Omega r - v_{\phi,\delta}\|/\sqrt{\nu\Omega}$
Re_ϕ	rotational Reynolds number $= \rho_{ref}\Omega b^2/\mu_{ref}$
Ro	Rossby number $= v_{z,in}/\Omega a$
s	streamwise coordinate in planes of constant ϕ or θ, or spacing between discs or cones
s_c	minimum rim seal clearance
t	time
T	temperature or time dimension
T^*	rotary stagnation temperature $= I/C_p$
$\boldsymbol{u}$	velocity vector in rotating coordinates $= (u_r, u_\theta, u_z)$
U	magnitude of velocity vector $= \|\boldsymbol{u}\| = \left(u_r^2 + u_\theta^2 + u_z^2\right)^{1/2}$
$\boldsymbol{v}$	velocity vector in stationary coordinates $= (v_r, v_\phi, v_z)$
V	magnitude of velocity vector $= \|\boldsymbol{v}\| = \left(v_r^2 + v_\phi^2 + v_z^2\right)^{1/2}$ or volume
W	work
x_{ing}	ingress as fraction of the shortfall between the supplied and minimum sealing flow
z	axial coordinate
Z	viscous work and dissipation
α	angle of preswirl nozzle to circumferential direction
β	coefficient of thermal expansion
γ	ratio of specific heats $= C_p/C_v$
δ	boundary layer thickness
δ^*	scaled boundary layer thickness in integral equation $= Re_\phi^{2/(n+3)} \sin(\varepsilon)^{(n+1)/(n+3)} \breve{\gamma}\delta/b$
ε	cone half angle, turbulent energy dissipation or rim seal effectiveness
θ	angular coordinate in rotating system
κ	thermal conductivity or von Kármán constant
λ	mass flow parameter $= C_W/Re_\phi^{0.8}$ or $C_W/[Re_\phi/\sin\varepsilon]^{0.8}$
λ_r	local mass flow parameter $= \lambda(r/b)^{-2.6}$
μ	viscosity
ν	kinematic viscosity $= \mu/\rho$

ξ	non-dimensional boundary layer coordinate $= z\sqrt{\Omega/\nu}$ or $n\sqrt{\Omega\sin(\varepsilon)/\nu}$
ρ	density
τ	shear stress component
ϕ	angular coordinate in stationary system
Φ	frictional dissipation
ω	frequency or rate of energy dissipation per unit turbulent kinetic energy
Ω	angular rotational speed

Subscripts

a	inner radius
ann	annulus or main gas path
b	outer radius or location of bolts
c	associated with rim seal
eff	effective
ext	external to the cavity
in	inlet
min	minimum to avoid ingestion
n	component normal to the surface
r, ϕ, z	radial, circumferential and axial components in a stationary system
r, θ, z	radial, circumferential and axial components in a rotating system
ref	reference conditions
rel	in the rotating frame of reference
s	streamwise component or rotating flow structure
sh	shroud
t	total (or stagnation) property or turbulent
δ	boundary layer edge
∞	freestream
0	at the surface

Superscripts

$+$	distance normalised by “wall unit”, e.g. $n^+ = n\sqrt{\rho\tau_0}/\mu$
$'$	turbulent fluctuation
$\bar{}$	mean value
$\breve{}$	non-dimensional
$\tilde{}$	mass or mass flow weighted average

Abbreviations

CFD	computational fluid dynamics
DNS	direct numerical simulation
FE	finite element
FEA	finite element analysis
HP	high-pressure

HPC	high-pressure compressor
HPT	high-pressure turbine
IP	intermediate-pressure
IPC	intermediate-pressure compressor
IPT	intermediate-pressure turbine
LES	large eddy simulation
LP	low-pressure
LPC	low-pressure compressor
LPT	low-pressure turbine
RANS	Reynolds-averaged Navier-Stokes
RNG	renormalisation group
RSM	Reynolds stress model
SA	Spalart-Allmaras
SST	shear stress transport
WMLES	wall modelled large-eddy simulation
WRLES	wall resolved large-eddy simulation
WMURANS	wall modelled unsteady Reynolds-averaged Navier-Stokes

1

Introduction

The main aim of this book is to help practising engineers and researchers understand, apply and advance methods for estimating flow and heat transfer in rotating disc systems. While this subject has much in common with other areas of fluid mechanics, rotating flows exhibit some unique, and sometimes counter-intuitive, features. Combined with their practical significance, this has attracted considerable interest from researchers for more than a century.

The content is particularly relevant to turbomachinery internal air systems, reflecting the motivation for much research in recent decades. Cavities between rotating discs and between rotating discs and stationary discs are an important element of such systems. Examples are shown in Figures 1.1 and 1.2. Both figures relate to Rolls-Royce three-shaft (or three-spool) turbofan aeroengines. HP, IP and LP denote high-pressure, intermediate-pressure and low-pressure compressors (C) or turbines (T). The first figure highlights the rotating blades, stationary vanes, rotating discs and rotating shafts (connecting the compressors and turbines), with contours of temperature at a particular operating condition. The blade and vanes reflect the positive axial temperature gradient of the air as it is compressed and a negative axial gradient in the turbine as the hot combustion gases expand. Cooling of the discs and shafts balances heat transfer from the mainstream gas and gives generally positive radial temperature gradients. This figure is from a finite element 'whole-engine thermomechanical model', as described by Dixon *et al.* (2004). Such finite element analyses (FEA) often assume axisymmetry with approximate treatment of 3D features. Transient analyses are conducted to obtain component temperatures, movements and stresses through a flight cycle. These models may contain around 1000 'boundaries' for which aerothermal boundary conditions are required throughout the flight cycle.

Figure 1.2 shows a central section of an engine core containing part of the HPC, the burner (or combustor), the HPT and the IPT. The white areas correspond to engine components, including discs, blades and shafts, and the main gas path through the compressor, burner and turbines. The colours or shades in this picture show the source of the air for the internal air system. For example, the highest pressure cooling air is fed from the HPC outlet, bypasses the burner and supplies cooling for the HPT disc front face and blades. Areas in other colours or shades are supplied with lower-pressure air from further upstream in the compressors. The bearing chamber will contain a mixture of oil and air.

The internal air system plays a vital role in modern gas turbines, serving to dissipate heat generated by shaft work (known as windage) and heat conducted to the discs, deliver air for turbine blade cooling, maintain the required axial pressure loading on the bearings, prevent hot gas ingestion from the main annulus overheating the turbine discs and isolate the oil system. Running clearances between rotating and stationary components, and the variation of clearances through a flight or operating cycle, are also affected or controlled by the air system. These depend on a combination

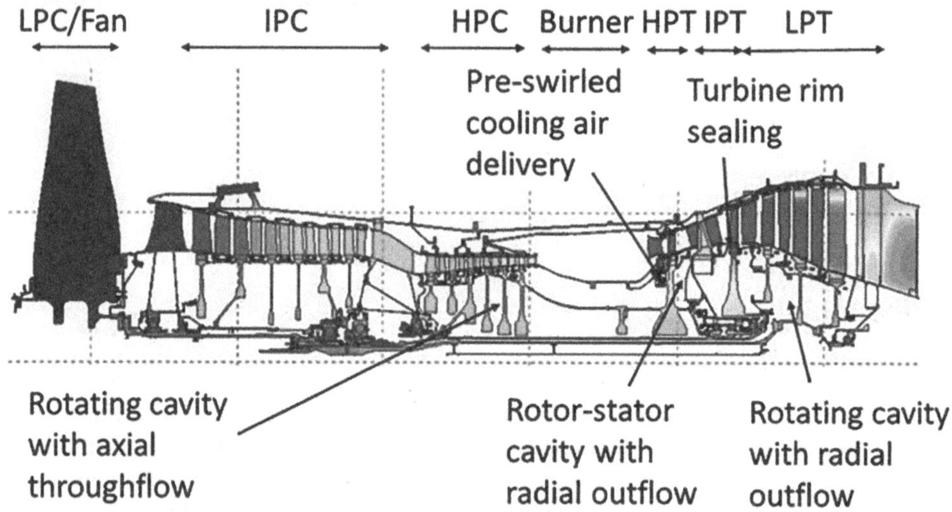

Figure 1.1 Whole-engine thermomechanical model. *Source:* Following Dixon *et al.* (2004).

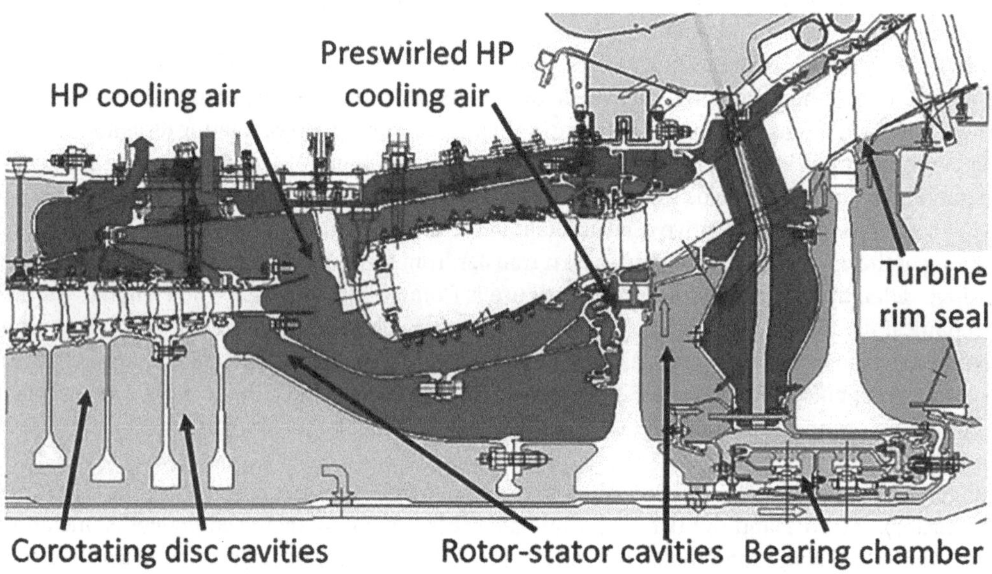

Figure 1.2 Section of the internal air system, showing the HP and IP sections for a three-shaft engine. *Source:* courtesy Rolls-Royce plc.

of thermal and centrifugal expansion of components and pressure loads. With clearances affecting compressor and turbine efficiencies, and up to 25% of compressor air being used for cooling and sealing, the air system has a significant impact on engine efficiency. Considerable attention must be given to the internal air system and its interaction with the main annulus flow.

Some rotating disc cavities are indicated in Figures 1.1 and 1.2. In multistage axial compressors and turbines, such as the HPC and LPT in Figure 1.1, cavities are formed between co-rotating discs. In the compressors shown, these are vented by air flowing axially through the gap between the disc

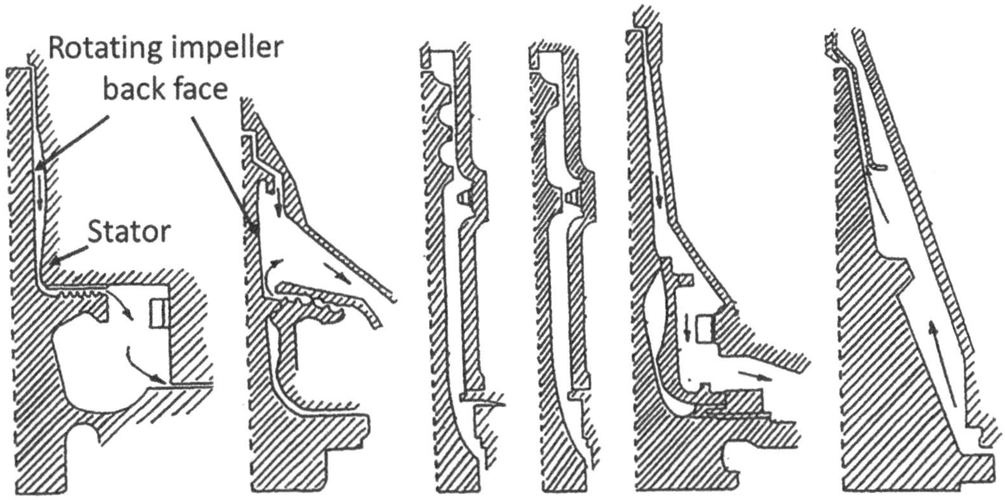

Figure 1.3 Typical impeller rear face rotor-stator cavities for centrifugal compressors. *Source:* Hart and Turner (1994) / with permission of American Society of Mechanical Engineers (ASME).

hubs and the IP shaft. This air subsequently flows radially outwards through co-rotating LPT disc cavities and re-joins the main annulus flow through turbine rim seals. These seals (or gaps) are formed between the rotating blade platforms and stator vane platforms that define the inner annulus line (or radial boundary) for the main gas path. In the HP and IP turbines in both figures, rotor-stator disc cavities are formed between the rotating discs and static components. Such co-rotating and rotor-stator cavities are common in turbomachinery and may vary considerably in geometry and in the nature of the throughflow. Cavities between differentially co-rotating or counter-rotating discs occur in some engines at the interface between compressors and turbines rotating at different speeds.

Rotor-stator disc cavities are also common and of considerable interest in other rotating machines, including centrifugal pumps, compressors and turbines. Variations in geometry that can occur are illustrated in Figure 1.3. This shows typical impeller rear face rotor-stator cavities for centrifugal compressors, as considered by Hart and Turner (1994), and was based on a survey of small gas turbine engines powering helicopters, business jets and light aircraft. Hart and Turner state that imposed radial inflows or outflows are typically in the range of 0–1% of the compressor main flow. Geometries vary to accommodate labyrinth seals, stiffening rings and balancing features, and the air flow must provide adequate cooling and acceptable axial pressure loads for the structure.

It is common industrial practice to model internal air systems with a 1D flow network comprised of multiple elements relating pressure changes and mass flow rates through the various components. For example, discharge coefficient correlations may be used for seals, and frictional pressure loss coefficients may be specified for duct flows with further allowance for pressure drops through bends and expansions or contractions. The network equations are solved numerically to provide pressure and mass flow data throughout the system. In some cases, the flow network model may be coupled to a thermal model, such as that illustrated in Figure 1.1, so that better estimates of seal clearances and temperature-dependent fluid properties are obtained. This may also improve thermal modelling through an engine acceleration or deceleration as seal clearances and the air system flow rates change. The book by Sultanian (2018) gives a good account of network modelling for internal air systems with guidance and illustration of how rotating disc cavity flow characteristics may be estimated and incorporated into the network. Such treatments inevitably involve some simplification and engineering judgement, so adjustments are often required to match engine test data.

Many experimental and theoretical research studies have provided correlations or formulae that may be used in flow network and thermal modelling. An early example is von Kármán's (1921) analysis of the flow due to a disc rotating in a quiescent environment. As will be discussed in Chapter 5, this solution continues to provide a benchmark and basis for engineering estimates. Further examples are given in books by Dorfman (1963), Owen and Rogers (1989, 1995), Shevchuk (2009), Childs (2011) and Sultanian (2018).

Interest and research in rotating flows is not confined to engineering applications. On earth and other planets, large-scale motions in the atmosphere and oceans are affected by rotation. Early research in geophysical fluid dynamics, often treating the flow as a small perturbation in a rotating frame of reference, has relevance to engineering flows. Ekman's (1905) laminar flow boundary layer solution is shown later to apply directly in some engine conditions and has inspired consideration of turbulent Ekman layers. Simple co-rotating disc cavities have often been used in research to give insights into geophysical flows, and flow features similar to those encountered in engineering have been identified. Examples and descriptions of the linear mathematical theory are given in Greenspan's (1968) book.

Chapters 2–6 of this book describe analysis, computational methods and solutions available for the prediction of rotating disc flow and heat transfer. Examples are given to illustrate the underlying flow mechanisms that are relevant in a wide range of applications and to show the strengths and weaknesses of the methods. Comparisons with experimental data indicate the degree to which methods have been validated and the uncertainty in their use for engineering estimates. Chapters 7 and 8 then focus on more specific applications to rotor-stator cavities and co-rotating disc cavities and include relatively complex examples. Chapter 9 considers the interaction of a rotor-stator cavity flow with the surrounding flow when the gap between the rotor and stator allows the external flow to be ingested. This relates to the turbine rim seals shown in Figures 1.1 and 1.2. Chapters 6–9 include results from relatively recent research using advanced computation and experimental methods to provide insights into the more complex flow phenomena.

References

Childs, P. R. N., 2011. Rotating flow, Elsevier.

Dixon, J. A., Verdicchio, J. A., Benito, D., Karl, A., and Tham, K. M., 2004. Recent developments in gas turbine component temperature prediction methods, using computational fluid dynamics and optimization tools, in conjunction with more conventional finite element analysis techniques. Proc. Inst. Mech. Eng. Part A: J. Power Energy, 218, 241–155. https://doi.org/10.1243/0957650041200641.

Dorfman, L. A., 1963. Hydrodynamic resistance and the heat loss of rotating solids, First English edition, London: Oliver and Boyd.

Ekman, V. W., 1905. On the influence of the earth's rotation on ocean-currents. Ark. Mat. Astr. Fys., 2, 1–53.

Greenspan, H. P., 1968. The theory of rotating fluids, Cambridge University Press.

Hart, K. J. and Turner, A. B., 1994. Influence of radial inflow on rotor-stator cavity pressure distributions. Proceedings of the ASME 1994 International Gas Turbine and Aeroengine Congress and Exposition. Volume 1: Turbomachinery, Paper 94-GT-106, https://doi.org/10.1115/94-GT-106.

Owen, J. M. and Rogers, R. H., 1989. Flow and heat transfer in rotating disc systems, Vol. 1: Rotor-stator systems, Taunton, UK: Research Studies Press.

Owen, J. M. and Rogers, R. H., 1995. Flow and heat transfer in rotating disc systems, Vol. 2: Rotating cavities, Taunton, UK: Research Studies Press.

Shevchuk, I. V., 2009. Convective heat and mass transfer in rotating disc systems, Springer.

Sultanian, B. K., 2018. Gas turbines: internal flow systems modeling, Cambridge University Press.

von Kármán, T., 1921. Über laminare und turbulente Reibung. Z. Angew. Math. Mech., 1, 233–252. https://doi.org/10.1002/zamm.19210010401.

Part I

Theory and Modelling Methods

2

Essential Theory

The theoretical basis of continuum fluid dynamics and thermodynamics has developed over many years in tandem with experimental studies. Many advances that were groundbreaking at the time, such as Newton's laws of motion (c. 1686), Euler's inviscid flow equations (mid 1700s), the Navier–Stokes viscous flow equations (early 1800s), the first and second laws of thermodynamics (mid to late 1800s) and Reynolds averaging (c. 1895), are now accepted wisdom. It is assumed that readers will have prior knowledge of these fundamental results, as taught in Mechanical and Aerospace Engineering degree programmes. The aim here is to introduce the theory and nature of rotating flows, and to indicate the origin of equations that will be used in other parts of the book. The equations given are written in the natural cylindrical coordinate system for the problems considered, with little use of vector operator or summation convention notation. This form of the equations relates most easily to their control volume derivation in cylindrical coordinates and the underlying physical assumptions to which they directly relate. Alternatively, the governing equations can be derived in Cartesian coordinates and transformed to the cylindrical coordinate system used here.

While the theory presented is essential to the applications and examples given later, it will not be essential reading for all. Readers may prefer to refer to parts of this chapter when they are referenced later in the book. Nor is the chapter a complete account of the background theory, much of which is more fully covered and explained in general introductory fluid dynamics textbooks. For example, the isentropic gas flow relations are used without derivation as these are independent of the coordinate system and are widely covered in introductory texts and undergraduate degree programmes.

2.1 Mass, Momentum and Energy Balances

Equations expressing mass, momentum and energy balances form the basis for analysis in fluid mechanics and heat transfer, and these equations can take many forms. The choice of form is subject to individual preference and purpose, and different approaches have different merits. Here, emphasis is placed on physical interpretation rather than mathematical elegance and generality, aiming to explain the flow physics through both physically intuitive arguments and mathematics. A cylindrical coordinate system (r, ϕ, z) is adopted with the z-axis corresponding to the axis of rotation or machine axis. This coordinate system, and its relation to Cartesian coordinates (x, y, z), is shown in Figure 2.1, along with a basic control volume δV containing fluid of mass δm. The velocity vector in the cylindrical system is $\boldsymbol{v} = (v_r, v_\phi, v_z)$.

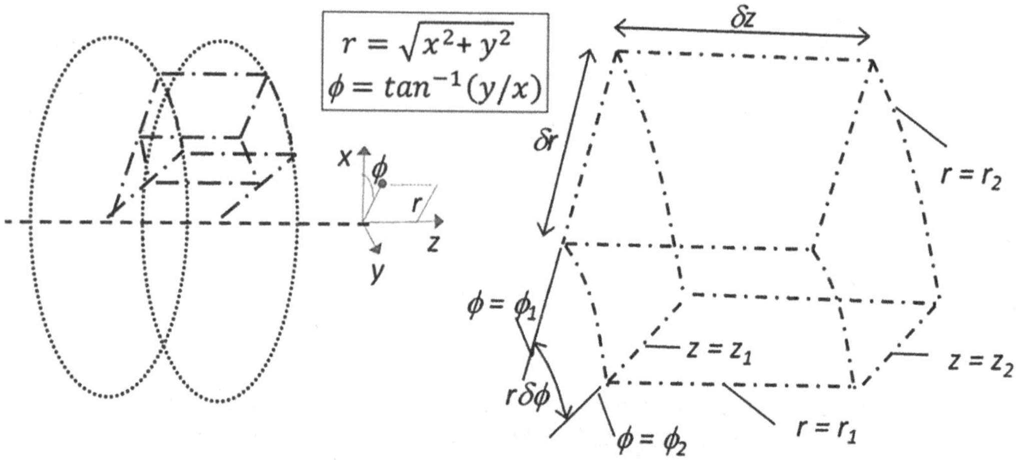

Figure 2.1 Cylindrical coordinate system and elemental control volume.

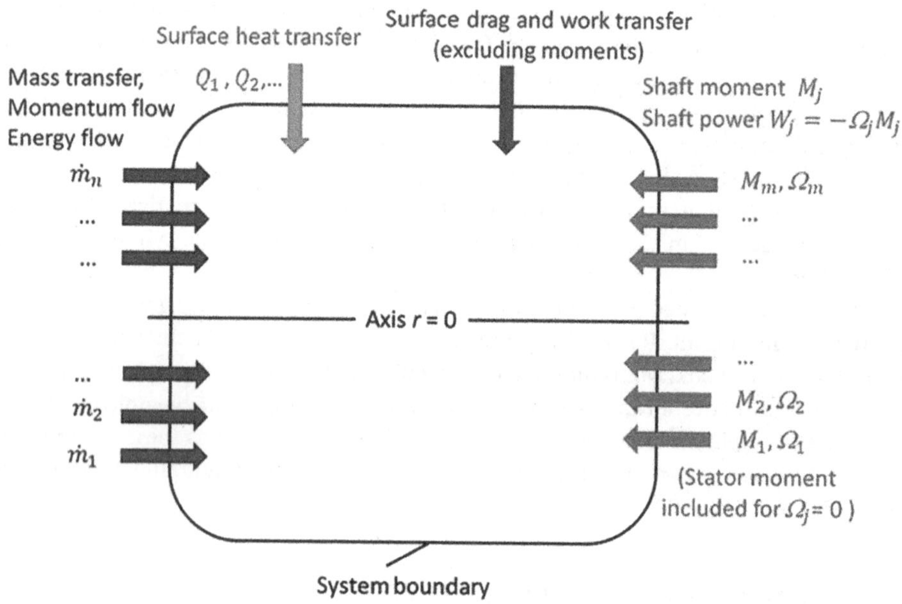

Figure 2.2 Open system mass, momentum and energy fluxes.

The control volume in Figure 2.1 may be used in deriving the governing differential equations for the flow, which assume the fluid behaves as a continuum with the molecular structure accounted for through bulk properties. It is also often useful to consider overall mass, momentum and energy balances for a given system. Figure 2.2 illustrates a general open system that includes fluid flow through multiple inlets and outlets, heat transfer and work transfer as shaft power through a number of rotors. When angular momentum is considered, the axis of rotation must be defined, and it is convenient to separate shaft moment and work from other drag and surface work transfer contributions. Mass, momentum and energy equations for the open system and in differential form are discussed below. Only single-phase, non-reacting flows are considered.

Attention focuses on the ideal cases of fluids with constant density ρ (as representative of many liquids and low-speed gas flows) and perfect gases (for which pressure $p = \rho RT$ where R is the gas constant and T denotes temperature) with Newton's law of viscosity. Gravitational effects are neglected, assuming that for the flows of interest, changes in potential energy can be neglected in comparison to changes in kinetic energy or, equivalently, that the high Froude number limit of a negligible gravitational field applies.

2.1.1 Mass Conservation

Mass conservation or continuity can simply be expressed as

> *The rate of change of mass in a volume with time is equal to the difference between the total inflow and outflow mass flow rates.*

For the open system in Figure 2.2, this gives

$$\frac{dm}{dt} = \sum_{inlet\ \&\ outlets\ i} \dot{m}_i \tag{2.1.1}$$

where t denotes time, m is the mass of fluid in the volume and $\dot{m}_i$ is the mass flow rate through the inlet or outlet i (and is positive for an inflow). The fluid mass and flow rates equate to volume integrals of the density ρ and surface integrals of the dot product of the velocity and surface area normal times density, as below

$$m = \int_V \rho dV, \quad \dot{m}_i = \int_{S_i} \rho \boldsymbol{v}.\boldsymbol{dS}_i. \tag{2.1.2}$$

Considering each of the 6 faces of the control volume or element δV in Figure 2.1 as an inlet or outlet and applying equation 2.1.1:

$$\begin{aligned}\delta V\frac{\partial\rho}{\partial t} &= (\rho r\delta\phi\delta z v_r)_{r=r_1} - (\rho r\delta\phi\delta z v_r)_{r=r_2} + (\rho\delta r\delta z v_\phi)_{\phi=\phi_1}\\ &- (\rho\delta r\delta z v_\phi)_{\phi=\phi_2} + (\rho r\delta\phi\delta r v_z)_{z=z_1} - (\rho r\delta\phi\delta r v_z)_{z=z_2}.\end{aligned} \tag{2.1.3}$$

Dividing this equation by $\delta V = r\delta\phi\delta r\delta z$ and taking the limit as $\delta\phi$, δr and δz tend to zero, gives the continuity equation below:

$$\frac{\partial\rho}{\partial t} + \frac{\partial(\rho r v_r)}{r\partial r} + \frac{\partial(\rho v_\phi)}{r\partial\phi} + \frac{\partial(\rho v_z)}{\partial z} = \frac{\partial\rho}{\partial t} + \boldsymbol{div}(\rho\boldsymbol{v}) = 0 \tag{2.1.4}$$

For any axisymmetric flow of an incompressible fluid or for steady, axisymmetric flow of a compressible fluid, the derivatives with respect to t and ϕ in equation 2.1.4 reduce to zero and the axisymmetric stream function ψ may be defined as follows:

$$\rho_{ref}(\psi - \psi_{ref}) = -\int_{z_{ref}}^{z} \rho r v_r dz + \int_{r_{ref}}^{r} \rho r v_z dr \tag{2.1.5}$$

where *ref* denotes reference values, and the integral is taken along any curve in the $r-z$ plane connecting (r_{ref}, z_{ref}) to (r, z). The radial and axial velocity components are then given by

$$\frac{\rho}{\rho_{ref}}v_r = -\frac{1}{r}\frac{\partial\psi}{\partial z}, \qquad \frac{\rho}{\rho_{ref}}v_z = \frac{1}{r}\frac{\partial\psi}{\partial r}. \tag{2.1.6}$$

From 2.1.5, the mass flow rate across the surface formed by the rotation of an arbitrary curve between two points in the $r-z$ plane is $2\pi\rho_{ref}$ times the difference in stream function between the two points. It follows that lines $\psi = constant$ are everywhere parallel to the vector $(v_r, 0, v_z)$ and represent streamlines in the $r-z$ plane. Contours of ψ (at equally spaced values) give a good visualisation of axisymmetric flows. Flow vectors align with the streamlines and the product of ρr times velocity magnitude in the $r-z$ plane is inversely proportional to the distance between streamlines.

2.1.2 Angular Momentum

The principle of conservation of angular momentum is fundamental to understanding rotating flow. At any point in a fluid, the angular momentum per unit mass (or specific angular momentum) is rv_ϕ and the conservation principle may be stated as follows.

If the net external moment (or torque) acting on a system about a given axis is zero, then the total angular momentum of the system about that axis remains constant.

Note that this statement emphasises the significance of the choice of axis and, in contrast to linear momentum, the angular momentum depends on this choice. This principle can be derived directly from Newton's second law regarding linear momentum, which also gives the following result.

The rate of change of angular momentum of a fluid or material element with time is equal to the net moment (or torque) acting on the element.

Considering an open system in which the fluid is bounded by stationary and rotating surfaces, as in Figure 2.2, the angular momentum balance is

Rate of change of AM in system = Inflow of AM – Outflow of AM + Rotor moment + Stator moment.

Here AM denotes angular momentum, and the moments are taken as positive if they tend to increase the angular momentum. For the open system in Figure 2.2, this can be expressed as follows:

$$\frac{d\{m(rv_\phi)_{bulk}\}}{dt} = \sum_{\text{inlet \& outlets } i} \dot{m}_i\,(\widetilde{rv_\phi})_i + \sum_{\text{all surfaces } j} M_j \tag{2.1.7}$$

Here the subscript *bulk* denotes a mass weighted volume average, and the tilde ˜ denotes a mass flow weighted surface average with the following definitions (for an arbitrary quantity ψ):

$$\psi_{bulk} = \int_V \rho\psi \, dV/m \qquad \widetilde{\psi_i} = \int_{S_i} \rho\psi \, \boldsymbol{v}.\boldsymbol{dS}_i/\dot{m}_i \tag{2.1.8}$$

Moments M_j are associated with pressure and shear stress forces as in the following equation:

$$M_j = \int_{S_j} rp\,\boldsymbol{i}_\phi.\boldsymbol{dS}_j + \int_{S_j} r\tau_\phi \, dS_j \tag{2.1.9}$$

where p and τ_ϕ denote static pressure and the circumferential component of surface shear stress. The vector dot product of the unit vector $\boldsymbol{i}_\phi$ and the surface area normal $\boldsymbol{dS}_j$ gives the elemental area projected onto an $r-z$ plane.

Applying the angular momentum balance to the control volume δV in Figure 2.1 and taking the limit $\delta\phi, \delta r, \delta z \to 0$ gives the following differential equation:

$$\frac{\partial(\rho rv_\phi)}{\partial t} + \frac{\partial(\rho r^2 v_r v_\phi)}{r\partial r} + \frac{\partial\left(\rho rv_\phi^2\right)}{r\partial\phi} + \frac{\partial(\rho rv_z v_\phi)}{\partial z} = -\frac{\partial p}{\partial\phi} + \frac{\partial(r^2\tau_{r\phi})}{r\partial r} + \frac{\partial(r\tau_{\phi\phi})}{r\partial\phi} + \frac{\partial(r\tau_{z\phi})}{\partial z} \tag{2.1.10}$$

The left-hand side terms account for the change of angular momentum in the volume and the advection through the element surfaces. The right-hand side accounts for the moments associated with pressure forces and viscous forces acting on the element surfaces. Here τ_{ij} denotes the j component of viscous stress on a plane with outward normal in the i direction.

Equation 2.1.10 is written in conservation form, as derived directly from the angular momentum balance. Expanding the left-hand side and using the continuity equation 2.1.4 gives a non-conservation form:

$$\frac{\rho\partial(rv_\phi)}{\partial t}+\frac{\rho v_r\partial(rv_\phi)}{\partial r}+\frac{\rho v_\phi\partial(rv_\phi)}{r\partial\phi}+\frac{\rho v_z\partial(rv_\phi)}{\partial z}=-\frac{\partial p}{\partial\phi}+\frac{\partial(r^2\tau_{r\phi})}{r\partial r}+\frac{\partial(r\tau_{\phi\phi})}{r\partial\phi}+\frac{\partial(r\tau_{z\phi})}{\partial z} \tag{2.1.11}$$

For a steady flow, the left-hand side is proportional to the gradient of rv_ϕ in the flow direction. It is apparent from this equation that in steady flow with no forces acting in the circumferential direction, the angular momentum per unit mass rv_ϕ is constant along streamlines.

2.1.3 Axial and Radial Momentum

The balance of the axial components of momentum and force for the open system in Figure 2.2 can be written as follows:

$$\frac{d\{mv_{z,bulk}\}}{dt}=\sum_{\text{inlet \& outlets } i}\dot{m}_i\,\tilde{v}_{z,i}+\sum_{\text{all surfaces } j}F_{z,j} \tag{2.1.12}$$

Here $F_{z,j}$ is the axial force component on surface j.

A similar equation can also be derived for radial momentum but this is not used here in this form. The balance of radial forces is generally dependent on the internal flow distribution. As discussed below, the circumferential velocity v_ϕ and radial pressure gradient may dominate.

Applying a linear momentum balance to the control volume δV in Figure 2.1 and taking the limit $\delta\phi, \delta r, \delta z \to 0$ gives the conservation form of the differential radial and axial momentum equations:

$$\frac{\partial(\rho v_r)}{\partial t}+\frac{\partial\left(\rho r v_r^2\right)}{r\partial r}+\frac{\partial(\rho v_\phi v_r)}{r\partial\phi}+\frac{\partial(\rho v_z v_r)}{\partial z}-\frac{\rho v_\phi^2}{r}=-\frac{\partial p}{\partial r}+\frac{\partial(r\tau_{rr})}{r\partial r}+\frac{\partial\tau_{\phi r}}{r\partial\phi}+\frac{\partial\tau_{zr}}{\partial z}-\frac{\tau_{\phi\phi}}{r} \tag{2.1.13}$$

$$\frac{\partial(\rho v_z)}{\partial t}+\frac{\partial(\rho r v_r v_z)}{r\partial r}+\frac{\partial(\rho v_\phi v_z)}{r\partial\phi}+\frac{\partial\left(\rho v_z^2\right)}{\partial z}=-\frac{\partial p}{\partial z}+\frac{\partial(r\tau_{rz})}{r\partial r}+\frac{\partial\tau_{\phi z}}{r\partial\phi}+\frac{\partial\tau_{zz}}{\partial z} \tag{2.1.14}$$

The centrifugal force term $\rho v_\phi^2/r$ and stress component $\tau_{\phi\phi}$ appear in the radial momentum equation due to the circumferential velocity component v_ϕ and normal stress component at the two control volume faces $\phi = \phi_1, \phi_2$ having components in the direction of the radial vector through the centre of the volume. Unlike linear and angular momentum, the radial momentum per unit mass (that is v_r) is not conserved along steady flow streamlines in the absence of forces. It is also apparent that in an inviscid flow with $v_r = v_z = 0$, $v_\phi \neq 0$, a radial pressure gradient must balance the centrifugal force. Defining a rotationally-dominated flow as one in which the circumferential velocity dominates (with $v_\phi \gg v_r, v_z$), the inviscid radial momentum equation has the form

$$\frac{\partial p}{\partial r}=\frac{\rho v_\phi^2}{r}. \tag{2.1.15}$$

This is known as the radial equilibrium equation.

2.1.4 Energy Conservation

The first law of thermodynamics (expressing conservation of energy) applied to the control volume in Figure 2.2 gives the equation

$$\frac{d\{mE_{bulk}\}}{dt} = \sum_{\text{inlet \& outlets } i} \dot{m}_i \widetilde{h_{t,i}} + \sum_{\text{all surfaces } j} \{Q_j - W_j - W_{rz,j}\}. \tag{2.1.16}$$

Here E and h_t denote total energy and total (or stagnation) enthalpy, and Q_j and W_j are heat transfers to the system and shaft work outputs from the system (or windage) associated with each surface. $W_{rz,j}$ denotes work associated with radial and axial components of surface force and velocity. For control volume surfaces with $v_r = v_z = 0$, the term $W_{rz,j} = 0$. Note that in this form of the equation, the work required to force the mass into or out of the control volume (also known as the flow work) is accounted for by use of enthalpy and should not be included in other terms. The total energy E and total enthalpy h_t are related to the internal energy e and enthalpy $h = e + p/\rho$ as below:

$$E = e + \frac{\left(v_r^2 + v_\phi^2 + v_z^2\right)}{2}, \quad h_t = h + \left(v_r^2 + v_\phi^2 + v_z^2\right)/2 \tag{2.1.17}$$

For steady flow, the left-hand side of equation 2.1.16 is zero, and this reduces to a simplification of the widely used steady flow energy equation, as derived in classical thermodynamics. An important result is that total enthalpy is conserved in steady flow with no work or heat transfer.

Assuming the only heat transfer is due to conduction governed by Fourier's law, equation 2.1.16 applied to the control volume δV in Figure 2.1 in the limit $\delta\phi$, δr, $\delta z \to 0$ gives the following differential equation for conservation of energy:

$$\frac{\partial}{\partial t}(\rho E) + \frac{\partial}{r\partial r}(\rho r v_r h_t) + \frac{\partial}{r\partial \phi}(\rho v_\phi h_t) + \frac{\partial}{\partial z}(\rho v_z h_t) = \frac{\partial}{r\partial r}\left(\frac{r\kappa\partial T}{\partial r}\right) + \frac{\partial}{r\partial\phi}\left(\frac{\kappa\partial T}{r\partial\phi}\right) + \frac{\partial}{\partial z}\left(\frac{\kappa\partial T}{\partial z}\right) + Z \tag{2.1.18a}$$

$$Z = \frac{\partial}{r\partial r} r(\tau_{rr}v_r + \tau_{r\phi}v_\phi + \tau_{rz}v_z) + \frac{\partial}{r\partial\phi}(\tau_{\phi r}v_r + \tau_{\phi\phi}v_\phi + \tau_{\phi z}v_z) + \frac{\partial}{\partial z}(\tau_{zr}v_r + \tau_{z\phi}v_\phi + \tau_{zz}v_z) \tag{2.1.18b}$$

where Z gives the viscous work and viscous energy dissipation (whereby kinetic energy is converted to thermal energy), and κ denotes thermal conductivity.

The derivatives of internal energy e and enthalpy h with respect to temperature give the specific heats of the fluid at constant volume (C_v) and constant pressure (C_p). If C_v and C_p are constant, it can be assumed $e = C_v T$ and $h = C_p T$. For the ideal constant density fluid, there is no distinction between these specific heats, so the subscripts for specific heat (C) may be dropped.

An alternative form of the energy equation may be derived using the relationship between specific kinetic energy K, mechanical work and viscous dissipation. Multiplying each of the momentum balance equations 2.1.10, 2.1.13 and 2.1.14 by their respective velocity components, dividing the angular momentum equation by r, and summing the three equations gives

$$\frac{\partial(\rho K)}{\partial t} + \frac{\partial(\rho r v_r K)}{r\partial r} + \frac{\partial(\rho v_\phi K)}{r\partial\phi} + \frac{\partial(\rho v_z K)}{\partial z} = -\left(\frac{v_r\partial p}{\partial r} + \frac{v_\phi\partial p}{r\partial\phi} + \frac{v_z\partial p}{\partial z}\right) + \frac{v_r\partial(r\tau_{rr})}{r\partial r} + \frac{v_\phi\partial(r\tau_{r\phi})}{r\partial r} + \frac{v_z\partial(r\tau_{rz})}{r\partial r} + \frac{v_r\partial\tau_{\phi r}}{r\partial\phi} + \frac{v_\phi\partial\tau_{\phi\phi}}{r\partial\phi} + \frac{v_z\partial\tau_{\phi z}}{r\partial\phi} + \frac{v_r\partial\tau_{zr}}{\partial z} + \frac{v_\phi\partial\tau_{z\phi}}{\partial z} + \frac{v_z\partial\tau_{zz}}{\partial z} - \frac{v_r\tau_{\phi\phi}}{r} + \frac{v_\phi\tau_{r\phi}}{r}. \tag{2.1.19}$$

where $K = \left(v_r^2 + v_\phi^2 + v_z^2\right)/2$.

Introducing the viscous dissipation Φ given by

$$\Phi = \tau_{rr}\frac{\partial v_r}{\partial r} + \tau_{r\phi}\frac{\partial v_\phi}{\partial r} + \tau_{rz}\frac{\partial v_z}{\partial r} + \tau_{\phi r}\frac{\partial v_r}{r\partial\phi} + \tau_{\phi\phi}\frac{\partial v_\phi}{r\partial\phi} + \tau_{\phi z}\frac{\partial v_z}{r\partial\phi} + \tau_{zr}\frac{\partial v_r}{\partial z}$$
$$+ \tau_{z\phi}\frac{\partial v_\phi}{\partial z} + \tau_{zz}\frac{\partial v_z}{\partial z} + \frac{v_r\tau_{\phi\phi}}{r} - \frac{v_\phi\tau_{r\phi}}{r} \qquad (2.1.20)$$

equation 2.1.19 may be rewritten as

$$\frac{\partial(\rho K)}{\partial t} + \frac{\partial(\rho r v_r K)}{r\partial r} + \frac{\partial(\rho v_\phi K)}{r\partial\phi} + \frac{\partial(\rho v_z K)}{\partial z} = -\left(\frac{v_r\partial p}{\partial r} + \frac{v_\phi\partial p}{r\partial\phi} + \frac{v_z\partial p}{\partial z}\right) + Z - \Phi. \qquad (2.1.21)$$

Subtracting this equation, known as the mechanical energy equation, from equation 2.1.18a gives

$$\frac{\partial(\rho e)}{\partial t} + \frac{\partial(\rho r v_r h)}{r\partial r} + \frac{\partial(\rho v_\phi h)}{r\partial\phi} + \frac{\partial(\rho v_z h)}{\partial z} = \frac{v_r\partial p}{\partial r} + \frac{v_\phi\partial p}{r\partial\phi} + \frac{v_z\partial p}{\partial z} + \frac{\partial}{r\partial r}\left(\frac{r\kappa\partial T}{\partial r}\right)$$
$$+ \frac{\partial}{r\partial\phi}\left(\frac{\kappa\partial T}{r\partial\phi}\right) + \frac{\partial}{\partial z}\left(\frac{\kappa\partial T}{\partial z}\right) + \Phi. \qquad (2.1.22a)$$

or, using the thermodynamic relationship $e = h - p/\rho$,

$$\frac{\partial(\rho h)}{\partial t} + \frac{\partial(\rho r v_r h)}{r\partial r} + \frac{\partial(\rho v_\phi h)}{r\partial\phi} + \frac{\partial(\rho v_z h)}{\partial z} = \frac{\partial p}{\partial t} + \frac{v_r\partial p}{\partial r} + \frac{v_\phi\partial p}{r\partial\phi} + \frac{v_z\partial p}{\partial z} + \frac{\partial}{r\partial r}\left(\frac{r\kappa\partial T}{\partial r}\right)$$
$$+ \frac{\partial}{r\partial\phi}\left(\frac{\kappa\partial T}{r\partial\phi}\right) + \frac{\partial}{\partial z}\left(\frac{\kappa\partial T}{\partial z}\right) + \Phi. \qquad (2.1.22b)$$

In these forms of the energy equation, viscous work terms that redistribute the kinetic energy of the fluid are (indirectly) included through pressure and velocity terms that do not include the fluid viscosity. The viscous dissipation Φ gives the rate of conversion of kinetic energy to internal energy.

Consideration of equation 2.1.18a or 2.1.22a indicates a difference between the energy equation for *unsteady* constant density flow with $C_v = C_p$ and compressible gas flow (with $e = C_vT$ and $h = C_pT$) due to the first term and the difference between C_v and C_p for a gas. For a constant volume of quiescent fluid heated slowly through the walls so that its temperature is approximately uniform, integration of equation 2.1.22a over the volume shows that the temperature rises at a rate Q/mC_v, where Q is the rate of heat transfer and m is the mass of fluid. This is as expected from the definition of C_v. For constant pressure heating, the work done by the fluid movement required to accommodate a change of volume must be accounted for, and equation 2.1.22b shows that the temperature rises at rate Q/mC_p. Equation 2.1.22b further indicates that as pressure effects in the energy balance become negligible the constant density and perfect gas energy equations become equivalent, even with unsteady flow.

Further consideration of equation 2.1.22a, 2.1.22b, with the relations $e = h - p/\rho$, $\partial e/\partial T = C_v$, $\partial h/\partial T = C_p$, shows that for constant density fluids, the pressure terms in these equations must be negligible for the approximation $C_v = C_p$ to apply. Removing these terms results in a form of the energy equation often given for constant density fluids. Note, however, that if changes in density with temperature are considered, the exact energy equation includes terms in pressure. This is discussed in many textbooks, including that by Batchelor (1967). Note further that, through equation 2.1.21, kinetic energy scales with the pressure terms in 2.1.22a and 2.1.22b, and so kinetic energy must also be negligible compared to thermal energy in the limit $C_v = C_p$ (also as expected). Drawing from the results of the second law of thermodynamics, Batchelor derives a form of the energy equation in which the pressure terms in 2.1.22b are multiplied by βT where β is the coefficient of thermal expansion for the fluid. Hence, for a perfect gas, with $\beta = 1/T$, equation 2.1.22b is reproduced. For more general fluids, the second law imposes some restrictions on the specific heats. Provided kinetic energy is negligible in the energy equation when incompressibility and $C_v = C_p$ are assumed, there is no conflict with the equations above.

2.1.5 Total Temperature and Total Pressure

The total (or stagnation) temperature T_t and total (or stagnation) pressure p_t may be defined as the temperature and pressure of an element of fluid brought to rest in a thermodynamically reversible process without work or heat transfer. Such processes are isentropic. As noted above, for steady flow in the absence of work and heat transfer, the total enthalpy h_t is conserved. Thus the enthalpy of the stagnated fluid will equal the total enthalpy of the fluid in motion. For a fluid with constant specific heat at constant pressure C_p, the total temperature $T_t = h_t/C_p$ will also be conserved.

For a ***constant density fluid***, thermodynamic considerations show that the internal energy is unchanged in the stagnation process (since there can be no compressive work on a fluid element). Hence, noting $h = e + p/\rho$, the total pressure is given by $p_t - p = (h_t - h)\rho = \rho(v_r^2 + v_\phi^2 + v_z^2)/2$ it follows from the conservation of h_t in steady processes with no work or heat transfer that p_t is conserved *if* the process is also isentropic. Thus in steady, isentropic, constant density flow with no work or heat transfer

$$T_t = \frac{h_t}{C_p} = T + \frac{\left(v_r^2 + v_\phi^2 + v_z^2\right)}{2C_p} = \textit{constant along streamlines} \tag{2.1.23a}$$

$$p_t = p + \frac{\rho\left(v_r^2 + v_\phi^2 + v_z^2\right)}{2} = \textit{constant along streamlines.} \tag{2.1.23b}$$

Equation 2.1.23b is a form of Bernoulli's equation, and may be derived directly from the inviscid momentum equations without reference to the energy equation. The kinetic energy term in 2.1.23a is retained for consistency with low-speed gas flow but will be negligible for many constant density flows.

For a ***compressible perfect gas flow***, the assumption of isentropic flow leads to the relationship $p_t \propto T_t^{\gamma/(\gamma-1)}$ where $\gamma = C_p/C_v$ is the ratio of specific heats of the gas at constant pressure and constant volume. This and other isentropic gas flow relationships are discussed in many textbooks and can be derived for steady, inviscid flow from the governing equations given above. Applying this result to the stagnation process gives the total pressure as $p_t = p(T_t/T)^{\gamma/(\gamma-1)}$. In steady isentropic flow with no work or heat transfer, it follows that both T_t and p_t are conserved. Thus in steady, isentropic, compressible flow with no work or heat transfer

$$T_t = \frac{h_t}{C_p} = T + \frac{\left(v_r^2 + v_\phi^2 + v_z^2\right)}{2C_p} = \textit{constant along streamlines} \tag{2.1.24a}$$

$$p_t = p\left(\frac{T_t}{T}\right)^{\gamma/(\gamma-1)} = \textit{constant along streamlines.} \tag{2.1.24b}$$

It may further be shown that, for relatively small changes in pressure, equation 2.1.24b reduces to the constant density flow relationship equation 2.1.23b. A flow Mach number may be defined as $M = V/\sqrt{(\gamma RT)}$ where $(V = \sqrt{\left(v_r^2 + v_\phi^2 + v_z^2\right)}$ is the gas speed and $\sqrt{(\gamma RT)}$ is the speed of sound in a perfect gas. In practice, gas flow with $M < 0.2$ or 0.3 is often treated as having constant density.

2.1.6 Euler Work Equation and Rothalpy

Considering the open system of Figure 2.2, it is assumed here that the only work transfer is due to rotation of the surfaces (rather than, for example, by linear movement as in a piston/cylinder). The work can then be related to the angular moments, noting that W is positive for work output and moment M is positive for angular momentum input, as

$$W_j = -\Omega_j M_j. \tag{2.1.25}$$

Consider steady, adiabatic flow in an open system in which all surfaces are either stationary or rotate with angular speed Ω. Multiplying equation 2.1.7 by Ω and combining with equations 2.1.16 and 2.1.25 gives the following relations.

$$\sum_{\textit{inlet \& outlets } i} \Omega \dot{m}_i \left(\widetilde{rv_\phi}\right)_i = \sum_{\textit{rotors } j} W_j - \sum_{\textit{stators } k} \Omega M_k \tag{2.1.26}$$

$$\sum_{\textit{inlet \& outlets } i} \dot{m}_i \left\{\widetilde{h_{t,i}} - \Omega \left(\widetilde{rv_\phi}\right)_i\right\} = \sum_{\textit{stators } k} \Omega M_k \tag{2.1.27}$$

The second of these equations applied to a system with just one inlet and one outlet, and zero or negligible stator moments, gives the power or work equation

$$Power = \dot{m}_{in}(\widetilde{h_{t,in}} - \widetilde{h_{t,out}}) = \dot{m}_{in}\Omega[\left(\widetilde{rv_\phi}\right)_{in} - \left(\widetilde{rv_\phi}\right)_{out}]. \tag{2.1.28}$$

Equations 2.1.26, 2.1.27 and variations of these are often referred to as Euler's turbine or turbomachinery equation in recognition of Euler's early work in relating moment to angular momentum change. The term Euler work equation is used here (as in other texts) for equation 2.1.28.

A consequence of equation 2.1.27. is that for a rotating system, with no stationary boundaries, the rothalpy defined as $I = h_t - \Omega r v_\phi$ is a conserved property. The term on the right-hand side of equation 2.1.27 can be interpreted as work done in a rotating frame of reference. It follows that for steady flow with no heat or work transfer in a rotating system, mass flow weighted averages of rothalpy will be conserved. (Recall also that we are neglecting gravitational or other body forces.) Rothalpy is further discussed in Section 2.2 below.

2.1.7 Navier–Stokes Equations

For a Newtonian fluid with the coefficient of bulk viscosity assumed to be zero, also known as a Stokes' fluid, the viscous stresses in the cylindrical coordinate system are given by

$$\tau_{rr} = 2\mu\left(\frac{\partial v_r}{\partial r} - \frac{1}{3}\mathbf{div}(v)\right), \qquad \tau_{\phi\phi} = 2\mu\left(\frac{\partial v_\phi}{r\partial\phi} + \frac{v_r}{r} - \frac{1}{3}\mathbf{div}(v)\right),$$

$$\tau_{zz} = 2\mu\left(\frac{\partial v_z}{\partial z} - \frac{1}{3}\mathbf{div}(v)\right), \qquad \tau_{r\phi} = \tau_{\phi r} = \mu\left(\frac{r\partial(v_\phi/r)}{\partial r} + \frac{\partial v_r}{r\partial\phi}\right),$$

$$\tau_{rz} = \tau_{zr} = \mu\left(\frac{\partial v_r}{\partial z} + \frac{\partial v_z}{\partial r}\right), \qquad \tau_{\phi z} = \tau_{z\phi} = \mu\left(\frac{\partial v_\phi}{\partial z} + \frac{\partial v_z}{r\partial\phi}\right). \tag{2.1.29}$$

where μ is the fluid viscosity. The expression for $\tau_{r\phi}$ reflects the fact that shear stresses in a forced vortex $v_\phi \propto r$, $v_r = v_z = 0$ are zero. The viscous stresses also allow a steady, constant density free vortex solution $rv_{\phi,ann} = constant$, in which the net viscous stress terms for a control volume sum to zero.

A complete set of governing equations for the flow is given by the above viscous stress model and relations for fluid properties in combination with the mass, momentum and energy equations 2.1.4, 2.1.10, 2.1.13, 2.1.14 and 2.1.18a, reproduced below in conservation form. The momentum equations are known as the Navier–Stokes equations:

$$\frac{\partial\rho}{\partial t}+\frac{\partial(\rho r v_r)}{r\partial r}+\frac{\partial(\rho v_\phi)}{r\partial\phi}+\frac{\partial(\rho v_z)}{\partial z}=\frac{\partial\rho}{\partial t}+\mathbf{div}(\rho \boldsymbol{v})=0 \tag{2.1.30}$$

$$\frac{\partial(\rho v_r)}{\partial t}+\frac{\partial\left(\rho r v_r^2\right)}{r\partial r}+\frac{\partial(\rho v_\phi v_r)}{r\partial\phi}+\frac{\partial(\rho v_z v_r)}{\partial z}-\frac{\rho v_\phi^2}{r}=-\frac{\partial p}{\partial r}+\frac{\partial(r\tau_{rr})}{r\partial r}+\frac{\partial\tau_{\phi r}}{r\partial\phi}+\frac{\partial\tau_{zr}}{\partial z}-\frac{\tau_{\phi\phi}}{r} \tag{2.1.31}$$

$$\frac{\partial(\rho r v_\phi)}{\partial t}+\frac{\partial(\rho r^2 v_r v_\phi)}{r\partial r}+\frac{\partial\left(\rho r v_\phi^2\right)}{r\partial\phi}+\frac{\partial(\rho r v_z v_\phi)}{\partial z}=-\frac{\partial p}{\partial\phi}+\frac{\partial(r^2\tau_{r\phi})}{r\partial r}+\frac{\partial(r\tau_{\phi\phi})}{r\partial\phi}+\frac{\partial(r\tau_{z\phi})}{\partial z} \tag{2.1.32}$$

$$\frac{\partial(\rho v_z)}{\partial t}+\frac{\partial(\rho r v_r v_z)}{r\partial r}+\frac{\partial(\rho v_\phi v_z)}{r\partial\phi}+\frac{\partial\left(\rho v_z^2\right)}{\partial z}=-\frac{\partial p}{\partial z}+\frac{\partial(r\tau_{rz})}{r\partial r}+\frac{\partial\tau_{\phi z}}{r\partial\phi}+\frac{\partial\tau_{zz}}{\partial z} \tag{2.1.33}$$

$$\frac{\partial(\rho E)}{\partial t}+\frac{\partial(\rho r v_r h_t)}{r\partial r}+\frac{\partial(\rho v_\phi h_t)}{r\partial\phi}+\frac{\partial(\rho v_z h_t)}{\partial z}=\frac{\partial}{r\partial r}\left(\frac{r\kappa\partial T}{\partial r}\right)+\frac{\partial}{r\partial\phi}\left(\frac{\kappa\partial T}{r\partial\phi}\right)+\frac{\partial}{\partial z}\left(\frac{\kappa\partial T}{\partial z}\right)+Z \tag{2.1.34}$$

with viscous work and dissipation term

$$Z=\frac{\partial}{r\partial r}r(\tau_{rr}v_r+\tau_{r\phi}v_\phi+\tau_{rz}v_z)+\frac{\partial}{r\partial\phi}(\tau_{\phi r}v_r+\tau_{\phi\phi}v_\phi+\tau_{\phi z}v_z)+\frac{\partial}{\partial z}(\tau_{zr}v_r+\tau_{z\phi}v_\phi+\tau_{zz}v_z). \tag{2.1.35}$$

As previously described for the angular momentum equation, non-conservation forms of these equations can be derived using the continuity equation in combination with the advective terms on the left-hand side of equations 2.1.31–2.1.34. If the fluid viscosity μ is assumed constant, the following set of equations may be deduced. For the energy equation, the form given by equation 2.1.22a is used below:

$$\frac{\partial\rho}{\partial t}+\mathbf{div}(\rho\boldsymbol{v})=0 \tag{2.1.36}$$

$$\rho\left\{\frac{\partial v_r}{\partial t}+\frac{v_r\partial v_r}{\partial r}+\frac{v_\phi\partial v_r}{r\partial\phi}+\frac{v_z\partial v_r}{\partial z}-\frac{v_\phi^2}{r}\right\}$$
$$=-\frac{\partial p}{\partial r}+\mu\left[\frac{\partial}{r\partial r}\left(\frac{r\partial v_r}{\partial r}\right)+\frac{\partial^2 v_r}{r^2\partial\phi^2}+\frac{\partial^2 v_r}{\partial z^2}-\frac{2\partial v_\phi}{r^2\partial\phi}-\frac{v_r}{r^2}+\frac{1}{3}\frac{\partial}{\partial r}\mathbf{div}(\boldsymbol{v})\right] \tag{2.1.37}$$

$$\rho\left\{\frac{\partial v_\phi}{\partial t}+\frac{v_r\partial v_\phi}{\partial r}+\frac{v_\phi\partial v_\phi}{r\partial\phi}+\frac{v_z\partial v_\phi}{\partial z}+\frac{v_r v_\phi}{r}\right\}$$
$$=-\frac{\partial p}{r\partial\phi}+\mu\left[\frac{\partial}{r\partial r}\left(\frac{r\partial v_\phi}{\partial r}\right)+\frac{\partial^2 v_\phi}{r^2\partial\phi^2}+\frac{\partial^2 v_\phi}{\partial z^2}+\frac{2\partial v_r}{r^2\partial\phi}-\frac{v_\phi}{r^2}+\frac{1}{3}\frac{\partial}{r\partial\phi}\mathbf{div}(\boldsymbol{v})\right] \tag{2.1.38}$$

$$\rho\left\{\frac{\partial v_z}{\partial t}+\frac{v_r\partial v_z}{\partial r}+\frac{v_\phi\partial v_z}{r\partial\phi}+\frac{v_z\partial v_z}{\partial z}\right\}$$
$$=-\frac{\partial p}{\partial z}+\mu\left[\frac{\partial}{r\partial r}\left(\frac{r\partial v_z}{\partial r}\right)+\frac{\partial^2 v_z}{r^2\partial\phi^2}+\frac{\partial^2 v_z}{\partial z^2}+\frac{1}{3}\frac{\partial}{\partial z}\mathbf{div}(\boldsymbol{v})\right] \tag{2.1.39}$$

$$\frac{\partial(\rho e)}{\partial t}+\frac{\partial(\rho r v_r h)}{r\partial r}+\frac{\partial(\rho v_\phi h)}{r\partial\phi}+\frac{\partial(\rho v_z h)}{\partial z}$$

$$= \frac{v_r \partial p}{\partial r} + \frac{v_\phi \partial p}{r \partial \phi} + \frac{v_z \partial p}{\partial z} + \frac{\partial}{r \partial r}\left(\frac{r \kappa \partial T}{\partial r}\right) + \frac{\partial}{r \partial \phi}\left(\frac{\kappa \partial T}{r \partial \phi}\right) + \frac{\partial}{\partial z}\left(\frac{\kappa \partial T}{\partial z}\right) + \Phi. \quad (2.1.40)$$

$$\Phi = 2\mu \left[\left(\frac{\partial v_r}{\partial r}\right)^2 + \left(\frac{\partial v_\phi}{r \partial \phi} + \frac{v_r}{r}\right)^2 + \left(\frac{\partial v_z}{\partial z}\right)^2\right]$$
$$+ \mu \left[\left(\frac{\partial v_r}{r \partial \phi} + \frac{\partial v_\phi}{\partial r} - \frac{v_\phi}{r}\right)^2 + \left(\frac{\partial v_\phi}{\partial z} + \frac{\partial v_z}{r \partial \phi}\right)^2 + \left(\frac{\partial v_z}{\partial r} + \frac{\partial v_r}{\partial z}\right)^2\right] - \frac{2}{3}\mu(\mathbf{div}(\boldsymbol{v}))^2 \quad (2.1.41)$$

For a constant density fluid, the continuity equation simplifies to $\mathbf{div}(\boldsymbol{v}) = \mathbf{0}$, the divergence terms in the viscous parts of the other equations vanish, and with $C_v = C_p$ the pressure terms may be omitted and e replaced with h in equation 2.1.40. With constant fluid properties, the continuity and momentum equations are completely decoupled from the energy equation, and the flow solution is independent of temperature.

The governing equations presented have a wealth of solutions. Initial and boundary conditions are needed to restrict these solutions to particular problems. A problem can be described as mathematically well posed if the equations together with the boundary and initial conditions have a unique solution. Conditions can be overprescribed, in which case they cannot all be satisfied, or underprescribed, in which case there will be multiple possible solutions. Appropriate boundary conditions depend on the type of boundary. For example, on inlets, velocity and temperature, or total pressure, total temperature and flow angle may be specified. Zero streamwise velocity and temperature gradients and static pressure are often assumed at outlets. On walls no-penetration, no-slip conditions ($\boldsymbol{v} = \boldsymbol{v}_{\mathbf{wall}}$) usually apply, with temperature or heat flux also specified. In compressible flow, solutions may be affected by reflection of pressure waves at inlet or exit boundaries, and special treatments have been developed to avoid or minimise such reflections in numerical solutions.

It is usually assumed that for steady or periodic flow, the solution will not depend on the initial conditions. However, sometimes hysteresis occurs in which more than one steady state is possible, depending on the system history. Such cases do occur in fluid dynamics and initial conditions, giving the flow and energy fields at the start of a solution, are then required for a full specification of the problem.

2.1.8 Vorticity and Q-Criterion

Visualisation of turbulent flow simulations often uses contours of vorticity components or the Q-criterion. The vorticity is a vector describing the spinning motion of a fluid element as observed from a point travelling with the element. In cylindrical coordinates it is defined as follows:

$$vorticity = \left(\frac{\partial v_r}{r \partial \phi} - \frac{\partial v_\phi}{\partial z}\right)\boldsymbol{i}_r + \left(\frac{\partial v_r}{\partial z} - \frac{\partial v_z}{\partial r}\right)\boldsymbol{i}_\phi + \left(\frac{\partial (r v_\phi)}{r \partial r} - \frac{\partial v_r}{r \partial \phi}\right)\boldsymbol{i}_z \quad (2.1.42)$$

An equation for vorticity can be derived from the Navier–Stokes equations and has been used as a replacement for the momentum equations in some analyses.

The Q-criterion was introduced by Hunt *et al.* (1988). Values correspond to half the difference between the square of the vorticity magnitude and the square of the rate of strain. These two quantities were shown to drive pressure fluctuations by Bradshaw and Koh (1981). Contours of the positive Q-criterion are effective in identifying vortices.

2.2 Rotating Coordinate Systems

It is sometimes advantageous to conduct analysis or computations in a rotating coordinate system. For example, for a rotating non-axisymmetric geometry, the boundary conditions may be steady in a rotating frame of reference but time-dependent in a stationary system. Some flows can also be considered a small perturbation from the forced vortex (or 'solid body rotation') $v_r = v_z = 0$, $v_\phi = \Omega r$. Converting to coordinates (r, θ, z) rotating about the z-axis with constant angular speed Ω, the velocity is denoted $\boldsymbol{u}$ and the forced vortex is $u_r = u_\theta = u_z = 0$, simplifying analysis.

2.2.1 Governing Differential Equations

The differential mass and momentum equations may be obtained by transforming equations 2.1.30–2.1.33 from the stationary to rotating coordinates. Noting that $\phi = \theta + \Omega t$, $v_\phi = u_\theta + \Omega r$, $\frac{\partial}{\partial \phi} = \frac{\partial}{\partial \theta}$, and $\frac{\partial}{\partial t}$ in the stationary system equates to $\frac{\partial}{\partial t} - \Omega \frac{\partial}{\partial \theta}$ in the rotating system, the following equations may be deduced:

$$\frac{\partial \rho}{\partial t} + \frac{\partial(\rho r u_r)}{r \partial r} + \frac{\partial(\rho u_\theta)}{r \partial \theta} + \frac{\partial(\rho u_z)}{\partial z} = \frac{\partial \rho}{\partial t} + \mathbf{div}(\rho \boldsymbol{u}) = 0 \tag{2.2.1}$$

$$\begin{aligned} &\frac{\partial(\rho u_r)}{\partial t} + \frac{\partial\left(\rho r u_r^2\right)}{r \partial r} + \frac{\partial(\rho u_\theta u_r)}{r \partial \theta} + \frac{\partial(\rho u_z u_r)}{\partial z} - \frac{\rho u_\theta^2}{r} - 2\rho \Omega u_\theta - \rho \Omega^2 r \\ &= -\frac{\partial p}{\partial r} + \frac{\partial(r \tau_{rr})}{r \partial r} + \frac{\partial \tau_{\theta r}}{r \partial \theta} + \frac{\partial \tau_{zr}}{\partial z} - \frac{\tau_{\theta\theta}}{r} \end{aligned} \tag{2.2.2}$$

$$\begin{aligned} &\frac{\partial(\rho r u_\theta)}{\partial t} + \frac{\partial(\rho r^2 u_r u_\theta)}{r \partial r} + \frac{\partial\left(\rho r u_\theta^2\right)}{r \partial \theta} + \frac{\partial(\rho r u_z u_\theta)}{\partial z} + 2\rho \Omega r u_r \\ &= -\frac{\partial p}{\partial \theta} + \frac{\partial(r^2 \tau_{r\theta})}{r \partial r} + \frac{\partial(r \tau_{\theta\theta})}{r \partial \theta} + \frac{\partial(r \tau_{z\theta})}{\partial z} \end{aligned} \tag{2.2.3}$$

$$\frac{\partial(\rho u_z)}{\partial t} + \frac{\partial(\rho r u_r u_z)}{r \partial r} + \frac{\partial(\rho u_\theta u_z)}{r \partial \theta} + \frac{\partial\left(\rho u_z^2\right)}{\partial z} = -\frac{\partial p}{\partial z} + \frac{\partial(r \tau_{rz})}{r \partial r} + \frac{\partial \tau_{\theta z}}{r \partial \theta} + \frac{\partial \tau_{zz}}{\partial z} \tag{2.2.4}$$

with viscous stresses

$$\begin{aligned} &\tau_{rr} = 2\mu \left(\frac{\partial u_r}{\partial r} - \frac{1}{3} \mathbf{div}(\boldsymbol{u}) \right), \qquad \tau_{\theta\theta} = 2\mu \left(\frac{\partial u_\theta}{r \partial \theta} + \frac{u_r}{r} - \frac{1}{3} \mathbf{div}(\boldsymbol{u}) \right), \\ &\tau_{zz} = 2\mu \left(\frac{\partial u_z}{\partial z} - \frac{1}{3} \mathbf{div}(\boldsymbol{u}) \right), \qquad \tau_{r\theta} = \tau_{\theta r} = \mu \left(\frac{r \partial(u_\theta / r)}{\partial r} + \frac{\partial u_r}{r \partial \theta} \right), \\ &\tau_{rz} = \tau_{zr} = \mu \left(\frac{\partial u_r}{\partial z} + \frac{\partial u_z}{\partial r} \right), \qquad \tau_{\theta z} = \tau_{z\theta} = \mu \left(\frac{\partial u_\theta}{\partial z} + \frac{\partial u_z}{r \partial \theta} \right). \end{aligned} \tag{2.2.5}$$

The continuity and axial momentum equations, and the viscous stresses, have exactly the same form as in the stationary system. The radial and angular momentum equations contain new terms representing the centrifugal and Coriolis forces associated with the rotating frame of reference. In the radial momentum equation the centrifugal acceleration term v_ϕ^2/r in the stationary frame has become $u_\theta^2/r + 2\Omega u_\theta + \Omega^2 r$ in the rotating frame. The term $-\rho\Omega^2 r$ is the centrifugal force (per unit volume) associated with the reference frame and the term $-2\Omega u_\theta$ is the radial component of the Coriolis force. As apparent in equation 2.2.3, the Coriolis force also has a component $2\rho\Omega r u_r$ in the circumferential direction, and acts at right angles to the motion and the rotation axis. This arises from the tendency of a parcel of fluid to conserve angular momentum. A practical demonstration of the Coriolis force may be experienced on a playground roundabout. Moving a leg or arm radially outwards while rotating, one can feel the Coriolis force in the circumferential direction, to one's left or right, depending on the direction of rotation. This is a restoring force that resists the radial displacement.

For constant density flow it is common to introduce a reduced pressure $p_{red} = p - \rho\Omega^2 r^2/2$. With p_{red} replacing p throughout in the above equations, the term $\rho\Omega^2 r$ is eliminated from equation 2.2.2, and p_{red} gives the perturbation from the solid body rotation solution. For compressible gas flow (at sufficiently high Mach number or imposed temperature difference), density varies significantly. The pressure distribution in the solid body rotation case will depend on the Mach number and thermal boundary conditions and is not so straightforward to determine.

The energy equation is obtained by the transformation of equations 2.1.34 and 2.1.35 with the use of equation 2.2.3. The right-hand side of equation 2.2.3 multiplied by Ω relates to the work (in a stationary system) on a rotating fluid element required to maintain the rotation. Similar terms arise in the coordinate transformation of the energy equation, and equation 2.2.3 may be used to eliminate these. Writing the transformed energy equation in terms of rothalpy, as introduced in subsection 2.1.6 gives

$$\frac{\partial(\rho E_{rot})}{\partial t} + \frac{\partial(\rho r u_r I)}{r\partial r} + \frac{\partial(\rho u_\theta I)}{r\partial\theta} + \frac{\partial(\rho u_z I)}{\partial z} = \frac{\partial}{r\partial r}\left(\frac{r\kappa\partial T}{\partial r}\right) + \frac{\partial}{r\partial\theta}\left(\frac{\kappa\partial T}{r\partial\theta}\right) + \frac{\partial}{\partial z}\left(\frac{\kappa\partial T}{\partial z}\right) + Z_{rot} \tag{2.2.6}$$

with viscous work and dissipation

$$Z_{rot} = \frac{\partial}{r\partial r} r(\tau_{rr}u_r + \tau_{r\theta}u_\theta + \tau_{rz}u_z) + \frac{\partial}{r\partial\theta}(\tau_{\theta r}u_r + \tau_{\theta\theta}u_\theta + \tau_{\theta z}u_z) + \frac{\partial}{\partial z}(\tau_{zr}u_r + \tau_{z\theta}u_\theta + \tau_{zz}u_z) \tag{2.2.7}$$

and the total energy in the rotating system

$$E_{rot} = e + \frac{\left(u_r^2 + u_\theta^2 + u_z^2 - \Omega^2 r^2\right)}{2} = e + I - h. \tag{2.2.8}$$

The rothalpy is related to enthalpy, relative total enthalpy and total enthalpy as follows:

$$I = h + \frac{\left(u_r^2 + u_\theta^2 + u_z^2 - \Omega^2 r^2\right)}{2} = h_{t,rel} - \frac{\Omega^2 r^2}{2} = h_t - \Omega r v_\phi \tag{2.2.9}$$

Rothalpy is comparable to total enthalpy in a stationary system. It is conserved along streamlines in steady, inviscid, adiabatic flow and conserved on a mass-averaged basis in steady, adiabatic systems with no work done in the rotating frame. Equation 2.2.6 for rothalpy is similar to the energy equation in a stationary system, with I replacing h_t. For $\Omega = 0$ the correspondence between I and h_t is exact.

2.2.2 Relative Total Temperature and Pressure

The relative total temperature $T_{t,rel}$ and relative total pressure $p_{t,rel}$ may be defined as the temperature and pressure of an element of fluid brought to rest in the rotating system in a thermodynamically reversible process without work or heat transfer. These quantities are not conserved in the same way as T_t and p_t in a stationary system but naturally arise when measurements are taken in the rotating system. Local values of static and relative total pressure, temperature and density relate to each other and to the relative flow velocity in the same way as for static and total properties in a stationary system. In flow at constant radius, the relative total pressure and temperature do have similar conservation properties to their stationary coordinate system counterpart. This follows from consideration of the conservation of rothalpy, as below.

For ***constant density flow***, the relative total reduced pressure may be defined as $p_{red,t} = p_{red} + \rho\left(u_r^2 + u_\theta^2 + u_z^2\right)/2$. From the momentum equations 2.2.2–2.2.4 it can then be shown that $p_{red,t}$ is constant along streamlines in steady, inviscid flow in the rotating coordinate system. As discussed below, $p_{red,t}$ may also be referred to as the rotary stagnation pressure.

2.2.3 Rotary Stagnation Temperature and Pressure

The rotary stagnation temperature T^* is defined as the rothalpy divided by specific heat at constant pressure. The rotary stagnation pressure p^* may be defined as the pressure resulting from a steady, isentropic process without work transfer taking the fluid from its initial state to a state with temperature T^*. For $\Omega = 0$, T^* and p^* are equal to the total temperature and pressure T_t and p_t. In rotating systems, T^* and p^* have similar conservation properties to T_t and p_t in a stationary system.

For a ***constant density fluid***, the internal energy is unchanged in an isentropic process (since there can be no compressive work). Hence, recalling $h = e + p/\rho$, the rotary stagnation pressure is given by $p^* = p + \rho(C_pT^* - h) = p + \rho\left(u_r^2 + u_\theta^2 + u_z^2 - \Omega^2r^2\right)/2$. It follows from the conservation of rothalpy $I = C_pT^*$ discussed above that p^* is conserved in isentropic flow with no work transfer. Thus in steady, isentropic flow, with constant density and with no work or heat transfer (in the rotating frame), the following relations apply:

$$T^* = \frac{I}{C_p} = T + \frac{\left(u_r^2 + u_\theta^2 + u_z^2 - \Omega^2r^2\right)}{2C_p} = \textit{constant along streamlines} \tag{2.2.10a}$$

$$p^* = p + \frac{\rho\left(u_r^2 + u_\theta^2 + u_z^2 - \Omega^2r^2\right)}{2} = \textit{constant along streamlines} \tag{2.2.10b}$$

The rotary stagnation pressure for constant density flow is identical to the total reduced pressure discussed above, with the alternative terms arising from different application areas.

For a ***compressible perfect gas flow***, the isentropic flow relationships give the rotary stagnation pressure as $p^* = p(T^*/T)^{\gamma/(\gamma-1)}$. In steady isentropic flow with no work or heat transfer (in the rotating system), it follows that both T^* and p^* are conserved. Thus in isentropic, compressible flow with no work or heat transfer

$$T^* = \frac{I}{C_p} = T + \frac{\left(u_r^2 + u_\theta^2 + u_z^2 - \Omega^2r^2\right)}{2C_p} = T_{t,rel} - \frac{\Omega^2r^2}{2C_p} = T_t - \frac{v_\phi\Omega r}{C_p}$$

$$= \textit{constant along streamlines} \tag{2.2.11a}$$

$$p^* = p\left(\frac{T^*}{T}\right)^{\gamma/(\gamma-1)} = \textit{constant along streamlines.} \tag{2.2.11b}$$

For an isentropic perfect gas flow with no work transfer in the rotating system (associated with forces on the stator), $p^*/T^{*\gamma/(\gamma-1)}$ is constant in direct correspondence with total pressure and total temperature in isentropic flow in a stationary system.

For relatively small changes in pressure, equation 2.2.11b reduces to the constant density flow relationship equation 2.2.10b. In a gas flow with $M_{rel} \ll 1$, the effects of the relative motion on density are negligible. Here the relative flow Mach number is defined as $M_{rel} = U/\sqrt{(\gamma RT)}$ where $U = \sqrt{\left(u_r^2 + u_\theta^2 + u_z^2\right)}$ is the gas speed and $\sqrt{(\gamma RT)}$ is the speed of sound.

In a forced vortex with zero relative velocity, negative radial gradients in T^* can be established as there is no flow in the radial direction and the associated positive density gradient is stabilising.

However, positive temperature gradients will destabilise the vortex. A forced vortex in which T^* and p^* are constant might be described as an isentropic forced vortex. In such a vortex, the density of fluid displaced radially will adjust to that of the neighbouring fluid at the new radius so that there is no buoyancy effect.

2.3 Dimensional Analysis

2.3.1 Nondimensional Governing Equations

Considering the flow around a rotating body and specifying characteristic dimensional parameters as the rotor radius b, rotor speed Ω, reference properties ρ_{ref}, p_{ref}, T_{ref}, μ_{ref}, κ_{ref}, C_p and a representative temperature difference ΔT_{ref}, the following nondimensional variables are defined:

$$
\begin{aligned}
&\check{t} = \Omega t, && \check{r} = r/b, && \check{z} = z/b, \\
&\check{\rho} = \rho/\rho_{ref}, && \check{\mu} = \mu/\mu_{ref}, && \check{\kappa} = \kappa/\kappa_{ref}, \\
&\check{u}_i = u_i/\Omega b, && \check{v}_i = v_i/\Omega b, && \check{p} = (p - p_{ref})/\rho_{ref}\Omega^2 b^2, \\
&\check{\tau}_{ij} = \tau_{ij}/\mu_{ref}\Omega, && \check{E}_{rot} = E_{rot}/C_p\Delta T_{ref}, && \check{I} = I/C_p\Delta T_{ref}, \\
&\check{T} = \frac{T - T_{ref}}{\Delta T_{ref}}, && \check{Z}_{rot} = Z_{rot}/\mu_{ref}\Omega^2, && \check{h} = \frac{h}{C_p\Delta T_{ref}}.
\end{aligned}
\tag{2.3.1}
$$

The governing equations in a rotating system from Section 2.2.1 may then be written in non-dimensional form, as below:

$$
\frac{\partial \check{\rho}}{\partial \check{t}} + \frac{\partial(\check{\rho}\check{r}\check{u}_r)}{\check{r}\partial \check{r}} + \frac{\partial(\check{\rho}\check{u}_\theta)}{\check{r}\partial\theta} + \frac{\partial(\check{\rho}\check{u}_z)}{\partial\check{z}} = 0 \tag{2.3.2}
$$

$$
\begin{aligned}
&\frac{\partial(\check{\rho}\check{u}_r)}{\partial\check{t}} + \frac{\partial\left(\check{\rho}\check{r}\check{u}_r^2\right)}{\check{r}\partial\check{r}} + \frac{\partial(\check{\rho}\check{u}_\theta\check{u}_r)}{\check{r}\partial\theta} + \frac{\partial(\check{\rho}\check{u}_z\check{u}_r)}{\partial\check{z}} - \frac{\check{\rho}\check{u}_\theta^2}{\check{r}} - 2\check{\rho}\check{u}_\theta - \check{\rho}\check{r} \\
&= -\frac{\partial\check{p}}{\partial\check{r}} + \frac{1}{Re_\phi}\left[\frac{\partial(\check{r}\check{\tau}_{rr})}{\check{r}\partial\check{r}} + \frac{\partial\check{\tau}_{\theta r}}{\check{r}\partial\theta} + \frac{\partial\check{\tau}_{zr}}{\partial\check{z}} - \frac{\check{\tau}_{\theta\theta}}{\check{r}}\right]
\end{aligned}
\tag{2.3.3}
$$

$$
\begin{aligned}
&\frac{\partial(\check{\rho}\check{r}\check{u}_\theta)}{\partial\check{t}} + \frac{\partial(\check{\rho}\check{r}^2\check{u}_r\check{u}_\theta)}{\check{r}\partial\check{r}} + \frac{\partial\left(\check{\rho}\check{r}\check{u}_\theta^2\right)}{\check{r}\partial\theta} + \frac{\partial(\check{\rho}\check{r}\check{u}_z\check{u}_\theta)}{\partial\check{z}} + 2\check{\rho}\check{r}\check{u}_r \\
&= -\frac{\partial\check{p}}{\partial\theta} + \frac{1}{Re_\phi}\left[\frac{\partial(\check{r}^2\check{\tau}_{r\theta})}{\check{r}\partial\check{r}} + \frac{\partial(\check{r}\check{\tau}_{\theta\theta})}{\check{r}\partial\theta} + \frac{\partial(\check{r}\check{\tau}_{z\theta})}{\partial\check{z}}\right]
\end{aligned}
\tag{2.3.4}
$$

$$
\frac{\partial(\check{\rho}\check{u}_z)}{\partial\check{t}} + \frac{\partial(\check{\rho}\check{r}\check{u}_r\check{u}_z)}{\check{r}\partial\check{r}} + \frac{\partial(\check{\rho}\check{u}_\theta u_z)}{\check{r}\partial\theta} + \frac{\partial\left(\check{\rho}\check{u}_z^2\right)}{\partial\check{z}} = -\frac{\partial\check{p}}{\partial\check{z}} + \frac{1}{Re_\phi}\left[\frac{\partial(\check{r}\check{\tau}_{rz})}{\check{r}\partial\check{r}} + \frac{\partial\check{\tau}_{\theta z}}{\check{r}\partial\theta} + \frac{\partial\check{\tau}_{zz}}{\partial\check{z}}\right] \tag{2.3.5}
$$

$$
\begin{aligned}
&\frac{\partial(\check{\rho}\check{E}_{rot})}{\partial\check{t}} + \frac{\partial(\check{\rho}\,\check{r}\check{u}_r\check{I})}{\check{r}\partial\check{r}} + \frac{\partial(\check{\rho}\,\check{u}_\theta\,\check{I})}{\check{r}\partial\theta} + \frac{\partial(\check{\rho}\check{u}_z\check{I})}{\partial\check{z}} \\
&= \frac{1}{Pr\,Re_\phi}\left[\frac{\partial}{\check{r}\partial\check{r}}\left(\frac{\check{r}\check{\kappa}\partial\check{T}}{\partial\check{r}}\right) + \frac{\partial}{\check{r}\partial\theta}\left(\frac{\check{\kappa}\partial\check{T}}{\check{r}\partial\theta}\right) + \frac{\partial}{\partial\check{z}}\left(\frac{\check{\kappa}\partial\check{T}}{\partial\check{z}}\right)\right] + \frac{Ec}{Re_\phi}\check{Z}_{rot}
\end{aligned}
\tag{2.3.6}
$$

with

$$
\check{\tau}_{rr} = 2\check{\mu}\left(\frac{\partial\check{u}_r}{\partial\check{r}} - \frac{1}{3}\mathbf{d\check{i}v}(\check{\mathbf{u}})\right), \quad \check{\tau}_{\theta\theta} = 2\check{\mu}\left(\frac{\partial\check{u}_\theta}{\check{r}\partial\theta} + \frac{\check{u}_r}{\check{r}} - \frac{1}{3}\mathbf{d\check{i}v}(\check{\mathbf{u}})\right),
$$

$$\check{\tau}_{zz} = 2\check{\mu}\left(\frac{\partial \check{u}_z}{\partial \check{z}} - \frac{1}{3}\check{\mathbf{div}}(\check{\mathbf{u}})\right), \quad \tau_{r\theta} = \check{\tau}_{\theta r} = \check{\mu}\left(\frac{\check{r}\partial(\check{u}_\theta/\check{r})}{\partial \check{r}} + \frac{\partial \check{u}_r}{\check{r}\partial\theta}\right),$$

$$\check{\tau}_{rz} = \check{\tau}_{zr} = \check{\mu}\left(\frac{\partial \check{u}_r}{\partial \check{z}} + \frac{\partial \check{u}_z}{\partial \check{r}}\right), \quad \tau_{\theta z} = \check{\tau}_{z\theta} = \check{\mu}\left(\frac{\partial \check{u}_\theta}{\partial \check{z}} + \frac{\partial \check{u}_z}{\check{r}\partial\theta}\right) \tag{2.3.7}$$

$$\check{Z}_{rot} = \frac{\partial}{\check{r}\partial\check{r}}\check{r}(\check{\tau}_{rr}\check{u}_r + \check{\tau}_{r\theta}\check{u}_\theta + \check{\tau}_{rz}\check{u}_z) + \frac{\partial}{\check{r}\partial\theta}(\check{\tau}_{\theta r}\check{u}_r + \check{\tau}_{\theta\theta}\check{u}_\theta + \check{\tau}_{\theta z}\check{u}_z) + \frac{\partial}{\partial\check{z}}(\check{\tau}_{zr}\check{u}_r + \check{\tau}_{z\theta}\check{u}_\theta + \check{\tau}_{zz}\check{u}_z) \tag{2.3.8}$$

$$\check{E}_{rot} = \check{T}/\gamma + Ec\left(\check{u}_r^2 + \check{u}_\theta^2 + \check{u}_z^2 - \check{r}^2\right)/2 \tag{2.3.9}$$

$$\check{I} = \check{T} + \frac{Ec\left(\check{u}_r^2 + \check{u}_\theta^2 + \check{u}_z^2 - \check{r}^2\right)}{2} = \check{T}_{t,rel} - \frac{Ec\,\check{r}^2}{2} = \check{T}_t - Ec\,\check{r}\check{v}_\phi. \tag{2.3.10}$$

Consider first the flow with negligible variation in density, viscosity, specific heats and thermal conductivity. Disregarding, for the moment, the effect of boundary conditions, the balance of terms in the above governing equations then depends on just four nondimensional parameters: rotational Reynolds number Re_ϕ, Prandtl number Pr, Eckert number Ec and the ratio of specific heats γ. The latter (γ) is a fluid property (and has a limited effect at low Mach number, as discussed in Section 2.1.4). The other three parameters are defined and discussed below:

$$Re_\phi = \rho_{ref}\Omega b^2/\mu_{ref}, \qquad Pr = \mu_{ref}C_p/\kappa_{ref}, \qquad Ec = \Omega^2 b^2/C_p\Delta T_{ref}. \tag{2.3.11}$$

The Reynolds number represents the ratio of inertial force to viscous force in the momentum equations. As in other areas of fluid dynamics, high values of Reynolds number are generally associated with essentially inviscid flow away from boundaries and internal shear layers, with adjustment of the solution to satisfy boundary conditions at walls through thin, viscous boundary layers. Such flow can be laminar and steady, but as the Reynolds number increases, steady flow solutions become unstable to infinitesimal and/or finite disturbances, resulting in turbulent flow. High Reynolds number, turbulent flows are most common in the engineering applications considered in this book, although other regimes do occur.

In flows that are a small perturbation from solid body rotation (with $\check{U} = U/\Omega b \leq 1$) a slightly different interpretation of Re_ϕ applies. In this case Coriolis forces dominate over advective terms in the momentum equations. The inverse of Re_ϕ equates to an Ekman number, which is used, particularly in geophysics, to describe the ratio of viscous to Coriolis force. In these circumstances the rotor speed Ωb does not represent the velocity differences in the flow field, and an alternative definition of Reynolds number, using a representative relative velocity, could be appropriate.

The Prandtl number represents the ratio of momentum diffusivity to thermal diffusivity of the fluid. It appears in the energy equation 2.3.6 along with the Reynolds number. The product $PrRe_\phi$ represents the ratio of convection of energy to diffusion of energy, playing a similar role in the energy equation to Re_ϕ in the momentum equations. Depending whether $Pr < 1$ or $Pr > 1$, thermal boundary layers are expected to be thicker or thinner than the momentum (or kinetic) boundary layers, respectively. Exceptions to this may occur when the Coriolis force dominates the advective terms in the momentum equations. This leads to Ekman layer flow, as to be discussed in Chapter 3.

The Eckert number Ec relates to the ratio of kinetic energy to internal energy. In the energy equation 2.3.6, Ec appears as a multiple of the viscous dissipation and work terms. For Ec order $1/Pr$ or higher, viscous heating (from dissipation of kinetic energy) is likely to be significant or dominate in the boundary layers. In equations 2.3.9 and 2.3.10, Ec is a factor in terms relating to the static temperature change associated with radial convection in the rotating system (and work required to maintain system rotation). For $Ec \ll 1$, frictional heating, kinetic energy effects and work required to maintain system rotation are generally negligible.

For a perfect gas, fluid density $\rho = p/RT$, and this equation of state must be included in the governing equations. In nondimensional form, this gives

$$\check{\rho} = \frac{Ec\left[\gamma M_{ref}^2 \check{p} + 1\right]}{\left[(\gamma - 1)M_{ref}^2 \check{T} + Ec\right]} \text{ or } \check{\rho} = \frac{\gamma M_{ref}^2 \check{p} + 1}{(\Delta T_{ref}/T_{ref})\check{T} + 1} \text{ or } \check{\rho} = \frac{[\gamma/(\gamma - 1)](\Delta T_{ref}/T_{ref})Ec\,\check{p} + 1}{(\Delta T_{ref}/T_{ref})\check{T} + 1} \tag{2.3.12}$$

where $M_{ref} = \Omega b/\sqrt{(\gamma RT_{ref})}$. The first form of this equation shows that just one new nondimensional parameter, the Mach number M_{ref} is introduced. However, the role of the Eckert number, normally associated with friction and kinetic energy effects, is somewhat obscure. Noting the relationship between Eckert and Mach numbers, $Ec = M_{ref}^2(\gamma - 1)T_{ref}/\Delta T_{ref}$, the second equation includes the ratio $\Delta T_{ref}/T_{ref}$ while eliminating Ec. $\Delta T_{ref}/T_{ref}$ characterises the change in density caused by temperature differences, with $1/T_{ref}$ being the coefficient of expansion $\beta = \left(\rho_{ref}^{-1}[\partial\rho/\partial T]_{ref}\right)$ for a perfect gas at the reference condition. The third form retains the Eckert number but introduces $\Delta T_{ref}/T_{ref}$ as the new nondimensional, rather than M_{ref}.

Table 2.1 summarises the conditions under which effects involving frictional work, kinetic energy, rotating system work, and density variation can often be neglected. There are many situations where all these conditions are satisfied and so Mach and Eckert number effects can be neglected. However, all the effects listed are significant at some engine conditions. For problems in which frictional and kinetic energy effects can be neglected in the energy equation, it is usual to use $\beta\Delta T_{ref}$, a rotational Grashof number $Gr = \beta\Delta T_{ref}Re_\phi^2$, or a rotational Rayleigh number $Ra = PrGr$ to characterise the buoyancy force in the centrifugal force field associated with the effect of temperature on density. These parameters will be further discussed later.

It is important to note that the conditions in Table 2.1 give a rough guide only, as the balance of terms in the governing equations varies according to the case considered. For example, the value of Ec low enough for frictional dissipation to be neglected will depend on the velocity difference across boundary layers and the level of heat transfer of interest.

Table 2.1 Summary of conditions when compressibility, density variations and Eckert number effects are negligible.

Effect	**Conditions when negligible**
Density variation due to flow (compressibility)	$Ec\,\Delta T_{ref}/(\gamma - 1)T_{ref} = M_{ref}^2 \ll 1$
Density variation due to the imposed temperature difference	$\beta\Delta T_{ref} = \dfrac{(\gamma - 1)M_{ref}^2}{\gamma Ec} \ll 1$ where $\beta = 1/T_{ref}$ for a perfect gas
Rotating system work	$Ec \ll 1$
Frictional heating/energy dissipation	$Ec \ll 1$
Kinetic energy effects (in thermal energy balance)	$Ec \ll 1$

Notes

1) $Ec = \Omega^2 b^2/C_p\Delta T_{ref}$, $M_{ref} = \Omega b/\sqrt{\gamma RT_{ref}}$.
2) For a stationary frame with representative fluid velocity Ωb, the above conditions apply with the omission of rotating system work.

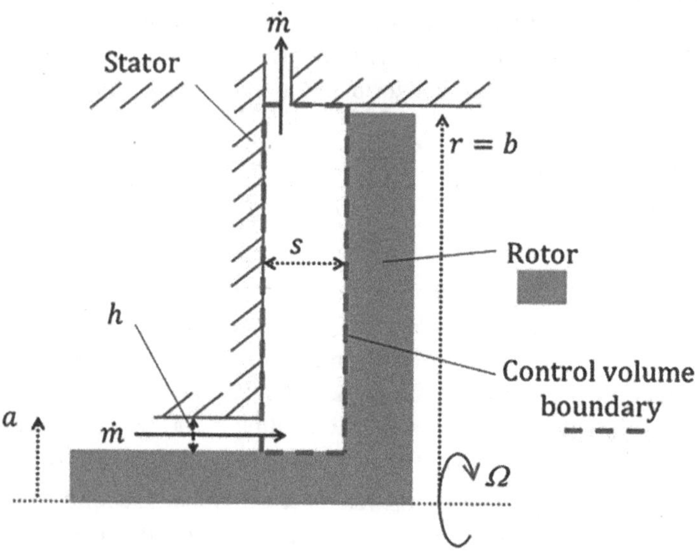

Figure 2.3 A simple rotor/stator disc cavity with radial outflow.

Boundary conditions, required to close the governing equations, and variable fluid properties introduce further dependencies in the nondimensional solution. For example, the inlet boundary condition in the simple rotor/stator disc cavity configuration of Figure 2.3 might be expressed for low-speed, incompressible flow as

$$\textit{inlet conditions:} \qquad \check{u}_r = 0, \qquad \check{u}_\theta = \check{v}_{\phi,in} - 1, \qquad \check{u}_z = \check{v}_{z,in} \tag{2.3.13}$$

Thus the nondimensional solution will depend on the inlet swirl $\check{v}_{\phi,in}$ (given in the absolute reference frame) and inlet axial velocity $\check{v}_{z,in}$. The latter parameter $\check{v}_{z,in}$ may be called a Rossby number or flow coefficient, with the difference in terminology arising from different application areas. An alternative to the velocity ratio $\check{v}_{z,in}$, which is popular in the turbomachinery internal air system literature is the nondimensional mass flow parameter $C_W = \dot{m}/\mu b$. This has the form of a Reynolds number based on the throughflow velocity, rather than the rotational speed as in Re_ϕ. A further mass flow parameter $\lambda = C_W/Re_\phi^{0.8}$ is often used in the air systems community. This is representative of the ratio of inlet flow to the flow entrained by a disc rotating in a quiescent environment and will be discussed further in Chapter 5.

2.3.2 Buckingham Pi Theorem

Use of the Buckingham Pi theorem is sometimes a more convenient way of identifying independent and dependent nondimensional parameters than reference to the governing equations and boundary conditions. The theorem is presented in many textbooks and may be stated as follows.

For a physical relationship $Q_1 = f(Q_2, Q_3, \ldots, Q_n)$ involving p fundamental dimensions we can define $n - p$ dimensionless ratios $\Pi_1, \Pi_2, \ldots, \Pi_{n-p}$ such that $\Pi_1 = \phi(\Pi_2, \Pi_3, \ldots, \Pi_{n-p})$.

Π_i are of the form $\Pi_1 = Q_1^a Q_2^b \ldots Q_n^?$ where $a, b, \ldots$ are rational numbers.

A general method of finding the nondimensionals Π_i is

Pick p of the n Q's as primary Q's. Each fundamental dimension must appear at least once in the primary Q's, the selected Q's should not be nondimensional, and the dimensions of each selection should be different.

The remaining $n-p$ Q's are then expressed as dimensionless ratios using the primary Q's. The resulting parameter sets are not unique but depend on the choice of variables and which of these are selected as primary variables. The constraints on the choice of primary variables given above are also not exhaustive, so adjustment of these is sometimes required. In practice, further adjustment of the nondimensional parameters is often undertaken, as in the example given below.

Consider the simple rotor/stator cavity with outflow in Figure 2.3, assuming that the rotor has a uniform temperature T_0, the stator is adiabatic, the inlet has uniform total temperature $T_{t,in}$ and circumferential velocity $v_{\phi.in}$, and that the perfect gas law applies with otherwise constant fluid properties. From physical principles, it might then be assumed that the mean surface heat flux to the rotor q_0 is a function of other parameters, as below:

$$q_0 = f(a, b, s, h, \dot{m}, \Omega, \mu, \kappa, C_p, \gamma, T_{t,in}, \rho_{t,in}, v_{\phi.in}, T_{t,in} - T_0) \tag{2.3.14}$$

where $\rho_{t,in} = p_{t,in}/RT_{t,in}$. Denoting fundamental dimensions length, time, mass and temperature by L, T, M and Θ, the 15 parameters in this equation have the dimensions below.

Mean rotor heat flux q_0:	M/T^3
Inner radius a:	L
Outer radius b:	L
Cavity width s:	L
Inlet height h:	L
Mass flow rate, $\dot{m}$:	M/T
Rotor angular speed Ω:	$1/T$
Fluid viscosity μ:	$M/(LT)$
Fluid thermal conductivity κ:	$ML/(T^3\Theta)$
Fluid specific heat at constant pressure C_p:	$L^2/(T^2\Theta)$
Fluid ratio of specific heats γ:	–
Inlet total temperature $T_{t,in}$:	Θ
Inlet density $\rho_{t,in}$:	M/L^3
Inlet circumferential velocity $v_{\phi.in}$:	L/T
Inlet-to-surface temperature difference $T_{t,in} - T_0$	Θ

Following the general method, four primary dimensional parameters, μ, $\rho_{t,in}$, b and $T_{t,in} - T_0$, are selected. Seeking exponents a, b, c and d such that $q_0\mu^a\rho_{t,in}^b b^c(T_{t,in} - T_0)^d$ is dimensionless requires

$$[M/T^3][M/(LT)]^a[M/L^3]^b[L]^c[\Theta]^d = 1$$

Collecting coefficients for each fundamental dimension then gives

$$M:\quad 1 + a + b = 0$$

$$T:\quad -3 - a = 0$$

$$L:\quad -a - 3b + c = 0$$

$$M:\quad d = 0$$

with the solution $a = -3$, $b = 2$, $c = 3$, $d = 0$. The nondimensional heat flux is then $\rho_{t,in}^2 b^3 q_0/\mu^3$. Repeating the procedure for all non-primary parameters shows the nondimensional heat flux to be a function of 10 nondimensional independent parameters, as below:

$$\frac{\rho_{t,in}^2 b^3 q_0}{\mu^3} = f\left(\frac{a}{b}, \frac{s}{b}, \frac{h}{b}, C_W, Re_\phi, \frac{\rho_{t,in}^2 b^2 \kappa (T_{t,in} - T_0)}{\mu^3}, \frac{\rho_{t,in}^2 b^2 C_p (T_{t,in} - T_0)}{\mu^2}, \gamma, \frac{T_{t,in}}{(T_{t,in} - T_0)}, \frac{\rho_{t,in} v_{\phi,in} b}{\mu}\right) \tag{2.3.15}$$

with $C_W = \dot{m}/\mu b$, $Re_\phi = \rho_{ref}\Omega b^2/\mu_{ref}$, as introduced in Section 2.3.1.

The nondimensional functional relationship above can be adapted to incorporate other familiar parameters. For example, the term on the left-hand side of equation 2.3.15 divided by any of the nondimensionals in the function on the right-hand side can still be written in the same functional form. Dividing $\rho_{t,in}^2 b^3 q_0/\mu^3$ by $\rho_{t,in}^2 b^2 \kappa (T_{t,in} - T_0)/\mu^3$ gives a Nusselt number Nu where

$$Nu = b q_0/\kappa (T_{t,in} - T_0) \tag{2.3.16}$$

With further manipulation, equation 2.3.15 gives

$$Nu = f\left(\frac{a}{b}, \frac{s}{b}, \frac{h}{b}, C_W, Re_\phi, Ec, Pr, \gamma, \beta(T_{t,in} - T_0), \frac{v_{\phi,in}}{\Omega b}\right) \tag{2.3.17}$$

Thus the nondimensional solution for rotor heat transfer can be regarded as a function of the independent parameters Re_ϕ, Ec, Pr, γ and $\beta(T_{t,in} - T_0)$ as expected from the governing equations, and a/b, s/b, h/b, C_W and $v_{\phi,in}/\Omega b$, as expected from the boundary conditions. For gas flow, the Eckert number Ec might be replaced by a Mach number. For constant density flow with $C_p = C_v$ the dependency on γ and $\beta(T_{t,in} - T_0)$ does not apply. Extension of the analysis to further dependent parameters is straightforward. For example, in constant density flow, the rotor moment M will be independent of the energy equation, and the moment coefficient $C_M = M/\frac{1}{2}\rho_{ref}\Omega^2 b^5$ will be a function of a/b, s/b, h/b, C_W, Re_ϕ and $v_{\phi,in}/\Omega b$.

It is clear that the choice of nondimensionals is arbitrary in many respects, but the number of independent parameters is fixed by the governing equations, boundary conditions and fluid properties. This can have important consequences for the interpretation and correlation of results. Understanding of the effects of the different parameters is crucial in many engineering applications.

2.4 Reynolds-Averaged Equations and Eddy Viscosity

While the governing equations given in Section 2.2.1 are valid for both laminar and turbulent flow, analytical solution is often impossible and numerical or computational solutions for turbulent flow are extremely computationally demanding. Turbulence length and time scales reduce as Reynolds number increases, prohibiting direct numerical solutions at high Reynolds numbers. Considering also that the main interest is often in the mean flow rather than the turbulent fluctuations, it is natural to break down the flow and temperature fields into mean and fluctuating components. With Reynolds decomposition, for example, the circumferential velocity component is written as $\overline{u}_\theta + \acute{u}_\theta$, with $\overline{u}_\theta$ and $\acute{u}_\theta$ being the mean and fluctuating components. The mean can be a time-average (covering the turbulent timescales) or an ensemble average. The mean component can be allowed to vary with time if this timescale is much greater than that for the turbulence. The Reynolds-averaged governing equations are obtained by time or ensemble averaging the equations written in terms of mean and fluctuating components. These equations give some insight into the nature of turbulent flow and form the basis of much of the computational fluid dynamics (CFD) discussed later.

Favre averaging is used on velocity components to simplify the treatment of the advective terms for compressible flow. For example, the mass-weighted average $\tilde{u}_\theta$ is defined as follows:

$$\tilde{u}_\theta = \int_{\Delta t} \rho u_\theta dt \Big/ \int_{\Delta t} \rho dt \tag{2.4.1}$$

with $u_\theta = \tilde{u}_\theta + \acute{u}_\theta$, compared to $u_\theta = \overline{u}_\theta + \acute{u}_\theta$ for Reynolds averaging, and Δt being a suitable time period. This treatment avoids the inclusion of 'extra terms' associated with density variations that occur with Reynolds averaging. However, these terms are negligible when the density variations associated with the turbulence are small. This is generally the case unless very highly compressible flow is considered. Here Reynolds-averaging is used with neglect of turbulent density fluctuations. Turbulent fluctuations in fluid viscosity and thermal conductivity are also assumed to be negligible.

Mean flow equations may be derived by representing $\boldsymbol{u}, p, T, h$ and I in terms of appropriate mean and fluctuating components, and then averaging the equations in Section 2.2.1 over a suitable time. The mass and momentum balances give the Reynolds-averaged equations below:

$$\frac{\partial \rho}{\partial t} + \frac{\partial(\rho r \overline{u}_r)}{r\partial r} + \frac{\partial(\rho \overline{u}_\theta)}{r\partial \theta} + \frac{\partial(\rho \overline{u}_z)}{\partial z} = 0 \tag{2.4.2}$$

$$\begin{aligned} &\frac{\partial(\rho \overline{u}_r)}{\partial t} + \frac{\partial\left(\rho r \overline{u}_r^2\right)}{r\partial r} + \frac{\partial(\rho \overline{u}_\theta \overline{u}_r)}{r\partial \theta} + \frac{\partial(\rho \overline{u}_z \overline{u}_r)}{\partial z} - \frac{\rho \overline{u}_\theta^2}{r} - 2\rho\Omega\overline{u}_\theta - \rho\Omega^2 r \\ &= -\frac{\partial \overline{p}}{\partial r} + \frac{\partial[r(\overline{\tau}_{rr} - \rho\overline{\acute{u}_r\acute{u}_r})]}{r\partial r} + \frac{\partial[\overline{\tau}_{\theta r} - \rho\overline{\acute{u}_\theta\acute{u}_r}]}{r\partial \theta} + \frac{\partial[\overline{\tau}_{zr} - \rho\overline{\acute{u}_z\acute{u}_r}]}{\partial z} - \frac{[\overline{\tau}_{\theta\theta} - \rho\overline{\acute{u}_\theta\acute{u}_\theta}]}{r} \end{aligned} \tag{2.4.3}$$

$$\begin{aligned} &\frac{\partial(\rho r \overline{u}_\theta)}{\partial t} + \frac{\partial(\rho r^2 \overline{u}_r \overline{u}_\theta)}{r\partial r} + \frac{\partial\left(\rho r \overline{u}_\theta^2\right)}{r\partial \theta} + \frac{\partial(\rho r \overline{u}_z \overline{u}_\theta)}{\partial z} + 2\rho\Omega r \overline{u}_r \\ &= -\frac{\partial \overline{p}}{\partial \theta} + \frac{\partial[r^2(\overline{\tau}_{r\theta} - \rho\overline{\acute{u}_r\acute{u}_\theta})]}{r\partial r} + \frac{\partial[r(\overline{\tau}_{\theta\theta} - \rho\overline{\acute{u}_\theta\acute{u}_\theta})]}{r\partial \theta} + \frac{\partial[r(\overline{\tau}_{z\theta} - \rho\overline{\acute{u}_z\acute{u}_\theta})]}{\partial z} \end{aligned} \tag{2.4.4}$$

$$\begin{aligned} &\frac{\partial(\rho \overline{u}_z)}{\partial t} + \frac{\partial(\rho r \overline{u}_r \overline{u}_z)}{r\partial r} + \frac{\partial(\rho \overline{u}_\theta \overline{u}_z)}{r\partial \theta} + \frac{\partial\left(\rho \overline{u}_z^2\right)}{\partial z} \\ &= -\frac{\partial \overline{p}}{\partial z} + \frac{\partial[r(\overline{\tau}_{rz} - \rho\overline{\acute{u}_r\acute{u}_z})]}{r\partial r} + \frac{\partial[\overline{\tau}_{\theta z} - \rho\overline{\acute{u}_\theta\acute{u}_z}]}{r\partial \theta} + \frac{\partial[\overline{\tau}_{zz} - \rho\overline{\acute{u}_z\acute{u}_z}]}{\partial z} \end{aligned} \tag{2.4.5}$$

with viscous stresses

$$\begin{aligned} &\overline{\tau}_{rr} = 2\mu\left(\frac{\partial \overline{u}_r}{\partial r} - \frac{1}{3}\mathbf{div}(\overline{\mathbf{u}})\right), \quad \overline{\tau}_{\theta\theta} = 2\mu\left(\frac{\partial \overline{u}_\theta}{r\partial \theta} + \frac{\overline{u}_r}{r} - \frac{1}{3}\mathbf{div}(\overline{\mathbf{u}})\right), \\ &\overline{\tau}_{zz} = 2\mu\left(\frac{\partial \overline{u}_z}{\partial z} - \frac{1}{3}\mathbf{div}(\overline{\mathbf{u}})\right), \quad \overline{\tau}_{r\theta} = \overline{\tau}_{\theta r} = \mu\left(\frac{r\partial(\overline{u}_\theta/r)}{\partial r} + \frac{\partial \overline{u}_r}{r\partial \theta}\right), \\ &\overline{\tau}_{rz} = \overline{\tau}_{zr} = \mu\left(\frac{\partial \overline{u}_r}{\partial z} + \frac{\partial \overline{u}_z}{\partial r}\right), \quad \overline{\tau}_{\theta z} = \overline{\tau}_{z\theta} = \mu\left(\frac{\partial \overline{u}_\theta}{\partial z} + \frac{\partial \overline{u}_z}{r\partial \theta}\right). \end{aligned} \tag{2.4.6}$$

The continuity and momentum equations for the mean flow have the same form as the original flow equations with the addition of terms involving the averages of products of velocity perturbations. These terms arise from the nonlinear advective terms and are known as Reynolds stresses. Their effects are in some ways analogous to the viscous terms, and their grouping with the viscous stress terms in the momentum equations emphasises this point.

The Reynolds-averaged energy equation includes terms involving the turbulent kinetic energy (per unit mass) $k = (\overline{\acute{u}_r\acute{u}_r} + \overline{\acute{u}_\theta\acute{u}_\theta} + \overline{\acute{u}_z\acute{u}_z})/2$ and its transport due to molecular diffusion and turbulence. These are often neglected and it is not unduly restrictive to do so in the present context. A more detailed description and discussion of the averaged energy equation is given, for example, by Wilcox (2006). With these approximations the mean energy equation may be written as follows:

$$\frac{\partial(\rho\overline{E}_{rot})}{\partial t} + \frac{\partial(\rho r\overline{u}_r\overline{I})}{r\partial r} + \frac{\partial(\rho\overline{u}_\theta\overline{I})}{r\partial\theta} + \frac{\partial(\rho\overline{u}_z\overline{I})}{\partial z} = \frac{\partial}{r\partial r}\left[r\left(\frac{\kappa\partial\overline{T}}{\partial r} - \rho\overline{\acute{u}_r\acute{h}}\right)\right] + \frac{\partial}{r\partial\theta}\left[\frac{\kappa\partial\overline{T}}{r\partial\theta} - \rho\overline{\acute{u}_\theta\acute{h}}\right] + \frac{\partial}{\partial z}\left[\frac{\kappa\partial\overline{T}}{\partial z} - \rho\overline{\acute{u}_z\acute{h}}\right] + Z_{rot} + Z_{t,rot} \quad (2.4.7)$$

with

$$Z_{rot} = \frac{\partial}{r\partial r}r(\overline{\tau}_{rr}\overline{u}_r + \overline{\tau}_{r\theta}\overline{u}_\theta + \overline{\tau}_{rz}\overline{u}_z) + \frac{\partial}{r\partial\theta}(\overline{\tau}_{\theta r}\overline{u}_r + \overline{\tau}_{\theta\theta}\overline{u}_\theta + \overline{\tau}_{\theta z}\overline{u}_z) + \frac{\partial}{\partial z}(\overline{\tau}_{zr}\overline{u}_r + \overline{\tau}_{z\theta}\overline{u}_\theta + \overline{\tau}_{zz}\overline{u}_z) \quad (2.4.8)$$

$$Z_{t,rot} = -\frac{\partial}{r\partial r}\rho r(\overline{\acute{u}_r\acute{u}_r}\overline{u}_r + \overline{\acute{u}_r\acute{u}_\theta}\overline{u}_\theta + \overline{\acute{u}_r\acute{u}_z}\overline{u}_z) - \frac{\partial}{r\partial\theta}\rho(\overline{\acute{u}_r\acute{u}_\theta}\overline{u}_r + \overline{\acute{u}_\theta\acute{u}_\theta}\overline{u}_\theta + \overline{\acute{u}_\theta\acute{u}_z}\overline{u}_z) - \frac{\partial}{\partial z}\rho(\overline{\acute{u}_r\acute{u}_z}\overline{u}_r + \overline{\acute{u}_\theta\acute{u}_z}\overline{u}_\theta + \overline{\acute{u}_z\acute{u}_z}\overline{u}_z) \quad (2.4.9)$$

$$\overline{E}_{rot} = \overline{e} + \frac{\left(\overline{u}_r^2 + \overline{u}_\theta^2 + \overline{u}_z^2 - \Omega^2 r^2\right)}{2}. \quad (2.4.10)$$

$$\overline{I} = \overline{h} + \frac{\left(\overline{u}_r^2 + \overline{u}_\theta^2 + \overline{u}_z^2 - \Omega^2 r^2\right)}{2} = \overline{h}_{t,rel} - \frac{\Omega^2 r^2}{2} = \overline{h}_t - \Omega r\overline{v}_\phi \quad (2.4.11)$$

As for the momentum equations, the mean energy equation is similar to the original equation, but with extra terms arising from products of fluctuations. The terms $\rho\overline{\acute{u}_i\acute{h}}$ are associated with turbulent diffusion of heat, and $Z_{t,rot}$ represents the work and energy dissipation associated with the Reynolds (or turbulent) stresses. Further equations may be derived for the mean mechanical energy $\overline{K}$ and the turbulent kinetic energy k, with the latter being frequently used in CFD.

The Reynolds-averaged equations above cannot be solved without assumptions regarding the turbulent stresses and diffusion. This is known as the closure problem. Although, for example, further equations can be derived for the Reynolds stresses from the governing equations, these introduce triple correlation terms $\overline{\acute{u}_i\acute{u}_j\acute{u}_k}$ that again prevents closure of the problem.

Many approaches to modelling turbulence make use of the Boussinesq eddy viscosity hypothesis. This draws an analogy with molecular dissipation and introduces a turbulent or eddy viscosity μ_t. In cylindrical coordinates the Reynolds stresses are then given by

$$-\rho\overline{\acute{u}_r\acute{u}_r} = 2\mu_t\left(\frac{\partial\overline{u}_r}{\partial r} - \frac{1}{3}\mathrm{div}(\overline{\mathbf{u}}) - \frac{1}{3}\rho k\right), \quad -\rho\overline{\acute{u}_\theta\acute{u}_\theta} = 2\mu_t\left(\frac{\partial\overline{u}_\theta}{r\partial\theta} + \frac{\overline{u}_r}{r} - \frac{1}{3}\mathrm{div}(\overline{\mathbf{u}}) - \frac{1}{3}\rho k\right),$$
$$-\rho\overline{\acute{u}_z\acute{u}_z} = 2\mu_t\left(\frac{\partial\overline{u}_z}{\partial z} - \frac{1}{3}\mathrm{div}(\overline{\mathbf{u}}) - \frac{1}{3}\rho k\right), \quad -\rho\overline{\acute{u}_r\acute{u}_\theta} = -\rho\overline{\acute{u}_\theta\acute{u}_r} = \mu_t\left(\frac{r\partial(\overline{u}_\theta/r)}{\partial r} + \frac{\partial\overline{u}_r}{r\partial\theta}\right),$$
$$-\rho\overline{\acute{u}_r\acute{u}_z} = -\rho\overline{\acute{u}_z\acute{u}_r} = \mu_t\left(\frac{\partial\overline{u}_r}{\partial z} + \frac{\partial\overline{u}_z}{\partial r}\right), \quad -\rho\overline{\acute{u}_\theta\acute{u}_z} = -\rho\overline{\acute{u}_z\acute{u}_\theta} = \mu_t\left(\frac{\partial\overline{u}_\theta}{\partial z} + \frac{\partial\overline{u}_z}{r\partial\theta}\right). \quad (2.4.12)$$

Note that the terms in turbulent energy k are needed to ensure the normal stresses sum to $-2\rho k$ in incompressible flow.

The eddy viscosity is a (supposed) property of the flow that has still to be determined. It is important to recognise that the Boussinesq hypothesis is known to have significant shortcomings. For example, its isotropic nature (with μ_t having no directional dependence) is inconsistent with measurements. Nevertheless, it has proven to be a very useful concept as will be shown in later discussions of CFD. The model is most suited to simple shear flows, that is flows dominated by shear in one direction.

In the energy equation, a similar analogy with molecular diffusion may be made to give the following gradient transport relations for turbulent heat diffusion, which introduce the turbulent thermal conductivity κ_t:

$$-\overline{\rho \acute{u}_r \acute{h}} = \kappa_t \frac{\partial \overline{T}}{\partial r} \qquad -\overline{\rho \acute{u}_\theta \acute{h}} = \frac{\kappa_t}{r}\frac{\partial \overline{T}}{\partial \theta} \qquad -\overline{\rho \acute{u}_z \acute{h}} = \kappa_t \frac{\partial \overline{T}}{\partial z} \tag{2.4.13}$$

Noting that $h = C_pT$ and introducing a turbulent Prandtl number $Pr_t = \mu_t C_p/\kappa_t$, these relations can be written in terms of the eddy viscosity μ_t. In practice, Pr_t is often assumed to be constant at ~0.9, but measurements show that variations in Pr_t can be significant.

2.5 Heat Transfer

The focus here is on convective heat transfer. This may be either forced convection associated with an imposed flow field, free or natural convection associated with buoyancy-generated flow, or mixed convection. Some general results regarding forced and free convection are given in the following subsections. Although surface radiative heat transfer can be significant in gas turbine internal air systems, it is not considered here.

2.5.1 Forced Convection

This subsection considers incompressible flow with uniform properties. In these conditions the velocity and pressure fields are independent of the fluid temperature and energy equation. For a given flow field the convective heat transfer is determined by the energy equation and the thermal boundary conditions. In stationary coordinates the energy equation, from Section 2.1, is as follows:

$$\frac{\partial(\rho E)}{\partial t} + \frac{\partial(\rho r v_r h_t)}{r\partial r} + \frac{\partial(\rho v_\phi h_t)}{r\partial \phi} + \frac{\partial(\rho v_z h_t)}{\partial z} = \frac{\partial}{r\partial r}\left(\frac{r\kappa\partial T}{\partial r}\right) + \frac{\partial}{r\partial \phi}\left(\frac{\kappa\partial T}{r\partial \phi}\right) + \frac{\partial}{\partial z}\left(\frac{\kappa\partial T}{\partial z}\right) + Z. \tag{2.5.1}$$

For small Eckert numbers, Z and the kinetic energy components in E and h_t will be negligible, as was discussed in Section 2.3. With $C_v = C_p$ equation 2.5.1 then reduces to

$$\frac{\partial(\rho C_p T)}{\partial t} + \frac{\partial(\rho r v_r C_p T)}{r\partial r} + \frac{\partial(\rho v_\phi C_p T)}{r\partial \phi} + \frac{\partial(\rho v_z C_p T)}{\partial z} = \frac{\partial}{r\partial r}\left(\frac{r\kappa\partial T}{\partial r}\right) + \frac{\partial}{r\partial \phi}\left(\frac{\kappa\partial T}{r\partial \phi}\right) + \frac{\partial}{\partial z}\left(\frac{\kappa\partial T}{\partial z}\right). \tag{2.5.2}$$

This equation is linear and homogeneous in temperature so that the principle of superposition applies. If $T_1, T_2, \ldots, T_n$ all satisfy equation 2.5.2 then any linear combination of the n solutions will also be a valid solution. This property can be used to construct solutions for arbitrary thermal boundary conditions from a set of known solutions. However, the practicality of this is often limited by the infinite possible variation in surface temperature distributions.

The linearity of the energy equation underpins the common use of convective heat transfer coefficients. For a two-temperature problem, such as that of Figure 2.3 discussed in Section 2.3.2, the solution $T - T_{0,1}$ to equation 2.5.2 will be proportional to $T_{t,in} - T_{0,1}$, where $T_{t,in}$ and $T_{0,1}$ are uniform inlet and rotor surface temperatures. Noting that the solution of 2.5.2 $T = T_{t,in} = constant$ corresponds to an adiabatic rotor ($q_{0,1} = 0$) with $T_{0,1} = T_{t,in}$, it follows that the local or mean rotor heat transfer $q_{0,1} = htc(T_{t,in} - T_{0,1})$ where the heat transfer coefficient htc is a function of position but independent of the temperature difference for the given geometry and flow field. In this example the stator was assumed adiabatic. If instead the stator had a uniform temperature $T_{0,2}$ then the heat flux to the rotor could be represented as $q_{0,1} = htc_1(T_{t,in} - T_{0,1}) + htc_2(T_{0,2} - T_{0,1})$, with htc_1 and htc_2 independent of $T_{0,1}$, $T_{0,2}$ and $T_{t,in}$.

Referring to Section 2.3.2 above, the relationship of htc to Nusselt number is $Nu = htc.\, L/\kappa$, where an arbitrary length scale L is used rather than the disc radius for the example considered earlier. Dimensional analysis has shown that Nu depends on a limited number of parameters for problems where the thermal boundary conditions can be represented by just two temperatures (or a single temperature difference). Empirical correlations for Nusselt numbers have been developed for various cases, and these may be applied to calculate heat transfer coefficients for engineering predictions. Care is required when the thermal boundary conditions do not match those in the original experiments. For example, consider estimating local heat transfer in flow through a pipe with varying surface temperature using available correlations that are based on experiments with uniform surface temperature. The results will depend on the choice of temperature difference in the definition of htc and Nu. In these circumstances it is usually better to use local values of temperature, such as the local surface temperature and local bulk fluid temperature. In the pipe flow example, the use of the inlet fluid temperature for T_{ref} could lead to large errors in long pipes.

Equation 2.5.2 neglects terms involving friction, kinetic energy and/or pressure work terms. These and, in a rotating system, the work associated with maintaining the system rotation, can be taken into account by using adiabatic surface temperatures. Suppose the solution of equation 2.5.1 when all solid surfaces are adiabatic (with inlet total temperature $T_{t,in}$) is T_{ad}. Surface temperatures may vary and are denoted $T_{0,ad}$. Consideration of equations 2.5.1 and 2.5.2 shows that the sum of T_{ad} (calculated with inlet total temperature $T_{t,in}$) and any solution of equation 2.5.2 will satisfy the full energy equation 2.5.1. Thus, selecting a solution of 2.5.2 so that the sum with T_{ad} satisfies boundary conditions for the two-temperature problem above (with specified $T_{t,in}$ and $T_{0,1}$) it follows that the rotor heat transfer $q_{0,1} = htc(T_{0,1,ad} - T_{0,1})$. Thus the heat transfer coefficient for higher Eckert numbers is equal to that of the limiting case $Ec \ll 1$ provided the adiabatic surface temperature is used in calculating heat flux. As with the application of empirical Nusselt number correlations, some care is required in estimating $T_{0,1,ad}$ for engineering applications. It is generally better to base estimates on local fluid temperature and velocity.

2.5.2 Centrifugal Free Convection

In free or natural convection, fluid motion arises from the density gradients caused by heat transfer from boundaries, and so the flow field cannot be decoupled from the thermal boundary conditions. Here, the gravitational force is neglected, so the convective flow arises from the centrifugal force term in the radial momentum equation. Cooler (denser) fluid tends to be centrifuged radially outwards, and hotter (less dense) fluid tends to move radially inwards.

Considering a closed, rotating container, solid body rotation $u_r = u_\theta = u_z = 0$, with $T = A + Bln(r/b)$, where A and B are constants, satisfies the governing equations given in Section 2.2 and the velocity boundary conditions. For the solution to be valid the temperatures on the boundaries must match the solution. In this solution the pressure adjusts to balance the centrifugal force term in the radial momentum equation 2.2.2, depending on temperature through the density with $\partial p/\partial r = \rho\Omega^2 r$. Any axial or circumferential temperature gradients will lead to corresponding density and pressure gradients and to fluid motion. This is analogous to convection in an enclosure under gravity, where horizontal temperature gradients produce fluid motion. Also in correspondence with gravitational flows, for sufficiently high (positive) radial temperature gradients, the solid body rotation solution becomes unstable, leading to centrifugal Rayleigh-Bénard convection.

Assuming the relative velocity $\boldsymbol{u}$ is small (compared to Ωr) in free convection, kinetic energy and viscous terms may be neglected in the energy equation 2.2.6. This gives

$$\frac{\partial(\rho C_v T)}{\partial t} + \frac{\partial(\rho r u_r[C_p T - 0.5\Omega^2 r^2])}{r\partial r} + \frac{\partial(\rho u_\theta C_p T)}{r\partial\theta} + \frac{\partial(\rho u_z C_p T)}{\partial z}$$

$$= \frac{\partial}{r\partial r}\left(\frac{r\kappa\partial T}{\partial r}\right) + \frac{\partial}{r\partial\theta}\left(\frac{\kappa\partial T}{r\partial\theta}\right) + \frac{\partial}{\partial z}\left(\frac{\kappa\partial T}{\partial z}\right). \tag{2.5.3}$$

As discussed in Section 2.3, the Nusselt number is dependent on γ and Pr, which characterise the fluid, and Re_ϕ, Ec and $\Delta T_{ref}/T_{ref}$. If the Eckert number $Ec = \Omega^2 b^2/C_p\Delta T_{ref}$ is small, the term involving $\Omega^2 r^2$ in equation 2.5.3, will be negligible and the energy equation reverts to the form of equation 2.5.2. For these conditions M_{rel} will also be small and, referring to Table 2.1, density variations can be characterised solely by $\beta\Delta T_{ref}$.

The Boussinesq approximation is often used in analysing free convection. This assumes that the fluid can be treated as incompressible in the equations, apart from in the body force term that drives the motion. With this assumption the continuity and momentum balance equations 2.2.1–2.2.4 become

$$\frac{\partial(r u_r)}{r\partial r} + \frac{\partial u_\theta}{r\partial\theta} + \frac{\partial u_z}{\partial z} = 0 \tag{2.5.4}$$

$$\frac{\partial(\rho u_r)}{\partial t} + \frac{\partial\left(\rho r u_r^2\right)}{r\partial r} + \frac{\partial(\rho u_\theta u_r)}{r\partial\theta} + \frac{\partial(\rho u_z u_r)}{\partial z} - \frac{\rho u_\theta^2}{r} - 2\rho\Omega u_\theta + \rho\beta(T - T_{ref})\Omega^2 r$$

$$= -\frac{\partial p_{red}}{\partial r} + \frac{\partial(r\tau_{rr})}{r\partial r} + \frac{\partial\tau_{\theta r}}{r\partial\theta} + \frac{\partial\tau_{zr}}{\partial z} - \frac{\tau_{\theta\theta}}{r} \tag{2.5.5}$$

$$\frac{\partial(\rho r u_\theta)}{\partial t} + \frac{\partial(\rho r^2 u_r u_\theta)}{r\partial r} + \frac{\partial\left(\rho r u_\theta^2\right)}{r\partial\theta} + \frac{\partial(\rho r u_z u_\theta)}{\partial z} + 2\rho\Omega r u_r$$

$$= -\frac{\partial p_{red}}{\partial\theta} + \frac{\partial(r^2\tau_{r\theta})}{r\partial r} + \frac{\partial(r\tau_{\theta\theta})}{r\partial\theta} + \frac{\partial(r\tau_{z\theta})}{\partial z} \tag{2.5.6}$$

$$\frac{\partial(\rho u_z)}{\partial t} + \frac{\partial(\rho r u_r u_z)}{r\partial r} + \frac{\partial(\rho u_\theta u_z)}{r\partial\theta} + \frac{\partial\left(\rho u_z^2\right)}{\partial z} = -\frac{\partial p_{red}}{\partial z} + \frac{\partial(r\tau_{rz})}{r\partial r} + \frac{\partial\tau_{\theta z}}{r\partial\theta} + \frac{\partial\tau_{zz}}{\partial z}. \tag{2.5.7}$$

where the reduced pressure p_{red} includes the leading contribution from the centrifugal term, ρ is the reference density in these equations, and for small enough $\Delta T_{ref}/T_{ref}$ the coefficient of thermal expansion $\beta = 1/T_{ref}$ for a perfect gas.

An important consequence of the Bousinesq approximation in gravitational convection is that it eliminates one independent nondimensional parameter for the problem. This follows from the fact that β and gravitational force (g) only occur as a product in the governing equations with the Bousinesq approximation. In this case the single parameter βg can replace the two parameters in the application of the Buckingham Pi method.

For rotational free convection, as shown in equation 2.5.5, β multiplies the centrifugal force term involving Ω^2, but Ω also occurs in the Coriolis terms. Thus the inclusion of the Coriolis terms prevents a reduction in the number of independent parameters. This is shown in the nondimensional governing equations below. Considering the form of the driving force, these take $\sqrt{\beta \Delta T_{ref}}\Omega b$ as an appropriate velocity scale, rather than Ωb as in Section 2.3.

Assuming $Ec \ll 1$ and that the Boussinesq approximation applies with μ and ρ constant, we take

$$\check{t} = \sqrt{\beta \Delta T_{ref}}\Omega t, \quad \check{r} = \frac{r}{b}, \quad \check{z} = \frac{z}{b},$$

$$\check{u}_i = u_i/(\Omega\sqrt{\beta \Delta T_{ref}} b), \quad \check{p} = (p_{red} - p_{ref})/(\rho \beta \Delta T_{ref} \Omega^2 b^2),$$

$$\check{\tau}_{ij} = \tau_{ij}/\mu\sqrt{\beta \Delta T_{ref}}\Omega, \quad \check{T} = \frac{T - T_{ref}}{\Delta T_{ref}}. \tag{2.5.8}$$

The following equations may then be derived:

$$\frac{\partial(\check{r}\check{u}_r)}{\check{r}\partial\check{r}} + \frac{\partial\check{u}_\theta}{\check{r}\partial\theta} + \frac{\partial\check{u}_z}{\partial\check{z}} = 0 \tag{2.5.9}$$

$$\frac{\partial\check{u}_r}{\partial\check{t}} + \frac{\partial\left(\check{r}\check{u}_r^2\right)}{\check{r}\partial\check{r}} + \frac{\partial(\check{u}_\theta\check{u}_r)}{\check{r}\partial\theta} + \frac{\partial(\check{u}_z\check{u}_r)}{\partial\check{z}} - \frac{\check{u}_\theta^2}{\check{r}} - \frac{2\check{u}_\theta}{\sqrt{\beta\Delta T_{ref}}} - \check{r}\check{T}$$
$$= -\frac{\partial\check{p}}{\partial\check{r}} + \frac{1}{\sqrt{Gr}}\left[\frac{\partial(\check{r}\check{\tau}_{rr})}{\check{r}\partial\check{r}} + \frac{\partial\check{\tau}_{\theta r}}{\check{r}\partial\theta} + \frac{\partial\check{\tau}_{zr}}{\partial\check{z}} - \frac{\check{\tau}_{\theta\theta}}{\check{r}}\right] \tag{2.5.10}$$

$$\frac{\partial(\check{r}\check{u}_\theta)}{\partial\check{t}} + \frac{\partial(\check{r}^2\check{u}_r\check{u}_\theta)}{\check{r}\partial\check{r}} + \frac{\partial\left(\check{r}\check{u}_\theta^2\right)}{\check{r}\partial\theta} + \frac{\partial(\check{r}\check{u}_z\check{u}_\theta)}{\partial\check{z}} + \frac{2\check{r}\check{u}_r}{\sqrt{\beta\Delta T_{ref}}}$$
$$= -\frac{\partial\check{p}}{\partial\theta} + \frac{1}{\sqrt{Gr}}\left[\frac{\partial(\check{r}^2\check{\tau}_{r\theta})}{\check{r}\partial\check{r}} + \frac{\partial(\check{r}\check{\tau}_{\theta\theta})}{\check{r}\partial\theta} + \frac{\partial(\check{r}\check{\tau}_{z\theta})}{\partial\check{z}}\right] \tag{2.5.11}$$

$$\frac{\partial\check{u}_z}{\partial\check{t}} + \frac{\partial(\check{r}\check{u}_r\check{u}_z)}{\check{r}\partial\check{r}} + \frac{\partial(\check{u}_\theta u_z)}{\check{r}\partial\theta} + \frac{\partial\left(\check{u}_z^2\right)}{\partial\check{z}} = -\frac{\partial\check{p}}{\partial\check{z}} + \frac{1}{\sqrt{Gr}}\left[\frac{\partial(\check{r}\check{\tau}_{rz})}{\check{r}\partial\check{r}} + \frac{\partial\check{\tau}_{\theta z}}{\check{r}\partial\theta} + \frac{\partial\check{\tau}_{zz}}{\partial\check{z}}\right] \tag{2.5.12}$$

$$\frac{\partial(\check{T})}{\partial\check{t}} + \frac{\partial(\check{r}\check{u}_r\check{T})}{\check{r}\partial\check{r}} + \frac{\partial(\check{u}_\theta\check{T})}{\check{r}\partial\theta} + \frac{\partial(\check{u}_z\check{T})}{\partial\check{z}}$$
$$= \frac{1}{Pr\sqrt{Gr}}\left[\frac{\partial}{\check{r}\partial\check{r}}\left(\frac{\check{r}\check{\kappa}\partial\check{T}}{\partial\check{r}}\right) + \frac{\partial}{\check{r}\partial\theta}\left(\frac{\check{\kappa}\partial\check{T}}{\check{r}\partial\theta}\right) + \frac{\partial}{\partial\check{z}}\left(\frac{\check{\kappa}\partial\check{T}}{\partial\check{z}}\right)\right] \tag{2.5.13}$$

with

$$\check{\tau}_{rr} = 2\check{\mu}\left(\frac{\partial\check{u}_r}{\partial\check{r}} - \frac{1}{3}\check{\text{div}}(\check{\boldsymbol{u}})\right), \quad \check{\tau}_{\theta\theta} = 2\check{\mu}\left(\frac{\partial\check{u}_\theta}{\check{r}\partial\theta} + \frac{\check{u}_r}{\check{r}} - \frac{1}{3}\check{\text{div}}(\check{\boldsymbol{u}})\right)$$

$$\check{\tau}_{zz} = 2\check{\mu}\left(\frac{\partial\check{u}_z}{\partial\check{z}} - \frac{1}{3}\check{\text{div}}(\check{\boldsymbol{u}})\right), \quad \check{\tau}_{r\theta} = \check{\tau}_{\theta r} = \check{\mu}\left(\frac{\check{r}\partial(\check{u}_\theta/\check{r})}{\partial\check{r}} + \frac{\partial\check{u}_r}{\check{r}\partial\theta}\right)$$

$$\check{\tau}_{rz} = \check{\tau}_{zr} = \check{\mu}\left(\frac{\partial\check{u}_r}{\partial\check{z}} + \frac{\partial\check{u}_z}{\partial\check{r}}\right), \quad \check{\tau}_{\theta z} = \check{\tau}_{z\theta} = \check{\mu}\left(\frac{\partial\check{u}_\theta}{\partial\check{z}} + \frac{\partial\check{u}_z}{\check{r}\partial\theta}\right) \tag{2.5.14}$$

These equations include the Grashof number $Gr = \beta\Delta T_{ref} Re_\phi^2$, as introduced earlier.

The energy equation 2.5.13 neglects the rotating system work. This may be significant when relative velocities are small and so frictional heating and kinetic energy in the rotating frame are still negligible. In these circumstances a suitable approximation to the energy equation for use with the Boussinesq approximation may be

$$\frac{\partial(\check{T}^*)}{\partial\check{t}} + \frac{\partial(\check{r}\check{u}_r\check{T}^*)}{\check{r}\partial\check{r}} + \frac{\partial(\check{u}_\theta\,\check{T}^*)}{\check{r}\partial\theta} + \frac{\partial(\check{u}_z\check{T}^*)}{\partial\check{z}}$$
$$= \frac{1}{Pr\sqrt{Gr}}\left[\frac{\partial}{\check{r}\partial\check{r}}\left(\frac{\check{r}\check{\kappa}\partial\check{T}^*}{\partial\check{r}}\right) + \frac{\partial}{\check{r}\partial\theta}\left(\frac{\check{\kappa}\partial\check{T}^*}{\check{r}\partial\theta}\right) + \frac{\partial}{\partial\check{z}}\left(\frac{\check{\kappa}\partial\check{T}^*}{\partial\check{z}}\right)\right] \tag{2.5.15}$$

with the nondimensional rotary stagnation temperature $\check{T}^* = (T^* - T_{ref})/\Delta T_{ref}$. This equation has the form of the incompressible flow energy equation with rotary stagnation temperature replacing static temperature. Pressure effects in the energy balance are assumed to be negligible, as discussed in Section 2.1.4, and the length scale for radial heat conduction is assumed $\ll b$. Using $\check{T}^*$ in place of $\check{T}$ in equation 2.5.10 allows for the density variation in an isentropic forced vortex (as discussed in Section 2.2.3). The Boussinesq approximation has been widely used in gravitational natural convection. Its validity has been examined in some detail by, for example, Spiegel and Veronis (1960) and Gray and Giorgini (1976). While there is evidence that some results from gravitational convection may read across to centrifugal convection, the limits of the Boussinesq approximation for applications considered here are less well investigated.

From the equations given above, it can be shown that the Nusselt number for a 'two temperature' enclosure problem will be a function of γ, Pr, $\beta\Delta T_{ref}$, Gr and geometrical parameters. The significance of the Coriolis terms compared to other inertial terms in equations 2.5.10 and 2.5.11 will be greatest at low values of $\beta\Delta T_{ref}$. This and the range of validity of the Boussinesq approximation are discussed further later in this book. In gravitational convection the dependency on $\beta\Delta T_{ref}$ is fully captured by the Grashof number (as there are no Coriolis terms), and the Boussinesq approximation is often assumed to apply for $\beta\Delta T_{ref}$ up to ~0.2.

2.5.3 Similarity Between Heat and Momentum Transfer

Convective heat transfer and viscous drag on surfaces are commonly associated with boundary layers or simple shear layers such as those that occur in pipe and channel flows. As discussed in Section 2.3 for laminar flow, momentum and thermal boundary layers are expected to have similar thickness for Prandtl numbers close to unity. In a turbulent boundary layer, molecular diffusion will dominate in the viscous sub-layer close to the wall, but turbulent diffusion dominates further away from the wall. It was noted in Section 2.4 that turbulent Prandtl numbers of ~0.9 were often assumed when using eddy viscosity models. This reflects the similarities between the transport of energy and momentum by turbulent eddies. Further similarities between heat and momentum transfer are evident in the governing equations.

Consider first the case of steady, 2D boundary layer flow over an isothermal flat plate for a uniform density fluid with neglect of kinetic energy and frictional work in the energy equation. The streamwise velocity will be much greater than the velocity component normal to the plate, and gradients normal to the plate will be much greater than those in the streamwise direction. The Reynolds-averaged boundary layer equations using the eddy viscosity hypothesis approximate to the following:

$$\frac{\partial\overline{v}_s}{\partial s} + \frac{\partial\overline{v}_n}{\partial n} = 0 \tag{2.5.16}$$

$$\frac{\partial \overline{p}}{\partial n} = 0 \tag{2.5.17}$$

$$\rho\frac{\partial\left(\overline{v}_s^{\,2}\right)}{\partial s} + \rho\frac{\partial(\overline{v}_n\overline{v}_s)}{\partial n} = -\frac{\partial \overline{p}}{\partial s} + \frac{\partial}{\partial n}\left(\frac{\mu_{eff}\partial\overline{v}_s}{\partial r}\right) \tag{2.5.18}$$

$$\rho C_p\frac{\partial(\overline{v}_s\overline{T})}{\partial s} + \rho C_p\frac{\partial(\overline{v}_n\overline{T})}{\partial n} = \frac{\partial}{\partial n}\left(\frac{\kappa_{eff}\partial\overline{T}}{\partial n}\right) \tag{2.5.19}$$

Here s and n are streamwise and surface normal coordinates, $\mu_{eff} = \mu_t + \mu$ and $\kappa_{eff} = \kappa_t + \kappa$. In the case of zero pressure gradient $\left(\partial\overline{p}/\partial s = 0\right)$ and an effective Prandtl number ($Pr_{eff} = \mu_{eff}C_p/\kappa_{eff}$) equal to one, the streamwise momentum and energy equations can be written as

$$\rho\frac{\partial(\overline{v}_s\psi)}{\partial s} + \rho\frac{\partial(\overline{v}_n\psi)}{\partial n} = \frac{\partial}{\partial n}\left(\frac{\mu_{eff}\partial\psi}{\partial n}\right) \tag{2.5.20}$$

with $\psi = \overline{v}_s$ or $\psi = \overline{T} - T_0$ and T_0 being the plate surface temperature. For boundary conditions $\overline{T} = T_0, \overline{v}_s = 0$ at $n = 0$ and conditions $\overline{T} = T_\infty, \overline{v}_s = v_\infty$ at $n = \infty$ it follows that $\overline{T} - T_0 = (T_\infty - T_0)\overline{v}_s/v_\infty$ is a solution to the energy equation. According to this solution, the heat flux normal to the surface is

$$q_0 = \kappa_{eff}\left[\frac{\partial\overline{T}}{\partial n}\right]_{n=0} = \mu_{eff}C_p\frac{(T_\infty - T_0)}{v_\infty}\left[\frac{\partial\overline{v}_s}{\partial n}\right]_{n=0} = \frac{C_p(T_\infty - T_0)\tau_0}{v_\infty} \tag{2.5.21}$$

where τ_0 is the surface shear stress. In terms of the Nusselt number $Nu = Lq_0/\kappa(T_\infty - T_0)$ and the friction factor $c_f = 2\tau_0/\rho\overline{v}_\infty^2$ this gives

$$Nu = \frac{\rho C_p L c_f \overline{v}_\infty}{2\kappa} = \frac{c_f Re}{2} \tag{2.5.22}$$

with $Re = \rho L\overline{v}_\infty/\mu$, since $Pr = 1$ and $\kappa_{eff} = \kappa = \mu C_p$ in the laminar sublayer close to the wall.

The result above is known as the Reynolds analogy for heat transfer, and is discussed further in many textbooks; see, for example, Schlichting (1979) and Rogers and Mayhew (1980). It can be extended to include frictional heating, giving the result that the adiabatic surface temperature for $Pr_{eff} = 1$ is $T_{0.ad} = T_\infty + v_\infty^2/2C_p$.

For axisymmetric rotating flow the circumferential pressure gradient $\partial\overline{p}/\partial\phi = 0$ and the Reynolds analogy applies to the angular momentum equation. This has been discussed by Dorfman (1963), Owen (1971) and Chew (1985). In this case s and n are taken to be streamwise and surface normal coordinates in the $r-z$ plane, and it is assumed that the boundary layer thickness is small compared to the radius (so that $\partial r/\partial n \approx 0$ in the diffusive terms). With the eddy viscosity hypothesis, the equations for $\overline{v}_\phi$ and $\overline{h}_t$ in steady, axisymmetric, boundary layer flow on a body of revolution are then approximated as follows:

$$\frac{\partial(\rho r^2\overline{v}_s\overline{v}_\phi)}{r\partial s} + \frac{\partial(\rho r^2\overline{v}_n\overline{v}_\phi)}{r\partial n} = \frac{\partial}{\partial n}\left[\mu_{eff}\left(\frac{\partial(r\overline{v}_\phi)}{\partial n}\right)\right] \tag{2.5.23}$$

$$\frac{\partial(\rho r\overline{v}_s\overline{h}_t)}{r\partial s} + \frac{\partial(\rho r\overline{v}_n\overline{h}_t)}{r\partial n} = \frac{\partial}{\partial n}\left[\frac{\mu_{eff}}{Pr_{eff}}\left(\frac{\partial\overline{h}_t}{\partial n}\right)\right] \tag{2.5.24}$$

For an effective Prandtl number of unity, the equations for $r\overline{v}_\phi$ and $\overline{h}_t$ have identical form. This form does not change when either variable is multiplied by a constant, or is subject to the addition of a constant. It follows that equations for a nondimensional circumferential velocity $G = \overline{v}_\phi/\Omega r$ and nondimensional total enthalpy $H = (\bar{h}_t - h_{ref})/cr^2$, where c is an arbitrary (dimensional) constant, are identical. The solutions for these two variables will be the same if they have the same boundary conditions.

For a body of revolution rotating at speed Ω, $G = 1$ on the body, and for H to also equal unity, it is required that $h_{t,0} - h_{ref} = C_p(T_0 - T_{ref}) + \Omega^2 r^2/2 = cr^2$, where T_0 is the surface temperature. Thus one requirement for $G = H$ is that the surface temperature has a quadratic form $T_0 - T_{ref} = (c - \Omega^2/2)r^2/C_p$. Assuming (for the moment) that other boundary conditions also correspond, and the above assumptions apply, so that $G = H$ everywhere, the surface heat flux is given by

$$q_0 = \left(\frac{\kappa_{eff}\partial \overline{T}}{\partial n}\right)_{n=0} = \left(\frac{cr}{\Omega}\frac{\mu_{eff}\partial \overline{v}_\phi}{\partial n} - \frac{\mu_{eff}\partial\left(\overline{v}_r^2 + \overline{v}_\theta^2 + \overline{v}_z^2\right)}{2\partial n}\right)_{n=0} = \left(\frac{c}{\Omega^2} - 1\right)\Omega r\,\tau_{\phi,0} \quad (2.5.25)$$

where $\tau_{\phi,0}$ is the circumferential component of surface shear stress. Thus, considering $q_0 = 0$, the adiabatic surface temperature is given by $T_{0,ad} - T_{ref} = \Omega^2 r^2/2C_p$. In terms of the fluid temperature and velocity at the edge of the boundary layer (T_δ and $v_{\phi,\delta}$) the analogy can be shown to give

$$T_{0,ad} - T_\delta = (\Omega r - v_{\phi,\delta})^2/2C_p \quad (2.5.26)$$

$$q_{s0} = \frac{C_p(T_{0,ad} - T_0)\tau_{\phi,0}}{(v_{\phi,\delta} - \Omega r)} \quad (2.5.27)$$

Noting that $v_{\phi,\delta} - \Omega r$ is a 'freestream' velocity relative to the rotating body, the correspondence between these results and those for a flat plate in zero pressure gradient is clear.

For a body rotating in a quiescent environment, the far-field boundary conditions give $G = H = 0$ at $n = \infty$ (with an appropriate choice of h_{ref}), and for the quadratic surface temperature distribution, the solutions for G and H match in the entire flow field. Further cases where boundary conditions for G and H match can be identified, but the required conditions are often restrictive. For engineering calculations, it is often more appropriate to use equations 2.5.26 and 2.5.27 as local approximations.

The effects of the Prandtl number have been considered by several authors, as reported in many textbooks. If, following practice for flat plate boundary layers, a recovery factor R_f is used to account for the effect of Pr on frictional heating, the adiabatic rotor surface temperature is given by

$$T_{0,ad} - T_\delta = R_f(\Omega r - v_{\phi,\delta})^2/2C_p. \quad (2.5.28)$$

R_f is commonly taken as $Pr^{1/2}$ for laminar flow and $Pr^{1/3}$ for turbulent flow.

Further assuming (as is also common practice) that the effects of Pr on the Nusselt number can be carried by a factor $\psi(Pr)$, equation 2.5.26 gives

$$\frac{rq_0}{\kappa(T_{0,ad} - T_0)} = \frac{\psi(Pr)}{2}\left[\frac{\rho(v_{\phi,\delta} - \Omega r)r}{\mu}\right]\left[\frac{2\tau_{\phi,0}}{\rho(v_{\phi,\delta} - \Omega r)^2}\right]. \quad (2.5.29)$$

Here the left-hand side gives a local Nusselt number, and the two terms in square brackets represent a local rotational Reynolds number and a circumferential friction factor. There is a clear correspondence with equation 2.5.21 for flat plate heat transfer. For air in laminar flow, Dorfman (1963) gives the function $\psi(Pr) \approx 0.87$ for heat transfer from a disc rotating in a quiescent environment, compared to $Pr^{1/3} = 0.89$. For turbulent flow Dorfman gives $\psi(Pr) \approx Pr^{0.6}$, as an approximation to the more elaborate expression, following von Kármán (1939), given below:

$$\psi = Pr\left[1 + 5\left(Pr - 1 + ln\left\{\frac{5Pr+1}{6}\right\}\right)\left(\frac{\tau_{\phi,0}}{\rho(v_{\phi,\delta} - \Omega r)^2}\right)^{1/2}\right]^{-1} \quad (2.5.30)$$

Modifications of the Reynolds analogy results to account for different surface temperature distributions have also been considered by various authors but are less widely used in practice. In engineering estimations, it is often assumed that equations 2.5.28 and 2.5.29 can be applied locally. The accuracy of this approach will depend on the degree of departure from the quadratic temperature distribution described above and the 'freestream' boundary conditions. The analogy may also be applied to a stationary surface in an axisymmetric rotating flow, but the stator surface temperature must be uniform for the analogy to be exact.

2.6 Rotating Waves and Fourier Analysis

Rotating flow structures or waves feature in many rotating flows and have been found to be important for some aspects of gas turbine engine performance. For example, rotating stall is a well-known and potentially very damaging phenomenon in compressors, and the interaction of rotating flow and vibration modes can lead to blade flutter. Rotating flow modes have more recently been recognised as important in some aspects of engine internal air systems. Time-resolved measurements and computer predictions able to capture these effects are increasingly used. This section gives some elementary results that are useful in considering rotating waves, and a very brief explanation of some commonly used analysis techniques.

Fourier analysis has its roots and motivation in Fourier series and is widely used in science and engineering applications where waves occur. Fourier's theorem states that any continuously smooth, periodic function can be represented as a sum of sine and cosine terms. For example, consider the pressure p as a function of the angular coordinate ϕ at a given radial and axial position on the surface of an axisymmetric body. If the flow around the body is steady in the stationary coordinate frame, then p is a periodic function of ϕ, and can be written as

$$p(\phi) = a_0 + \sum_{n=1}^{\infty}[(a_n \cos(n\phi) + b_n \sin(n\phi)] \quad (2.6.1a)$$

or, equivalently,

$$p(\phi) = a_0 + \sum_{n=1}^{\infty} c_n \sin(n\phi + \phi_n) \quad (2.6.1b)$$

a_n, b_n, c_n and ϕ_n are constants in this example but could, in general, be functions of t, r and z. Each value of n corresponds to a standing wave of wavelength $2\pi r/n$ and phase ϕ_n relative to the (arbitrary) origin $\phi = 0$.

Writing all variables in terms of the Fourier series in ϕ, substituting into the governing equations, and considering each value of n in the series, would give an infinite set of equations for the Fourier coefficients a_n and b_n. These equations are coupled through non-linear terms, most notably the advective terms, in the governing equations. However, in the limit of small perturbations, the equations for each Fourier mode are independent. This formulation is an example of harmonic analysis, which can be extended to consider waves in all three dimensions and time.

For the rotating coordinate system, the angular coordinate $\theta = \phi - \Omega t$ and equation 2.6.1b transforms to

$$p(\theta) = a_0 + \sum_{n=1}^{\infty} c_n \sin[n(\theta + \Omega t) + \phi_n] \tag{2.6.2}$$

A standing wave in the stationary system is a backwards-travelling wave in the rotating system with angular speed $-\Omega$. A sensor fixed in the rotating system will see a time-varying signal with a regular frequency $n\Omega/2\pi$ (which corresponds to angular frequency $n\Omega$).

A sinusoidal wave rotating with angular speed Ω_s in the stationary frame of reference may have the form

$$p(\phi) = a_0 + c_n \sin[n(\phi - \Omega_s t) + \phi_n] \textit{ in the stationary frame} \tag{2.6.3a}$$

$$\text{and } p(\theta) = a_0 + c_n \sin[n(\theta + \Omega t - \Omega_s t) + \phi_n] \textit{ in the rotating frame.} \tag{2.6.3b}$$

Thus, pressure measurements at points on stationary and rotating surfaces will show angular frequencies of $n\Omega_s$ and $n(\Omega - \Omega_s)$, respectively. If sensors are available in both rotating and stationary frames, the measured frequencies can be used to deduce Ω_s and n.

Suppose there are two rotating waves with circumferential wave numbers of n and m, and angular speeds and Ω_{s1} and Ω_{s2}. These may interact through the non-linear advective terms in the governing equations. Products of the form $\sin[n(\phi - \Omega_{s1}t) + \phi_n]\sin[m(\phi - \Omega_{s2}t) + \phi_m]$ will appear in these terms. Using standard trigonometric relationships these products transform to terms of the type $\sin[(n-m)\phi - (n\Omega_{s1} - m\Omega_{s2})t + \phi_{n-m}]$ and $\sin[(n+m)\phi - (n\Omega_{s1} + m\Omega_{s2})t + \phi_{n+m}]$. These correspond to circumferential wave numbers $n \pm m$, rotating with angular velocities $(n\Omega_{s1} \pm m\Omega_{s2})/(n+m)$. Thus measurements at points in the stationary frame may show frequencies at $n\Omega_{s1}$, $m\Omega_{s2}$, $n\Omega_{s1} + m\Omega_{s2}$ and $n\Omega_{s1} - m\Omega_{s2}$ depending on the strength and nature of the interaction.

Fourier analysis is widely used to examine the frequency content of time-varying data from experimental and computational results, and hence to give insight into the flow mechanisms and structures. For a pressure recording $p(t)$, the Fourier transform is defined as

$$\overline{p}(\omega) = \int_{-\infty}^{+\infty} p(t)e^{-i\omega t}\, dt \tag{2.6.4}$$

where ω is angular frequency and $i^2 = -1$. The complex variable $\overline{p}$ can be written in terms of its modulus $|\overline{p}|$ and phase ψ, taking the form $|\overline{p}(\omega)|e^{i\psi(\omega)}$. These relate to the amplitude and phase of waves at frequency ω. Plotting the modulus against frequency shows the dominant frequencies in the measurements. Examples are shown later in this book and can be found in many textbooks.

If measurements are available at different circumferential positions, these may be analysed to extract the time lag or phase difference between the signals using the cross-correlation function, and hence the circumferential wave number and speed of rotation of the wave can be deduced. Consider the example of a rotating flow structure in Figure 2.4. Probes 1 and 2 in the stationary frame are spaced at an angle α apart and will show an angular frequency of $N\Omega_s$ generated by a flow pattern with N lobes (or N nodal diameters) rotating at an angular speed Ω_s. The angle between each lobe (or repeating flow structure) is $\beta = 2\pi/N$. The probes will both record the same frequency with a phase lag $\Delta t_\alpha = \alpha/\Omega_s$, corresponding to the time it takes for a flow structure to travel between them. Measurements of the frequency and phase lag allow the number of lobes N and angular velocity Ω_s of the unsteady flow structure to be determined. Generally, the angle α between the two

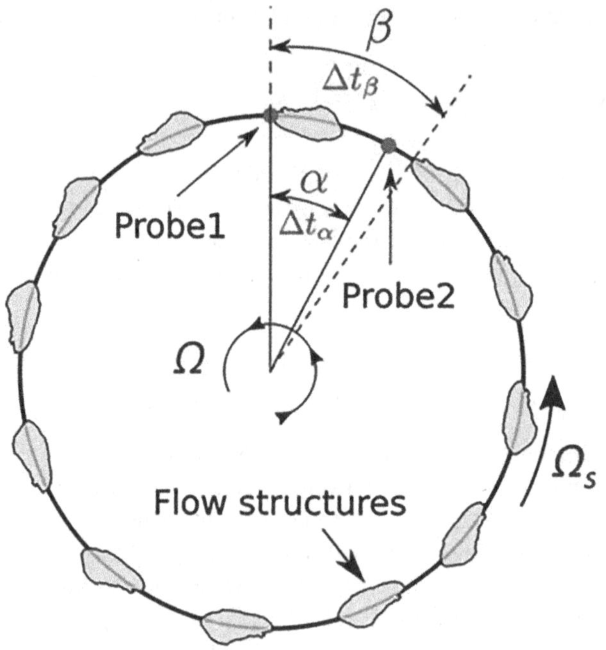

Figure 2.4 Illustration of a rotating flow structure. *Source:* Beard *et al.* (2017) / with permission of American Society of Mechanical Engineers (ASME).

sensors is not always smaller than the angle β between two pairs of perturbation cells (as shown in the figure). The determination of the phase difference is further complicated since the two sensors will show similar readings (and cross-correlation peaks) at all time intervals $n\alpha/\Omega_s$ where n is any integer, and flow features can be intermittent or modulating. Beard *et al.* (2017) give further details of procedures that can be used to analyse the data with application to experimental data.

Fourier analysis is also commonly used to analyse broadband fluctuations, with flow turbulence being a notable example. The power spectral density (PSD) of a signal $p(t)$, defined and discussed in many textbooks, is readily obtained in many software packages. The name comes from applications to measurement of voltage across a 1 ohm resistance, where PSD has units of power per frequency. Here the PSD of pressure has SI units of $\mathrm{Pa^2{\cdot}Hz^{-1}}$, and the PSD of velocity has units $\mathrm{m^2{\cdot}s^{-2}{\cdot}Hz^{-1}}$ or $\mathrm{J{\cdot}kg^{-1}{\cdot}Hz^{-1}}$. The PSD plotted against frequency shows the signal content for different ranges of frequency. For example, the PSD of velocity integrated between two values of frequency gives the specific kinetic energy in that frequency band.

2.7 Concluding Remark

This chapter has given a brief account of the theory underpinning the analysis of rotating flows. The material has much in common with more general introductions to fluid dynamics but focuses on aspects that have particular relevance to rotating flow and may not be explored in more general treatments. While the concepts of angular momentum, Coriolis and centrifugal force are relatively well known, rothalpy and rotary stagnation properties are less widely understood and so have been discussed in more detail. Dimensional analysis has been considered and some commonly used nondimensional parameters have been introduced. Some aspects of heat transfer particular to rotating flows have also been described, and methods of analysis of rotating flow structures have been introduced.

The equations and results given in this chapter will be used in the following chapters, which aim to give further insight into rotating flows. The next chapter considers analytical solutions of the governing equations for certain inviscid and laminar flows. Subsequent chapters look at more complex models and flow situations.

References

Batchelor, G. K., 1967. An introduction to fluid dynamics, Cambridge University Press.

Beard, P. F., Gao, F., Chana, K. S., and Chew, J. W., 2017. Unsteady flow phenomena in turbine rim seals. ASME J. Eng. Gas Turbines Power, 139(3), 032501. https://doi.org/10.1115/1.4034452.

Bradshaw, P. and Koh, Y. M., 1981. A note on Poisson's equation for pressure in a turbulent flow. Phys. Fluids, 24, 777. https://doi.org/10.1063/1.863442.

Chew, J. W., 1985. Effect of frictional heating and compressive work in rotating axisymmetric flow. ASME J. Heat Transfer, 107, 984–986. https://doi.org/10.1115/1.3247537.

Dorfman, L. A., 1963. Hydrodynamic resistance and the heat loss of rotating solids, First English edition, London: Oliver and Boyd.

Gray, D. D. and Giorgini, A., 1976. The validity of the Boussinesq approximation for liquids and gases. Int. J. Heat Mass Transfer, 19, 545–551. https://doi.org/10.1016/0017-9310(76)90168-X.

Hunt, J. C. R., Wray, A. A., and Moin, P., 1988. Eddies, streams, and convergence zones in turbulent flows. Center for Turbulence Research, Summer Program, 193–208. https://web.stanford.edu/group/ctr/Summer/201306111537.pdf.

Owen, J. M., 1971. The Reynolds analogy applied to the flow between a rotating and a stationary disc. Int. J. Heat Mass Transfer, 14, 451–460. https://doi.org/10.1016/0017-9310(71)90163-3.

Rogers, G. F. C. and Mayhew, Y. R., 1980. Engineering thermodynamics, work and heat transfer, Third edition, Longman Group limited.

Schlichting, H., 1979. Boundary-layer theory, Seventh edition, McGraw-Hill.

Spiegel, E. A. and Veronis, G., 1960. On the Boussinesq approximation for a compressible fluid. Astrophys. J., 131, 442–447. https://adsabs.harvard.edu/pdf/1960ApJ...131..442S.

von Kármán, T., 1939. The analogy between fluid friction and heat transfer. Trans. ASME, 61, 705–710. https://doi.org/10.1115/1.4021298.

Wilcox, D. C., 2006. Turbulence modelling for CFD, Third edition, DCW Industries Inc.

3

Analytical Solutions for Inviscid and Laminar Flow

This chapter gives some key results from analysis of the Navier–Stokes equations and their inviscid and boundary layer approximations. These are restricted to laminar and inviscid flows with turbulent flow being considered in the following chapters. Most of the solutions given were derived in the early and mid-1900s and are chosen for their simplicity and relevance to examples considered later. While a very considerable body of published work has built on the early analysis, no attempt is made to review the many interesting further developments.

3.1 Exact Solutions of the Navier–Stokes Equations

3.1.1 One-Dimensional Solutions (Free, Forced and Mixed Vortices)

It may be shown by substitution that the mass and momentum equations 2.1.36–2.1.39 for uniform density and viscosity flow are exactly satisfied by the following two one-dimensional solutions

$$v_r = \frac{A}{r}, \quad v_\phi = \frac{B}{r}, \quad v_z = 0, \quad p = -\frac{\rho}{2r^2}(A^2 + B^2) + C \tag{3.1.1}$$

$$v_r = 0, \quad v_\phi = Dr + \frac{B}{r}, \quad v_z = 0, \quad p = \rho\left(\frac{D^2r^2}{2} + 2DBln(r) - \frac{B^2}{2r^2}\right) + C \tag{3.1.2}$$

where A, B, C and D are constants.

Solution 3.1.1 describes a free vortex with radial flow. The pressure solution indicates a severe drop in pressure at low radii, with a singularity at the axis $r = 0$. This characteristic is exploited in flow control devices, such as vortex amplifiers and valves, where radially inwards flow with swirl is used to control flow rates, and is observed in natural phenomena such as hurricanes, cyclones and typhoons.

Another feature of the free vortex is that the radial pressure gradient is independent of the sign of the radial velocity. Hence, for a given value of rv_ϕ and pressure increase between $r = r_1$ and $r = r_2$, there may be two possible free vortex solutions. These correspond to (accelerating) radial inflow and (diffusing) outflow, with outflow occurring against a positive (adverse) pressure gradient. In practice, a change of flow direction is likely to change the vortex strength rv_ϕ. Situations where there are non-unique solutions can occur in engine internal air systems. In such circumstances, the flow will depend on other elements of the system and how the operating condition is approached. An example of a vortex system showing multiple solutions and hysteresis is given later in Section 5.3.3.

Solution 3.1.2 describes a mixed vortex. $B = 0$ gives a forced vortex and $D = 0$ gives a free or irrotational vortex. The forced vortex can also be described as solid body rotation. In 3.1.2, the radial pressure gradient simply balances the radial centrifugal force. Applying this solution to flow between cylinders at $r = r_1$ and $r = r_2$, rotating at speeds Ω_1 and Ω_2 gives

$$B = \frac{r_1^2 r_2^2(\Omega_1 - \Omega_2)}{\left(r_2^2 - r_1^2\right)}, \quad D = \frac{\left(\Omega_2 r_2^2 - \Omega_1 r_1^2\right)}{\left(r_2^2 - r_1^2\right)}. \tag{3.1.3}$$

Substituting for B and D in equation 3.1.2 gives the laminar Couette flow solution. Further discussions of laminar Couette flow are given by Dorfman (1963), including consideration of heat transfer, variable fluid properties and stability. Stability of flow between rotating cylinders has been subject to considerable research. Rayleigh's stability criterion for inviscid flow draws an analogy with stability of free convection under gravity, where a positive vertical temperature gradient (producing a negative density gradient) is unstable. Rayleigh's condition for stability of the mixed vortex 3.1.2 is

$$\frac{d(rv_\phi)^2}{dr} > 0. \tag{3.1.4}$$

This is derived by considering a small displacement of an element of fluid from its equilibrium condition, with conservation of angular momentum. For stability, it is required that the centrifugal force of the displaced element with a positive radial displacement is less than that of the surrounding fluid at that position.

Applied to Couette flow between rotating cylinders the stability condition gives

$$v_\phi\left(\Omega_2 r_2^2 - \Omega_1 r_1^2\right) > 0. \tag{3.1.5}$$

For counter-rotating cylinders, this condition shows that there will always be some region of the flow that is unstable (since v_ϕ will change sign at some point between the cylinders). For corotating cylinders, the condition reduces to $\Omega_2 r_2^2 > \Omega_1 r_1^2$. At low Reynolds numbers, instability is delayed by viscous effects, but Rayleigh's criterion and the viscous flow stability limit converge at higher Reynolds numbers. This was demonstrated by Taylor's (1923) analysis and experiments. As the flow becomes unstable, Taylor vortices form in the meridional plane. Examples of Taylor's calculated streamlines and flow visualisation for axisymmetric vortices are given in Figure 3.1 for conditions with $\Omega_1 > \Omega_2 > 0$.

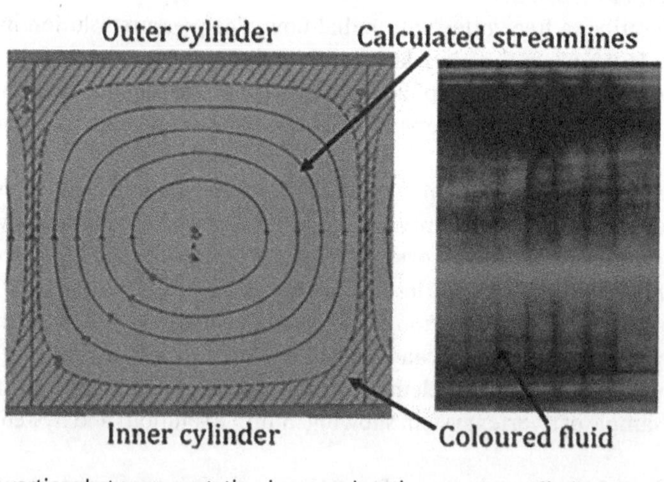

Figure 3.1 Taylor vortices between a rotating inner and stationary outer cylinders, showing streamlines in an azimuthal ($r - z$) plane and a photograph with dye injection seen through a transparent outer cylinder confirming the axisymmetric flow structure. *Source:* Taylor (1923) / with permission of The Royal Society.

With sufficiently high inner cylinder rotation, a wide variety of flow types including wavy vortices, spiralling vortices, turbulent Taylor vortices and featureless turbulence may occur between rotating cylinders. This is illustrated by Andereck *et al.*'s (1986) experimental study.

3.1.2 Axisymmetric Laminar Viscous Flow

Considering an infinite disc (at $z = 0$), some remarkable solutions can be obtained. Following von Kármán (1921), with allowance for a forced vortex far field, the velocity and pressure field are assumed to have the form

$$v_r = F(\xi)r\Omega, \quad v_\phi = G(\xi)r\Omega, \quad v_z = H(\xi)\sqrt{\nu\Omega}, \quad p = -P(\xi)\rho\nu\Omega + \frac{A\rho r^2\Omega^2}{2} \tag{3.1.6}$$

where

$$\xi = z\sqrt{\Omega/\nu},$$

A is constant, and $\nu = \mu/\rho$ is the kinematic viscosity. Substituting these relations into equations 2.1.36–2.1.39 gives the following set of ordinary differential equations:

$$2F + \frac{dH}{d\xi} = 0 \tag{3.1.7}$$

$$F^2 - G^2 + H\frac{dF}{d\xi} = \frac{d^2F}{d\xi^2} - A \tag{3.1.8}$$

$$2FG + H\frac{dG}{d\xi} = \frac{d^2G}{d\xi^2} \tag{3.1.9}$$

$$H\frac{dH}{d\xi} = \frac{dP}{d\xi} + \frac{d^2H}{d\xi^2} \tag{3.1.10}$$

Thus, the assumed form of the solution allows elimination of r from the governing equations and reveals a class of exact solutions of the Navier–Stokes equations.

Lingwood (1997) labelled this class of flow BEK after Bödewadt, Ekman and von Kármán and illustrated numerical solutions of equations 3.1.7–3.1.10 for a range of ratios of the far field-to-disc angular velocities Ω_∞/Ω_d. These show a gradual transition between three limiting cases that are considered below. For $\Omega_\infty/\Omega_d = 0$, the solution corresponds to von Kármán's free disc case described in Section 3.1.2.1. As $\Omega_\infty/\Omega_d \to 1$, the flow can be treated as a small perturbation from solid body rotation and the solution corresponds to the Ekman layer flow described in Section 3.3.2. For $\Omega_\infty/\Omega_d = \infty$, the solution corresponds to Bödewadt's case of flow rotating above a stationary disc described in Section 3.1.2.2.

3.1.2.1 The Free Rotating Disc

Von Kármán considered a free rotating disc, for which the disc rotates in a quiescent fluid. The boundary conditions for this case are $F(0) = H(0) = 0$, $G(0) = 1$, $F(\infty) = G(\infty) = 0$, and the constant $A = 0$. F, G and H may be found by numerical integration, or in terms of an expansion, of equations 3.1.7–3.1.9, and then 3.1.10 can be used to find the pressure distribution P. Tabulated values of F, G and H and further description of solution methods are given by Cochran (1934), Dorfman (1963), Owen and Rogers (1989) and Childs (2011). The solution is presented graphically in Figure 3.2, along with other illustrations visualising the spiralling, axisymmetric flow near the disc and showing the skewed nature of the boundary layer.

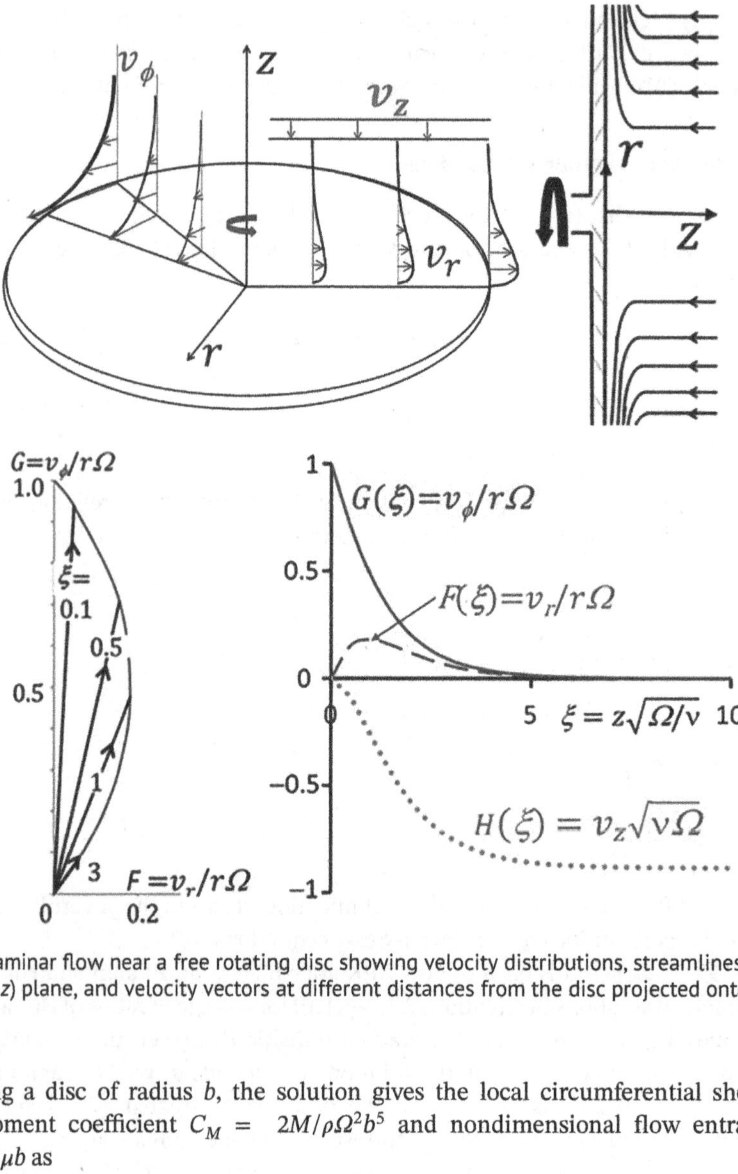

Figure 3.2 Laminar flow near a free rotating disc showing velocity distributions, streamlines in an azimuthal ($r-z$) plane, and velocity vectors at different distances from the disc projected onto the disc surface.

Considering a disc of radius b, the solution gives the local circumferential shear stress on the disc, moment coefficient $C_M = 2M/\rho\Omega^2 b^5$ and nondimensional flow entrainment rate $C_{W,fd} = \dot{m}_{fd}/\mu b$ as

$$\frac{\tau_{\phi,0}}{\rho\Omega^2 r^2} = -0.616\left(\frac{\rho\Omega r^2}{\mu}\right)^{-0.5}, \quad C_{M,fd} = 1.935Re_\phi^{-0.5}, \quad C_{W,fd} = 2.78Re_\phi^{0.5} \tag{3.1.11}$$

Note that here M and $\dot{m}_{fd}$ are for one side of the disc only. Applying the Reynolds analogy described in Section 2.5.3, gives the following result for heat transfer to a free disc with a surface temperature distribution $T_0 = T_\infty + cr^2$ and fluid with $Pr = 1$:

$$\frac{rq_0}{\kappa(T_{0,ad} - T_0)} = 0.616\left(\frac{\rho\Omega r^2}{\mu}\right)^{0.5} \tag{3.1.12}$$

The local heat transfer may be integrated over the disc to give the mean Nusselt number

$$Nu_{av} = \frac{bq_{0,av}}{\kappa(T_{0,ad,av} - T_{0,av})} = 0.616\left(\frac{\rho\Omega b^2}{\mu}\right)^{0.5} = 0.616Re_\phi^{0.5} \tag{3.1.13}$$

where subscript av denotes an area weighted average. As described by Dorfman, solutions to the energy equation for other disc temperature distributions can be derived (using the von Kármán flow field).

3.1.2.2 Rotating Flow Above a Stationary Disc

Bödewadt (1940) considered the flow due to rotation of the fluid (with angular velocity Ω) above a stationary infinite disc. The boundary conditions for this case are $F(0) = H(0) = G(0) = 0$, $F(\infty) = 0$, $G(\infty) = 1$ and the constant $A = 1$. Numerical solutions have been obtained by Nydahl as recorded by Schlichting (1979), Owen and Rogers (1989) and other authors. As illustrated in Figure 3.3, these show a thicker boundary layer than for the free rotating disc, with predominantly radial inflow. In addition, the velocity profiles show a damped oscillation with a distinct region of radial outflow from $\xi \approx 3$ to $\xi \approx 6.5$. As for the rotating disc, there is a uniform axial flow at large distances from the disc. This balances the radial inflow in the boundary layer. Considering a disc of radius b, the solution gives the local circumferential shear stress on the disc, moment coefficient and non-dimensional flow entrainment rate as

$$\frac{\tau_{\phi,0}}{\rho\Omega^2 r^2} = 0.773\left(\frac{\rho\Omega r^2}{\mu}\right)^{-0.5}, \quad C_{M,Böd} = -2.43Re_\phi^{-0.5}, \quad C_{W,Böd} = -4.24Re_\phi^{0.5} \tag{3.1.14}$$

The Reynolds analogy can be applied, but similarity of boundary conditions requires a quadratic radial temperature distribution in the main body of fluid.

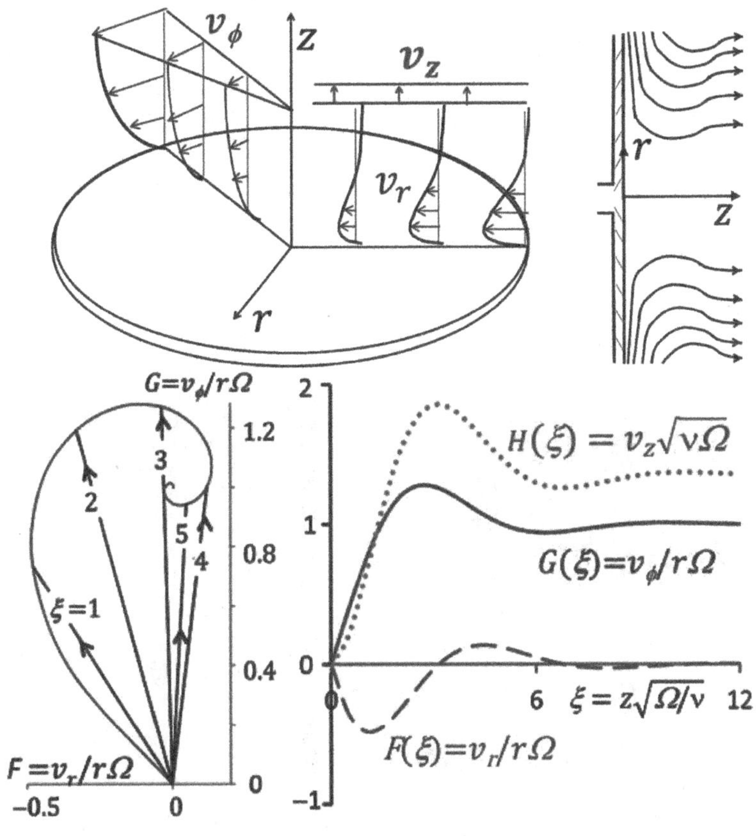

Figure 3.3 Laminar rotating flow near a stationary disc showing velocity distributions, streamlines in an azimuthal ($r-z$) plane, and velocity vectors at different distances from the disc projected onto the disc surface.

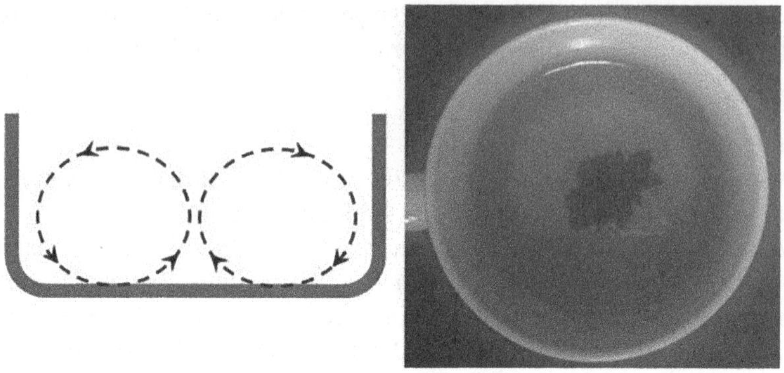

Figure 3.4 Teacup experiment showing flow circulation in the azimuthal plane and tea leaves settling at the centre as the rotation subsides. *Source:* Einstein (1954) / with permission of Springer Nature.

As discussed by Bowker (1988) and others, Einstein (1954) described a simple experiment providing evidence of Bödewadt flow. This involves the settling of tea leaves in a cup following vigorous stirring to create a swirling flow. Radial inflow in the boundary layer on the bottom of the cup creates a recirculation in the azimuthal plane. The tea leaves, which are heavier than the water, sink to the bottom where the boundary layer flow carries them to the centre bottom position. This is shown in the photograph in Figure 3.4 alongside Einstein's sketch of the flow circulation.

3.1.2.3 Flow Between Infinite Discs

Equations 3.1.7–3.1.10 can be solved numerically for arbitrary values of disc and far field angular velocities $G(0)$ and $G(\infty)$. Considering high Reynolds number flow between infinite discs (where

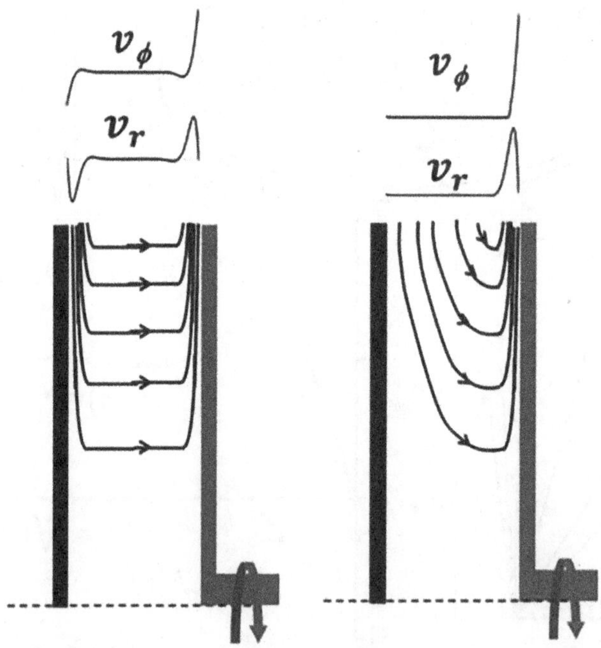

Figure 3.5 Streamlines and velocity profiles for Batchelor type (left) and Stewartson type (right) high Reynolds number flow between a rotating and a stationary disc.

Reynolds number is defined as $\rho\Omega s^2/\mu$, and s is the spacing between the discs), solutions can be obtained by combining solutions for the two discs with the same far field (or core) velocities. The rotation speeds of the two discs may be varied, and the core velocity for which the radial inflow and outflow in the two boundary layer balance gives a solution for the case of no net radial flow.

Streamlines for the flow between infinite stationary and rotating discs with no net radial flow are illustrated in Figure 3.5. As discussed by Batchelor (1951), there is a solution in which angular velocity in the core is constant and thin boundary layers form on the discs. However, as discussed by Stewartson (1953), an alternative solution is possible. This is illustrated in the right-hand figure. In this case, a weak radial inflow maintains a nonrotating flow outside the rotor boundary layer, and the flow near the rotating disc is approximated by von Kármán's solution.

Both types of flow shown in Figure 3.5 can occur in finite disc systems, depending on boundary conditions. The flow on the left is associated with closed rotor-stator disc cavities and is known as Batchelor type flow. The flow on the right is associated with open or ventilated disc cavities and is known as Stewartson type flow. The analysis can also be extended to consider cases when both discs rotate, as has been discussed by Batchelor and Stewartson. In practice, boundary conditions at the inner and outer radii must be considered and (for differentially rotating discs) the flow will become turbulent at high enough radii. When a net radial flow is imposed, the angular velocity in the core will vary with radius and the assumed form of the solution in equation 3.1.6 no longer applies.

3.2 Axisymmetric Laminar Boundary Layer Flow

With boundary layer approximations the family of exact solutions of the Navier–Stokes equations discussed in Section 3.1.2 can be extended to axisymmetric flow on more general bodies of revolution. This extension is described here. Further boundary layer solutions, known as Ekman layers, that additionally assume the flow to be a small perturbation from solid body rotation are given in Section 3.3.2.

Here, we use the stationary coordinate system (s, ϕ, n) where s and n are tangential and normal to the surface of the body in planes of constant angular coordinate ϕ. These coordinates are illustrated in Figure 3.6. It is assumed that the boundary layer thickness is small compared to the disc radius

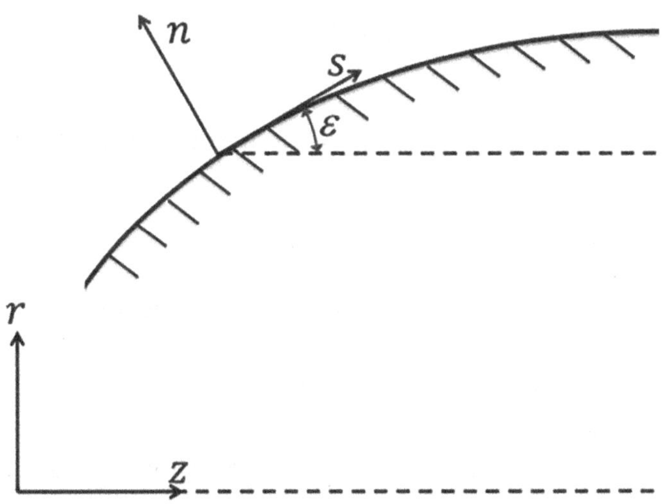

Figure 3.6 Boundary layer coordinates for an axisymmetric body.

and the surface slant angle $\varepsilon = \sin^{-1}(dr/ds)$ changes slowly with s (if at all). For constant values of slant angle, the body is conical and ε is the cone half angle.

With boundary layer assumptions, equations 2.1.36–2.1.39 for steady, axisymmetric, uniform density and viscosity flow give the following equations in the (s, n) coordinate system:

$$\frac{\partial(rv_s)}{\partial s} + \frac{\partial(rv_n)}{\partial n} = 0 \tag{3.2.1}$$

$$\rho\left\{v_s\frac{\partial v_s}{\partial s} + v_n\frac{\partial v_s}{\partial n} - \sin(\varepsilon)\frac{v_\phi^2}{r}\right\} = -\frac{\partial p}{\partial s} + \mu\frac{\partial^2 v_s}{\partial n^2} \tag{3.2.2}$$

$$\rho\left\{v_s\frac{\partial v_\phi}{\partial s} + v_n\frac{\partial v_\phi}{\partial n} + \sin(\varepsilon)\frac{v_s v_\phi}{r}\right\} = \mu\frac{\partial^2 v_\phi}{\partial n^2} \tag{3.2.3}$$

The assumed form of the solution for plane disc flow, given above as equation 3.1.6, is modified for the present case as follows:

$$v_s = F(\xi)r\Omega, \quad v_\phi = G(\xi)r\Omega, \quad v_n = H(\xi)\sqrt{\nu\Omega\sin(\varepsilon)}, \quad p = A\rho r^2\Omega^2/2 \tag{3.2.4}$$

where

$$\xi = n\sqrt{\Omega\sin(\varepsilon)/\nu}.$$

Substituting into equations 3.2.1–3.2.3 then gives the following ordinary differential equations, which are identical to those for the exact solution for the plane disc. (This correspondence of the cone and disc equations was noted by Wu, 1959.)

$$2F + \frac{dH}{d\xi} = 0 \tag{3.2.5}$$

$$F^2 - G^2 + H\frac{dF}{d\xi} = \frac{d^2F}{d\xi^2} - A \tag{3.2.6}$$

$$2FG + H\frac{dG}{d\xi} = \frac{d^2G}{d\xi^2} \tag{3.2.7}$$

With the assumed form of the solution, the circumferential shear stress on the body is given by

$$\frac{\tau_{\phi,s}}{\rho\Omega^2 r^2} = \left(\frac{\mu}{\rho\Omega r^2}\right)\left[\frac{\partial v_\phi}{\partial n}\right]_0 = \left(\frac{\nu}{\Omega r^2}\right)r\Omega\sqrt{\left(\frac{\Omega\sin\varepsilon}{\nu}\right)}\left[\frac{\partial G}{\partial \xi}\right]_0 = \sqrt{\left(\frac{\nu\sin\varepsilon}{\Omega r^2}\right)}\left[\frac{\partial G}{\partial \xi}\right]_0. \tag{3.2.8}$$

For a cone radius b, with $\varepsilon = constant$, the moment coefficient is obtained by integration over the surface to give

$$C_{M,cone} = \frac{2M}{\rho\Omega^2 b^5} = \left(\frac{\pi}{Re_\phi^{\,0.5}\sin^{0.5}\varepsilon}\right)\left[\frac{\partial G}{\partial \xi}\right]_0. \tag{3.2.9}$$

The flow entrained in the boundary layer at $r = b$ is obtained by integration across the layer, giving

$$C_{W,cone} = \frac{\dot{m}}{\mu b} = 2\pi\left(\frac{Re_\phi}{\sin\varepsilon}\right)^{0.5}\int_0^{\xi_\delta} F d\xi \tag{3.2.10}$$

where ξ_δ marks the boundary layer edge.

The results for a plane disc given in Section 3.1.2 are extended to cones with half angle ε as follows.

Free rotating cone in a quiescent environment

$$\frac{\tau_{\phi,0}}{\rho\Omega^2 r^2} = -0.616\left(\frac{\rho\Omega r^2}{\mu \sin\varepsilon}\right)^{-0.5}, \quad C_{M\,fcone} = 1.935(\sin\varepsilon\, Re_\phi)^{-0,5}, \quad C_{W\,fcone} = 2.78(\sin\varepsilon\, Re_\phi)^{0,5} \tag{3.2.11}$$

$$\frac{rq_0}{\kappa(T_{0,ad} - T_0)} = 0.616\left(\frac{\rho\Omega r^2 \sin\varepsilon}{\mu}\right)^{0.5}, \quad Nu_{av} = \frac{bq_{0,av}}{\kappa(T_{0,ad,av} - T_{0,av})} = 0.616(\sin\varepsilon\, Re_\phi)^{0,5}$$

The heat transfer results are obtained from the Reynolds analogy for a quadratic surface temperature distribution and $Pr = 1$. As described in Section 2.5.3, a correction may be applied for $Pr \neq 1$. Dorfman's (1963) free disc solutions for other surface temperature distributions may also be extended to the free cone.

Forced vortex flow above a stationary cone

$$\frac{\tau_{\phi,0}}{\rho\Omega^2 r^2} = 0.773\left(\frac{\rho\Omega r^2}{\mu \sin\varepsilon}\right)^{-0.5}, \quad C_{M,Bö,cone} = -2.43(\sin\varepsilon\, Re_\phi)^{-0,5}, \quad C_{W,Bö,cone} = -4.24(\sin\varepsilon\, Re_\phi)^{0,5} \tag{3.2.12}$$

Tien and Campbell (1963) demonstrated the correspondence of free disc and free cone flow experimentally through mass transfer experiments on a disc and for cones of half angles 75°, 60°, 45° and 30°. Through analogy between heat and mass transfer the Sherwood number characterising mass transfer was expected to be proportional to $(\sin\varepsilon\, Re_\phi)^{0,5}$. This was confirmed for $\sin\varepsilon\, Re_\phi$ from $\sim 4\times10^4$ to $\sim 2\times10^5$ for cone half angles of 75°, 60° and 45°. Results for the 30° half angle cone clearly departed from the other results at $\sin\varepsilon\, Re_\phi \approx 10^5$, possibly showing earlier transition to turbulence. The smaller cone angle also showed some differences from other mass transfer results for $Re_\phi < 10^5$, and differences in flow effects from the other cone angles were reported. Kreith *et al.*'s (1963) velocity measurements showed agreement with the laminar boundary layer solution for cone half angles ≥26.5°, with their experiments also showing earlier transition at low cone angles. Transition is discussed further in Chapter 4.

As for the disc flows, the von Kármán and Bödewadt solution methods can be adapted to find solutions for flow between infinite cones. With an inner rotating cone and outer stationary cone. Yamada and Ito (1975) deduced from measurements that for $\varepsilon \geq 60°$ there is correspondence between disc and cone flows, but for $\varepsilon \leq 45°$ differences may occur with Taylor type vortices being observed. The effects of cone angle on moment were most noticeable for larger cone spacings and lower rotational Reynolds numbers.

3.3 Further Approximate Solutions

Many workers have used linear approximations to the governing equations to study rotating flows, and this approach reveals some important results. The linear equations are obtained by assuming that the flow is a small perturbation from a known solution and neglecting terms that are second order or higher in small quantities. Taking the unperturbed fluid state to be solid body rotation, the linear equations for laminar flow with negligible variation in density and viscosity and $U/\Omega b \ll 1$ may be obtained from equations 2.2.1–2.2.5. Using the reduced pressure $p_{red} = p - \rho\Omega^2 r^2/2$, the equations may be written as follows:

$$\frac{\partial(ru_r)}{r\partial r} + \frac{\partial u_\theta}{r\partial\theta} + \frac{\partial u_z}{\partial z} = 0 \tag{3.3.1}$$

$$\frac{\partial u_r}{\partial t} - 2\Omega u_\theta = -\frac{\partial p_{red}}{\rho\,\partial r} + \nu\left[\frac{\partial}{r\partial r}\left(\frac{r\partial u_r}{\partial r}\right) + \frac{\partial^2 u_r}{r^2\partial\phi^2} + \frac{\partial^2 u_r}{\partial z^2} - \frac{2\partial u_\theta}{r^2\partial\theta} - \frac{u_r}{r^2}\right] \tag{3.3.2}$$

$$\frac{\partial u_\theta}{\partial t} + 2\Omega u_r = -\frac{\partial p_{red}}{\rho r\,\partial\theta} + \nu\left[\frac{\partial}{r\partial r}\left(\frac{r\partial u_\theta}{\partial r}\right) + \frac{\partial^2 u_\theta}{r^2\partial\theta^2} + \frac{\partial^2 u_\theta}{\partial z^2} + \frac{2\partial u_r}{r^2\partial\theta} - \frac{u_\theta}{r^2}\right] \tag{3.3.3}$$

$$\frac{\partial u_z}{\partial t} = -\frac{\partial p_{red}}{\rho\,\partial z} + \nu\left[\frac{\partial}{r\partial r}\left(\frac{r\partial u_z}{\partial r}\right) + \frac{\partial^2 u_z}{r^2\partial\theta^2} + \frac{\partial^2 u_z}{\partial z^2}\right] \tag{3.3.4}$$

3.3.1 Steady Inviscid Flow

For steady, inviscid conditions equations 3.3.2–3.3.4 reduce to

$$-2\Omega u_\theta = -\frac{\partial p_{red}}{\rho\,\partial r},\quad 2\Omega u_r = -\frac{\partial p_{red}}{\rho r\,\partial\theta},\quad 0 = -\frac{\partial p_{red}}{\rho\,\partial z} \tag{3.3.5}$$

Flow satisfying these equations, with Coriolis forces balancing the pressure gradients, is known as geostrophic. Multiplying these three equations by u_r, u_θ and u_z, respectively, and then summing gives

$$u_r\frac{\partial p_{red}}{\partial r} + u_\theta\frac{\partial p_{red}}{r\partial\theta} + u_z\frac{\partial p_{red}}{\partial z} = 0. \tag{3.3.6}$$

The left-hand side of this equation is the gradient of reduced pressure in the direction of the flow. It follows that streamlines are aligned with isobars in geostrophic flow. (This result also follows from equation 2.2.10b with $U/\Omega b \ll 1$.)

Taking the derivative of the first two equations of 3.3.5 with respect to z and using the third equation gives $\partial u_\theta/\partial z = \partial u_r/\partial z = 0$. Also, substituting for u_r and u_θ in the continuity equation 3.3.1 from 3.3.5 gives $\partial u_z/\partial z = 0$. Thus the velocities and pressure do not vary in the direction parallel to the axis of rotation. This is known as the Taylor–Proudman theorem. The effect can be demonstrated experimentally using a cylindrical bowl of water containing an obstacle on a rotating turntable. For example, the bowl is rotated steadily (so that a forced vortex develops), dye is injected at the free surface, and then the rotation rate is reduced slightly (so that the time derivatives in the momentum equations remain small). In accordance with the Taylor–Proudman theorem, the flow in the column above the obstacle remains fixed relative to the obstacle, and dye may be observed to flow around this 'Taylor column'. A schematic of the flow is shown in Figure 3.7.

For steady, axisymmetric flow (with $\partial/\partial\phi \cong 0$), the inviscid angular momentum equation in a stationary coordinate system is satisfied if either $rv_\phi = constant$ or $v_r = \partial v_\phi/\partial z = 0$. The first of these solutions is the free vortex discussed earlier. The second solution describes an arbitrary one dimensional vortex which is invariant in the axial direction (as suggested by the Taylor–Proudman theorem). With use of the inviscid radial momentum and continuity equations this gives

$$v_r = 0,\quad v_\phi = v_\phi(r),\quad v_z = v_z(r),\quad \frac{\partial p}{\partial r} = \frac{\rho v_\phi^2}{r}. \tag{3.3.7}$$

This inviscid, 1D, vortex solution does not depend on the assumption of small perturbations and is of practically significance in rotationally dominated flows. It will be further illustrated and used later. However, it should be noted that some radial distributions of v_ϕ will be subject to instability, and that viscous effects may become significant if the vortex extents over long enough axial distances.

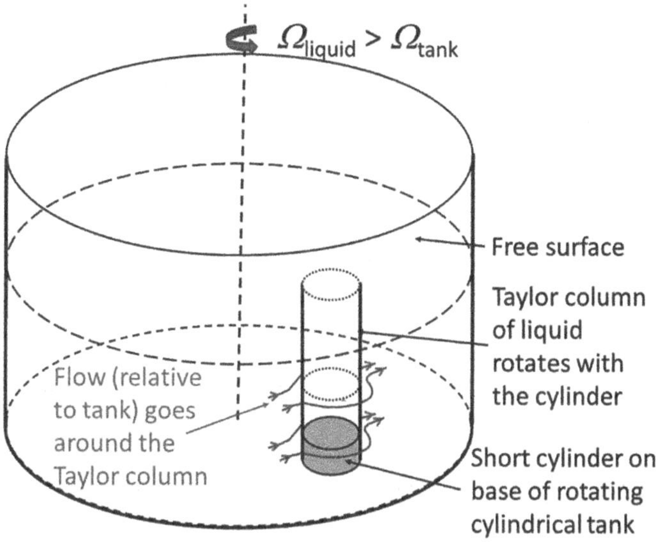

Figure 3.7 Illustration of the Taylor–Proudman theorem (viewed in the rotating frame of reference).

To account for buoyancy using the Boussinesq approximation (as in equation 2.5.5), the first of equation 3.3.5 becomes

$$-2\Omega u_\theta + \beta(T - T_{ref})\Omega^2 r = -\frac{\partial p_{red}}{\rho\,\partial r}. \tag{3.3.8}$$

From equation 2.5.3, the energy equation for steady, axisymmetric flow neglecting heat conduction in the radial and circumferential directions and taking $u_r = 0$ as derived above is

$$\rho u_z C_p \frac{\partial T}{\partial z} = \frac{\partial}{\partial z}\left(\frac{\kappa \partial T}{\partial z}\right). \tag{3.3.9}$$

Differentiating 3.3.8 with respect to z and using the last of equation 3.3.5 gives an equation for $\partial u_\theta/\partial z$. Integrating this and equation 3.3.9 between $z = 0$ and $z = s$ then gives

$$\frac{u_\theta}{\Omega r} = \frac{u_{\theta,z=0}(r)}{\Omega r} - \beta(T - T_{z=0})/2 \quad \text{and} \quad \frac{T - T_{z=0}}{T_{z=s} - T_{z=0}} = \frac{e^{\rho C_p u_z z/\kappa} - 1}{e^{\rho C_p u_z s/\kappa} - 1}. \tag{3.3.10}$$

where u_z is a function of r only. This solution is used in Section 3.3.3 in considering convective flow between infinite rotating discs at different temperatures.

3.3.2 Laminar Ekman Boundary Layers

Assuming boundary layer flow on an axisymmetric body and adopting an (s, n) coordinate system in planes of constant θ (as in Figure 3.6), the following equations may be obtained from 3.3.1–3.3.4:

$$\frac{\partial(ru_s)}{\partial s} + \frac{\partial(ru_n)}{\partial n} + \frac{\partial u_\theta}{\partial \theta} = 0 \tag{3.3.11}$$

$$-2\Omega u_\theta \sin\varepsilon = -\frac{\partial p_{red}}{\rho \partial s} + \nu\frac{\partial^2 u_s}{\partial n^2} \tag{3.3.12}$$

$$2\Omega u_s \sin\varepsilon = -\frac{\partial p_{red}}{\rho r \partial \theta} + \nu\frac{\partial^2 u_\theta}{\partial n^2} \tag{3.3.13}$$

$$-2\Omega u_n \cos\varepsilon = -\frac{\partial p_{red}}{\rho\partial n} \tag{3.3.14}$$

In these equations, the nonlinear advective terms are neglected, and $\Omega \sin\varepsilon$ is the component of the rotation vector in the surface normal (n) direction. Putting $\varepsilon = 90°$ recovers the equations for a plane disc. For flow in the earth's atmosphere and oceans ε corresponds to the latitude. Note also that, although the body is assumed axisymmetric, the flow can vary in the circumferential (θ) direction.

From equation 3.3.14, $\partial p_{red}/\partial n = 0$ is zero for $\varepsilon = 90°$. Assuming that the boundary layer is sufficiently thin, the normal pressure gradient can also be neglected for other values of ε. With this assumption, differentiating 3.3.13 three times with respect to n, and substituting from 3.3.12 then gives

$$\frac{\partial^5 u_\theta}{\partial \xi^5} + 4\frac{\partial u_\theta}{\partial \xi} = 0 \tag{3.3.15}$$

where the non-dimensional boundary layer coordinate $\xi = n\sqrt{\Omega \sin(\varepsilon)/\nu}$. The general solution to this equation is

$$u_\theta = Ae^{-\xi}\sin\xi + Be^{-\xi}\cos\xi + Ce^{\xi}\sin\xi + De^{\xi}\cos\xi + E \tag{3.3.16}$$

where A, B, C, D and E are constants of integration (that may depend on s and θ). From equation 3.3.13, it follows that

$$u_s = -Ae^{-\xi}\cos\xi + Be^{-\xi}\sin\xi + Ce^{\xi}\cos\xi - De^{\xi}\sin\xi + F \tag{3.3.17}$$

Considering equations 3.3.12 and 3.3.13 gives

$$E = \frac{1}{2\rho\Omega\sin(\varepsilon)}\frac{\partial p_{red}}{\partial s} \quad F = -\frac{1}{2\rho\Omega\sin(\varepsilon)}\frac{\partial p_{red}}{r\partial\theta}. \tag{3.3.18}$$

The remaining constants are determined from boundary conditions on the surface (at $\xi = 0$) and at the far field or boundary layer edge ($\xi \to \infty$). Restricting solutions to be bounded at the far field requires $C = D = 0$, with $E = u_{\theta,\delta}$ and $F = u_{s,\delta}$ then corresponding to the geostrophic flow. Assuming that the body rotates with the coordinate system at rate Ω, the no slip boundary condition $u_\theta = u_s = 0$ u_θ at $\xi = 0$ gives

$$u_\theta = u_{s,\delta}\, e^{-\xi}\sin\xi + u_{\theta,\delta}\,(1 - e^{-\xi}\cos\xi) \tag{3.3.19}$$

$$u_s = u_{s,\delta}(1 - e^{-\xi}\cos\xi) - u_{\theta,\delta}\, e^{-\xi}\sin\xi.$$

This solution may also be written in (x, y) coordinates, with x aligned with the freestream velocity, giving

$$u_x = u_{x,\delta}\,(1 - e^{-\xi}\cos\xi) \tag{3.3.20}$$

$$u_y = -u_{x,\delta}\, e^{-\xi}\sin\xi.$$

Ekman layers occur in the earth's oceans and atmosphere and were first investigated following observations of ice flows in the ocean at angles of 20°–40° from the wind direction. From 3.3.20, $\partial u_x/\partial\xi = +u_{x,\delta}$ and $\partial u_y/\partial\xi = -u_{x,\delta}$ at $\xi = 0$, showing that (in laminar flow) the surface shear stress and freestream velocity are at angle of 45°. For the ice flow at the ocean surface, the no-slip condition at $\xi = 0$ in the above solution is replaced by a specified shear stress (aligned with the wind direction). The velocity at the surface is then shown to be at 45° to the wind direction. Differences of this angle with the ice flow observations might be attributed to insufficient or varying ocean

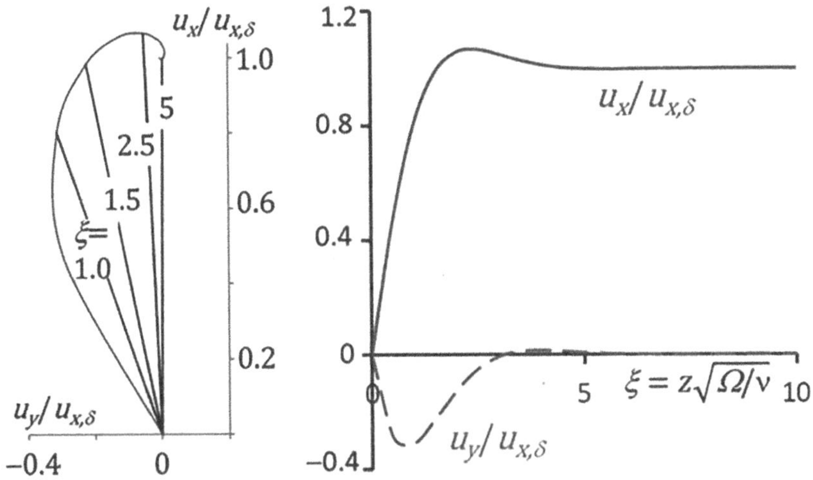

Figure 3.8 The Ekman spiral and velocity profiles.

depth and other departures from the theoretical ideal. The flow distribution in the boundary layer is called the Ekman spiral and is illustrated in Figure 3.8.

For the case when the freestream is described by the steady, axisymmetric vortex with $u_r = 0$ given in equation 3.3.7, $u_{s,\delta} \approx 0$ and the only radial transport is through the boundary layers. Integrating the second of equation 3.3.19 and applying the result at the radius $r = b$ gives the following expression for the radial mass flow in an Ekman layer:

$$\dot{m}_{EL} = -\pi\, r\rho\, u_{\theta,\delta} \Big/ \sqrt{\Omega \sin(\varepsilon)/\nu}. \tag{3.3.21}$$

This result is readily applied to the flow between infinite discs rotating at the same speed with a radial outflow. Assuming symmetry of flow on the two discs, the net radial outflow in the two boundary layers $\dot{m} = 2\dot{m}_{EL}$, and taking $\varepsilon = 90°$, the solution for the velocity in the inviscid core is

$$u_{r,\delta} = u_{z,\delta} = 0 \quad u_{\theta,\delta} = -C_W\left(2\pi Re_\phi^{0.5}\right)^{-1}(b/r)^2\Omega r \tag{3.3.22}$$

and the boundary layer solutions are given by equation 3.3.19. Note that the choice of representative radius b in the non-dimensional mass flow ($C_W = \dot{m}/\mu b$) and rotational Reynolds number ($Re_\phi = \Omega b^2/\nu$) is arbitrary as it cancels in the above equation.

The Ekman layer solution, free disc solution and inviscid, one-dimensional vortex are illustrated in the finite rotating cylindrical cavity with radial outflow shown in Figure 3.9, from Chew *et al.* (1984). Here, analytical solutions are compared with experiment and CFD (involving numerical solution of the steady, axisymmetric Navier–Stokes equations). The streamlines from the CFD and experimental flow visualisation show a source region, core flow, disc boundary layers and a sink region. Flow enters the cavity through the inner cylindrical boundary and is gradually entrained into the disc boundary layers. The radius at which all the flow is entrained marks the end of the source region. The disc boundary layer profile in the source region is close to the free disc solution, particularly at larger radii where the core velocity approaches zero and effects of the nonzero inner radius diminish. At the low flow rate shown in the figure, a rotating but otherwise quiescent core vortex develops, as described by equation 3.3.7. Although not directly shown in the figure, the core velocity agrees with equation 3.3.22 at higher radii where the radial velocity profiles in the boundary

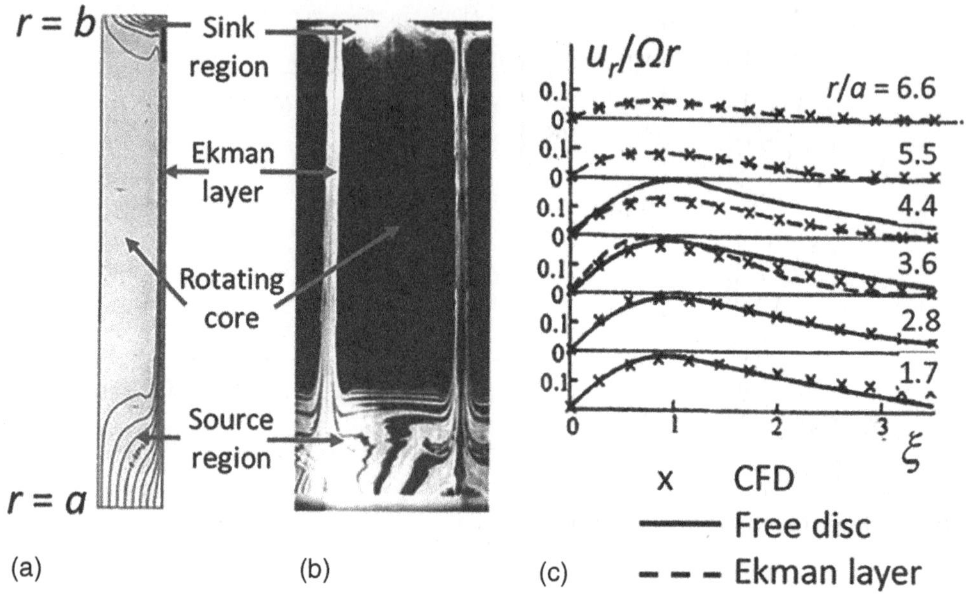

Figure 3.9 Radial outflow in a rotating cylindrical cavity, $Re_\phi = 2.5 \times 10^4$, $C_W = 79$, $a/b = 0.1$. (a) Streamlines from CFD for half the cavity. (b) Experimental flow visualisation (with reflections in the Perspex discs). (c) CFD and analytical solutions for the radial velocity in the disc boundary layer. *Source:* Chew, J. W, et al., (1984) / with permission of Cambridge University Press.

layers are shown to agree with the linear Ekman solution. As also illustrated by Owen *et al.* (1985), the linear Ekman layer solution for the core velocity agrees well with experimental measurements for a surprisingly large range of $u_{\theta,\delta}/\Omega r$. Flow from the Ekman layers feeds the outflow through the outer cylindrical surface via the sink region. The sink region here takes the form of a boundary layer and may be classed as a Stewartson (1957, 1966) layer.

The flow in Figure 3.9 can be classed as predominantly Batchelor-type, with Stewartson-type flow in the source region. If an open outer cylindrical boundary replaced the uniform sink in the CFD or discrete holes in the experiment, flow would be drawn into the cavity and Stewartson-type flow would develop throughout the cavity. If the discs were to rotate at different speeds, a weak axial velocity would develop in the rotating core, as allowed by equation 3.3.7 and shown for the rotor–stator configuration in the left of Figure 3.5.

3.3.3 Laminar Ekman Layer Heat Transfer

Applying the Reynolds analogy to the Ekman layer solution gives the following expression for the Nusselt number based on the local core-to-disc temperature difference (for $Pr = 1$ and negligible frictional heating):

$$Nu = \frac{rq_0}{\kappa(T_\delta - T_0)} = \sqrt{\frac{\Omega r^2 \sin(\varepsilon)}{\nu}}. \tag{3.3.23}$$

This assumes a quadratic temperature distribution on the disc and correspondence of the circumferential velocity and temperature solutions in the core. These conditions are rather restrictive and further analysis for specific cases provides more insight.

Disc heat transfer for radial outflow in a rotating cylindrical cavity, including the flow conditions shown in Figure 3.9, was considered by Chew (1985a). Assuming disc temperature distributions

$T_0 = T_{in} + cr^n$, the following expression was given for the disc Nusselt number for Ekman layer flow:

$$Nu = \frac{rq_0}{\kappa(T_{in} - T_0)} = \frac{nPrC_w b}{4\pi\, r}\left[1 - \frac{3(n-2)PrC_w b^2}{16\pi Re_\phi^{0.5} r^2}\right] - \frac{n^2 s}{2r}. \tag{3.3.24}$$

where s is the cavity width and the last term arises from heat conduction in the quiescent core. The Nusselt number here is based on the inlet-to-disc temperature difference. This result is based on an analytical solution with certain second and higher order terms neglected. The leading term in the solution corresponds to the heat flux required for the mean fluid temperature in the Ekman layers to change with radius at the same rate as the disc temperature.

The analytical solution giving equation 3.3.24 can be shown to agree with equation 3.3.23 for $n = 2, Pr = 1$ and with heat conduction in the core neglected. In these conditions, similarity of the boundary conditions for $v_\phi/\Omega r$ and $(T - T_{in})/cr^2$ holds. As might be expected from the velocity profiles in Figure 3.9, the free disc and Ekman layer heat transfer solutions (3.1.12 and 3.3.24) agree with numerical solutions of the axisymmetric, steady governing equations in appropriate regions of the flow.

If the two discs in a rotating cylindrical cavity are at different temperatures buoyancy effects may become significant and generate a weak axial flow in the rotating core, as it occurs for differentially rotating discs. Without a net radial flow, free convection occurs. Analytical solutions for axisymmetric flow between infinite discs are compared with CFD solutions by Chew (1985b). Some agreement between the analytical solution and measured heat transfer in a sealed rotating cavity with different disc temperatures has been shown. An example of the flow and temperature fields is given in Figure 3.10. Here, the thermal expansion coefficient β is taken as $1/T_{ref}$ with the hot disc temperature as the reference, and ΔT_{ref} is the difference between the disc temperatures. To first order in the analysis, the fluid temperature in the Ekman layers equals that on the adjacent disc and the two Ekman layer flows (given by equation 3.3.21) must balance. Equation 3.3.10 then gives the temperature and circumferential velocity variation across the core flow. The thermal boundary layer scale in this case is determined by the nonlinear convection and the heat conduction terms. This results in the thermal boundary layer being considerably thicker than the kinetic boundary layer. Similarity of $v_\phi/\Omega r$ and $(T - T_{in})/cr^2$ across the boundary layers is lost in this case as their boundary conditions differ.

The streamlines and radial velocity profiles in Figure 3.10 show radially outward and inward Ekman layer flow on the cold and hot discs, respectively. A weak axial flow convects heat from the hot to cold disc in the main part of the cavity. Away from the Ekman layers, the circumferential velocity profile has similar form to that for the temperature, as in equation 3.3.10. In this region, the change in velocity compensates for the change in density to maintain an equilibrium between the buoyancy-affected centrifugal force and the axially constant radial pressure gradient. The radial temperature gradient is close to zero in the main part of the cavity with the balance between axial convection and axial heat conduction controlling the axial temperature distribution. At higher Reynolds numbers, the axial convection will strengthen reducing axial temperature gradients further near the hot disc and increasing gradients near the cold disc.

It is important to note, as mentioned in Chapter 2, that positive radial temperature gradients can lead to centrifugal Rayleigh–Bénard instability in the quiescent core considerably changing the flow. In turbomachinery disc cavities, heat transfer from the (usually hot) cylindrical shroud is likely to be a strong effect. As will be illustrated later, this can result in large-scale unsteadiness and turbulence in the core flow producing a uniform mean core temperature, with unsteady laminar Ekman layers on the discs. Useful estimates of the mean disc heat transfer may then be found

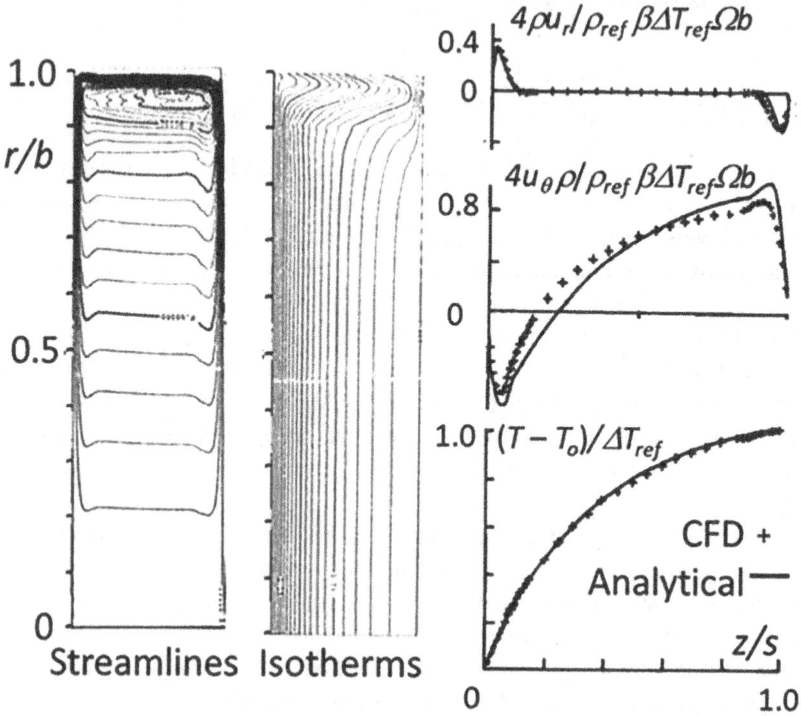

Figure 3.10 Steady, axisymmetric solutions for air flow in a rotating cavity between a cold isothermal disc (left), a hot isothermal disc (right) and a shroud with linear temperature distribution, $Re_\phi = 2 \times 10^4$, $\beta \Delta T_{ref} = 0.39$. Velocity and temperature profiles are at $r/b = 0.5$.

by assuming heat conduction across the Ekman layer thickness. In terms of Nusselt number this gives

$$Nu = \frac{rq_0}{\kappa(T_\delta - T_0)} = \frac{1}{\pi}\sqrt{\frac{\Omega r^2 \sin(\varepsilon)}{\nu}}. \tag{3.3.25}$$

3.3.4 Inertial Waves

The restoring nature of the Coriolis force leads to rotating flows being susceptible to the formation of waves. This may be illustrated mathematically by considering a small perturbation of a flow rotating as a solid body. This subject has received much attention from the atmospheric and oceanographic research communities as indicated, for example, by Greenspan (1968). In the natural cylindrical coordinates for the present problem, possible solutions may be found with an arbitrary number of sinusoidal circumferential lobes, radial variation given by Bessel functions, and sinusoidal variation in the axial direction. Frequency of these waves is limited to $<2\Omega$ in the rotating frame of reference. Thus, for a wave with N circumferential lobes, the angular speed of the pattern in the rotating frame will be $<2\Omega/N$, limiting the departure from the mean flow rotation speed.

The starting point for analysis is the inviscid, incompressible governing equations in a rotating system with the neglect of the nonlinear advective terms, as follows:

$$\frac{\partial(ru_r)}{r\partial r} + \frac{\partial u_\theta}{r\partial \theta} + \frac{\partial u_z}{\partial z} = 0 \tag{3.3.26}$$

$$\frac{\partial(\rho u_r)}{\partial t} - 2\rho\Omega u_\theta = -\frac{\partial p_{red}}{\partial r} \tag{3.3.27}$$

$$\frac{\partial(\rho r u_\theta)}{\partial t} + 2\rho\Omega r u_r = -\frac{\partial p_{red}}{\partial \theta} \tag{3.3.28}$$

$$\frac{\partial(\rho u_z)}{\partial t} = -\frac{\partial p_{red}}{\partial z} \tag{3.3.29}$$

Differentiating each of the momentum equations and eliminating time derivative terms through use of the continuity equation gives

$$\frac{\partial}{r\partial r}\left(r\frac{\partial p_{red}}{\partial r}\right) + \frac{\partial^2 p_{red}}{r^2\partial\theta^2} + \frac{\partial^2 p_{red}}{\partial z^2} = 2\rho\Omega\left(\frac{\partial(r u_\theta)}{r\partial r} - \frac{\partial u_r}{r\partial\theta}\right). \tag{3.3.30}$$

Also, differentiating the axial momentum equation and using the continuity equation to eliminate u_z gives the first part of equation 3.3.31. Using the radial and angular momentum equations then gives the second part of the equation:

$$\frac{\partial^2 p_{red}}{\partial z^2} = \rho\frac{\partial}{\partial t}\left(\frac{\partial(r u_r)}{r\partial r} + \frac{\partial u_\theta}{r\partial\theta}\right) = \frac{\rho}{2\Omega}\frac{\partial^2}{\partial t^2}\left(\frac{\partial u_r}{r\partial\theta} - \frac{\partial(r u_\theta)}{r\partial r}\right). \tag{3.3.31}$$

Combining results from equations 3.3.30 and 3.3.31 gives an equation for pressure:

$$\frac{\partial^2}{\partial t^2}\left[\frac{\partial}{r\partial r}\left(r\frac{\partial p_{red}}{\partial r}\right) + \frac{\partial^2 p_{red}}{r^2\partial\theta^2} + \frac{\partial^2 p_{red}}{\partial z^2}\right] + 4\Omega^2\frac{\partial^2 p_{red}}{\partial z^2} = 0. \tag{3.3.32}$$

It is convenient mathematically and standard practice in harmonic analysis to represent a solution by the real part of complex variables. Assuming a solution of the above equation of the form $p_{red} = \hat{p}_{red}(r)e^{i\omega t + ni\theta + miz}$ where $i^2 = -1$, substituting this into 3.3.32, and rearranging gives

$$\frac{r\partial}{\partial r}\left(r\frac{\partial \hat{p}_{red}}{\partial r}\right) - \left(1 - 4\frac{\Omega^2}{\omega^2}\right)m^2 r^2 \hat{p}_{red} - n^2\hat{p}_{red} = 0. \tag{3.3.33}$$

If n is an integer, as expected for a circumferentially periodic solution and $m \neq 0$, this equation has the general solution

$$\hat{p}_{red} = C_1 J_n(r^*) + C_2 Y_n(r^*) \quad \text{with} \quad r^* = m\sqrt{4\Omega^2/\omega^2 - 1}\,r \tag{3.3.34}$$

where J_n and Y_n are Bessel functions of the first and second kind and C_1 and C_2 are constants. Here, it is assumed that $\omega \neq 0$. The steady flow case $\omega = 0$ corresponds to the geostrophic flow discussed in Section 3.3.1.

In the limiting case where the radius r is large (compared to the radial length scale of the flow), equation 3.3.33 admits solutions of the form $\hat{p}_{red} \propto e^{lir}$, with the following result:

$$l^2 + (n/r)^2 + m^2 = 4\frac{m^2\Omega^2}{\omega^2} \tag{3.3.35}$$

For real values of l, m, n and ω in equation 3.3.35 (or for r^* to be real in 3.3.34) it is required that $-2\Omega \leq \omega \leq 2\Omega$.

Figure 3.11 illustrates a solution of the form 3.3.34 in an axial plane. The inertial wave illustrated has five high pressure/low pressure cell pairs and associated radial inflow and outflow regions distributed circumferentially, with an inner to outer radius ratio of ~0.65. The linear analysis allows many such solutions with variable radius ratios, cell numbers etc. In practice, these will be constrained by boundary conditions, viscous dissipation, nonlinear effects, and the assumption of a simple forced vortex base flow. Nevertheless, waves similar to that in the figure have been observed in CFD and experimental research. These are illustrated in later chapters.

In an azimuthal plane, the above solution has a form similar to that shown for Taylor vortices in Figure 3.1. Experiments indicate that as the inner cylinder rotation is increased above that of the outer cylinder axisymmetric vortices appear first and then develop into wavy and spiralling vortices

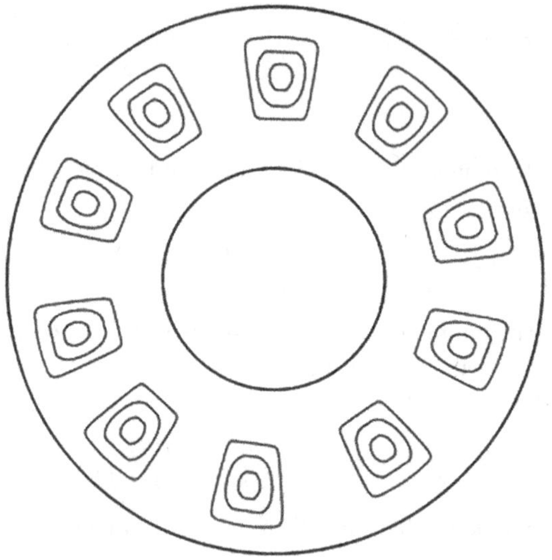

Figure 3.11 Schematic of rotating inertial wave pressure contours.

which would correspond to nonzero values of n and l in the above analysis. As Rayleigh's stability criterion assumes that an element of fluid conserves angular momentum as it moves radially, it might be expected that the axisymmetric mode would be most unstable. Asymmetrical disturbances will create circumferential pressure gradients tending to equalise the angular momentum of the disturbed element with that of the surrounding fluid, and hence stabilising the flow compared to the axisymmetric case. The axisymmetric centrifugal instability is stronger than the stabilising Coriolis force that resists axial gradients (as in the Taylor–Proudman theorem).

For axially uniform disturbances with $m = 0$ equation 3.3.33 still holds if $\omega \neq 0$ and has the solution

$$\hat{p}_{red} = C_1 r^n + C_2 r^{-n} \quad \text{for} \quad p_{red} = \hat{p}_{red}(r)e^{i\omega t + ni\theta} \tag{3.3.36}$$

On physical grounds, it might be expected that a periodic disturbance in the circumferential direction would decay with radial distance from its source. Thus, for example, a disturbance at an outer radius would propagate inwards with amplitude $p_{red} \propto r^n$. This may be related to the rotating waves observed in rim seals for rotor–stator cavities discussed in Chapter 9. For centrifugal Rayleigh–Bénard convection, as will be described in Section 4.7.1, buoyancy forces generate flow disturbances at both a cooled inner cylinder and a heated outer cylinder. Disturbances with short circumferential length scale (i.e. high n) may decay rapidly with radius but disturbances with length scales similar to the radial separation of the cylinders may combine to form alternating plumes of inflow and outflow. This is qualitatively consistent with observations from numerical simulations and experiments, notwithstanding the omission of buoyancy force in the above analysis. In this example, the instability arises when a fluid element moves radially maintaining its original temperature but attaining the angular momentum of the fluid at its new radius. Thus asymmetric disturbances are less stable than axisymmetric disturbances and flow can develop within the constraints of the Taylor–Proudman theorem.

3.3.5 Acoustic Waves

Extending the analysis for inertial waves above to allow for compressible flow gives some insight into acoustic waves. Here, the starting point for analysis is the inviscid flow equations in a rotating system, with the neglect of the nonlinear advective terms and allowance for density variation, as follows:

$$\frac{\partial \rho}{\partial t} + \frac{\partial(\rho r u_r)}{r\partial r} + \frac{\partial(\rho u_\theta)}{r\partial\theta} + \frac{\partial(\rho u_z)}{\partial z} = 0 \tag{3.3.37}$$

$$\frac{\partial(\rho u_r)}{\partial t} - 2\rho\Omega u_\theta - (\rho - \rho_{ref})\Omega^2 r = -\frac{\partial p_{red}}{\partial r} \tag{3.3.38}$$

$$\frac{\partial(\rho r u_\theta)}{\partial t} + 2\rho\Omega r u_r = -\frac{\partial p_{red}}{\partial\theta} \tag{3.3.39}$$

$$\frac{\partial(\rho u_z)}{\partial t} = -\frac{\partial p_{red}}{\partial z} \tag{3.3.40}$$

The flow is assumed to be a perturbation from a forced vortex with uniform density ρ_{ref}. Here, the reduced pressure $p_{red} = p - \rho_{ref}\Omega^2 r^2/2$.

Differentiating each of the momentum equations, summing equations and eliminating time derivatives of velocity terms through use of the continuity equation gives

$$\frac{\partial}{r\partial r}\left(r\frac{\partial p_{red}}{\partial r}\right) + \frac{\partial^2 p_{red}}{r^2\partial\theta^2} + \frac{\partial^2 p_{red}}{\partial z^2} - \frac{\partial^2\rho}{\partial t^2} = 2\Omega\left(\frac{\partial(r\rho u_\theta)}{r\partial r} - \frac{\partial\rho u_r}{r\partial\theta}\right) + \Omega^2\frac{\partial[(\rho - \rho_{ref})r^2]}{r\partial r}. \tag{3.3.41}$$

Also, differentiating the axial momentum equation and using the continuity equation to eliminate u_z gives the first part of equation 3.3.42. Using the radial and angular momentum equations then gives the second part of the equation:

$$\frac{\partial^2 p_{red}}{\partial z^2} = \frac{\partial}{\partial t}\left(\frac{\partial\rho}{\partial t} + \frac{\partial(r\rho u_r)}{r\partial r} + \frac{\partial\rho u_\theta}{r\partial\theta}\right) = \frac{\partial^2\rho}{\partial t^2} + \frac{1}{2\Omega}\frac{\partial^2}{\partial t^2}\left(\frac{\partial\rho u_r}{r\partial\theta} - \frac{\partial(r\rho u_\theta)}{r\partial r}\right) - \frac{\Omega}{2}\frac{\partial^2\rho}{\partial t\partial\theta}. \tag{3.3.42}$$

Combining results from equations 3.3.41 and 3.3.42 gives an equation relating derivatives of pressure and density:

$$\frac{\partial^2}{\partial t^2}\left[\frac{\partial}{r\partial r}\left(r\frac{\partial p_{red}}{\partial r}\right) + \frac{\partial^2 p_{red}}{r^2\partial\theta^2} + \frac{\partial^2 p_{red}}{\partial z^2}\right] - \frac{\partial^4\rho}{\partial t^4} + 4\Omega^2\frac{\partial^2 p_{red}}{\partial z^2} - 4\Omega^2\frac{\partial^2\rho}{\partial t^2}$$
$$+ 2\Omega^3\frac{\partial^2\rho}{\partial t\partial\theta} - \Omega^2\frac{\partial^3(\rho r^2)}{r\partial r\partial t^2} = 0. \tag{3.3.43}$$

Pressure and density gradients are related through the speed of sound c by the equation

$$\partial\rho/\partial p = 1/c^2. \tag{3.3.44}$$

With c assumed constant, using 3.3.44 in 3.3.43 gives the following equation for pressure:

$$\frac{\partial^2}{\partial t^2}\left[\frac{\partial}{r\partial r}\left(r\frac{\partial p_{red}}{\partial r}\right) - \frac{\Omega^2}{c^2}\left(r\frac{\partial p_{red}}{\partial r}\right) + \frac{\partial^2 p_{red}}{r^2\partial\theta^2} + \frac{\partial^2 p_{red}}{\partial z^2}\right] - \frac{\partial^4 p_{red}}{c^2\partial t^4} + 4\Omega^2\frac{\partial^2 p_{red}}{\partial z^2}$$
$$- 4\Omega^2\frac{\partial^2 p_{red}}{c^2\partial t^2} + 2\Omega^3\frac{\partial^2 p_{red}}{c^2\partial t\partial\theta} - \frac{2\Omega^2}{c^2}\frac{\partial^2 p_{red}}{\partial t^2} = 0. \tag{3.3.45}$$

Assuming a solution of the above equation of the form $p_{red} = \hat{p}_{red}(r)e^{i\omega t + in\theta + miz}$ where $i^2 = -1$, substituting this into 3.3.45, and rearranging gives

$$\frac{r\partial}{\partial r}\left(r\frac{\partial\hat{p}_{red}}{\partial r}\right) - \frac{\Omega^2 r^2}{c^2}\left(r\frac{\partial p_{red}}{\partial r}\right) + \frac{\omega^2}{c^2}r^2\hat{p}_{red} - \left(1 - 4\frac{\Omega^2}{\omega^2}\right)m^2 r^2\hat{p}_{red} - 2\frac{\Omega^2}{c^2}r^2\hat{p}_{red}$$
$$+ 2\frac{n\Omega^3}{\omega c^2}r^2\hat{p}_{red} - n^2\hat{p}_{red} = 0. \tag{3.3.46}$$

As expected, this reduces to equation 3.3.33 for waves in incompressible flow in the limit $c \to \infty$.

A further limiting case of interest is obtained when the terms arising from the Coriolis and centrifugal body forces are omitted. This gives the following equation, corresponding to the result for non-rotating flow in a cylindrical annulus:

$$\frac{r\partial}{\partial r}\left(r\frac{\partial \widehat{p}_{red}}{\partial r}\right) + \frac{\omega^2}{c^2}r^2\widehat{p}_{red} - m^2r^2\widehat{p}_{red} - n^2\widehat{p}_{red} = 0 \tag{3.3.47}$$

For integer n this equation has the general solution

$$\widehat{p}_{red} = C_1 J_n(r^*) + C_2 Y_n(r^*) \quad \text{with} \quad r^* = \sqrt{\omega^2/c^2 - m^2}r \tag{3.3.48}$$

where J_n and Y_n are Bessel functions of the first and second kind and C_1 and C_2 are constants. The condition for r^* to be real is that the wave speed in the (positive or negative) axial direction is less than or equal to the speed of sound. While the form of equation 3.3.48 follows that of 3.3.34 for inertial waves, the variation of density in the acoustic waves allows solutions for axially uniform waves (with $m = 0$). Considering, for example, axially uniform waves, the Coriolis and centrifugal force terms in equation 3.3.46 will be negligible for $n^2 \gg \Omega^2r^2/c^2$. This suggests that in many cases acoustic modes in the rotating system will be similar to those in nonrotating flow that have been studied in some depth in acoustics. Further details of nonrotating acoustic modes are given, for example, by Eversden (1991).

In the limiting case where the radius r is large (compared to the radial length scale of the flow) equations 3.3.46 and 3.3.47 admit solutions of the form $\widehat{p}_{red} \propto e^{ilr}$. For equation 3.3.47, the following equation relating wave frequency and length scales is obtained:

$$\omega^2/c^2 = l^2 + (n/r)^2 + m^2 \tag{3.3.49}$$

For axially and radially uniform acoustic waves, as might be expected in narrow, high inner-to-outer radius ratio cylindrical cavities, forward and backward circumferential travelling acoustic waves will have frequency nc/r and speed $\pm c$ in the rotating frame. In the stationary reference frame, this corresponds to frequencies $n\Omega \pm nc/r$ and speeds $v_\varphi \pm c$ where $v_\varphi = \Omega r$ is the mean fluid velocity.

Acoustic waves have been detected in computational and experimental studies related to rim sealing. Examples are discussed in Chapter 9, but their significance for this application is questionable, with inertial waves dominating in most studies. Coupling of disc cavity acoustic modes with structural vibration modes has been considered for centrifugal compressor vibration. A recent experimental investigation motivated by this application is reported by Unglaube and Brillert (2022). Interest in understanding of acoustic waves may well increase as advanced instrumentation and simulations yield more unsteady flow data. It may be noted that unsteady flow measurements often show unexplained effects and that identifying individual modes and their source is challenging.

3.4 Concluding Remark

The analytical results given above for laminar flow find rather limited direct application in turbomachinery disc flows but give important insight into some underlying flow mechanisms. In inviscid flow regions (away from the boundary layers) axisymmetric vortices may form in which (following the Taylor–Proudman theorem) axial gradients of flow variables are zero. Kinetic boundary layers on rotating discs, with a length scale $\sqrt{\nu/\Omega}$ in laminar flow, may dominate the radial transport of fluid. The boundary layer flow is radially outward if the absolute 'freestream' circumferential velocity is less than disc speed, and inwards if the freestream rotates faster than the disc. The relative magnitude of the boundary radial flows to the imposed radial flow and the angular momentum of the supplied flow are important in determining the nature of the flow.

The inviscid vortices are unstable in some conditions and the restoring nature of the Coriolis force may lead to the formation of inertial waves. Centrifugal and buoyancy instabilities lead to a wide variety of unsteady and 3D flows with various degrees of identifiable large-scale structures. In these cases, unsteady Ekman layers may form on the discs. These are dominated by viscous and Coriolis forces and react rapidly to the unsteady core or mainstream flow. Interest in unsteady flow effects, including inertial and acoustic waves, has increased in engineering research as advanced simulations and experiments have revealed complex flow features.

A major limitation of the solutions in this chapter is the assumption of laminar boundary layers. The next chapter discusses transition to turbulent flow, and Chapter 5 considers turbulent boundary layers.

References

Andereck, C. D., Liu, S. S., and Swinney, H. L., 1986. Flow regimes in a circular Couette system with independently rotating cylinders. J. Fluid Mech., 164, 155–183. https://doi.org/10.1017/S0022112086002513.

Batchelor, K., 1951. Note on a class of solutions of the Navier–Stokes equations representing steady rotationally-symmetric flow. Q. J. Mech. Appl. Math., 4, 29–41. https://doi.org/10.1093/qjmam/4.1.29.

Bödewadt, U. T., 1940. Die Drehströmung über festem Grund. Z. Angew. Math. Mech., 20, 241–253. https://doi.org/10.1002/zamm.19400200502.

Bowker, K. A., 1988. Albert Einstein and meandering rivers. Earth Sci. Hist., 7, 45–45. http://www.jstor.org/stable/24136868.

Chew, J. W., 1985a. Computation of forced laminar convection in rotating cavities. ASME J. Heat Transfer, 107, 277–282. https://doi.org/10.1115/1.3247411.

Chew, J. W., 1985b. Computation of convective laminar flow in rotating cavities. J. Fluid Mech., 153, 339–360. https://doi.org/10.1017/S0022112085001288.

Chew, J. W., Owen, J. M., and Pincombe, J. R., 1984. Numerical predictions for laminar source-sink flow in a rotating cylindrical cavity. J. Fluid Mech., 143, 451–466. https://doi.org/10.1017/S0022112084001439.

Childs, P. R. N., 2011. Rotating flow, Elsevier.

Cochran, W. G., 1934. The flow due to a rotating disc. Math. Proc. Cambridge Philos. Soc., 30, 365–375. https://doi.org/10.1017/S0305004100012561.

Dorfman, L. A., 1963. Hydrodynamic resistance and the heat loss of rotating solids, First English edition, London: Oliver and Boyd.

Einstein, A., 1954. The cause of the formation of meanders in the courses of rivers and of the so-called Baer's law, New York: Ideas and Opinions, Bonanza Books, pp. 249–253.

Eversden, W., 1991. Theoretical models for duct acoustic propagation and radiation. Available from NASA, Langley Research Center, Aeroacoustics of Flight Vehicles: Theory and Practice. Volume 2: Noise Control. https://ntrs.nasa.gov/citations/19920005564.

Greenspan, H. P., 1968. The theory of rotating fluids, Cambridge University Press.

Kreith, F., Ellis, D., and Giesing, J., 1963. An experimental investigation of the flow engendered by a rotating cone. Appl. Sci. Res., 11, 430–440. https://doi.org/10.1007/BF03184627.

Lingwood, R., 1997. Absolute instability of the Ekman layer and related rotating flows. J. Fluid Mech., 331, 405–428. https://doi.org/10.1017/S0022112096004144.

Owen, J. M., Pincombe, J. R., and Rogers, R. H., 1985. Source–sink flow inside a rotating cylindrical cavity. J. Fluid Mech., 155, 233–265. https://doi.org/10.1017/S0022112085001793.

Owen, J. M. and Rogers, R. H., 1989. Flow and heat transfer in rotating disc systems, Vol. 1: Rotor-stator systems, Taunton, UK: Research Studies Press.

Schlichting, H., 1979. Boundary-layer theory, Seventh edition, McGraw-Hill.

Stewartson, K., 1953. On the flow between two rotating coaxial disks. Math. Proc. Cambridge Philos. Soc., 49(2), 333–341. http://doi.org/10.1017/S0305004100028437.

Stewartson, K., 1957. On almost rigid rotations. J. Fluid Mech., 3, 17–26. http://doi.org/10.1017/S0022112057000452.

Stewartson, K., 1966. On almost rigid rotations. Part 2. J. Fluid Mech., 26, 131–144. http://doi.org/10.1017/S0022112066001137.

Taylor, G. I., 1923. Stability of a viscous liquid contained between two rotating cylinders. Philos. Trans. R. Soc., Ser. A, 223, 289–343. https://doi.org/10.1098/rsta.1923.0008.

Tien, C. and Campbell, D., 1963. Heat and mass transfer from rotating cones. J. Fluid Mech., 17(1), 105–112. http://doi.org/10.1017/S0022112063001142.

Unglaube, T. and Brillert, D., 2022. Acoustoelastic modes in rotor-cavity systems: An overview on frequency shift effects supported with measurements. Int. J. Turbomach. Propul. Power, 7, 15. https://doi.org/10.3390/ijtpp7020015.

von Kármán, T., 1921. Über laminare und turbulente Reibung. Z. Angew. Math. Mech., 1, 233–252. https://doi.org/10.1002/zamm.19210010401.

Wu, C. S., 1959. The three dimensional incompressible laminar boundary layer on a spinning cone. Appl. Sci. Res., 8, 140–146. https://doi.org/10.1007/BF00411744.

Yamada, Y. and Ito, M., 1975. On the frictional resistance of enclosed rotating cones. Bull. JSME, 18(123), 1026–1034. https://doi.org/10.1299/jsme1958.18.1026.

4

Laminar–Turbulent Transition

Before considering fully turbulent flow, it is appropriate to review under what conditions laminar and turbulent flow occur. Over the last century, many experimental, analytical and computational studies have considered the onset of turbulence in rotating flows. These build on research in other classes of flow including Reynolds' seminal studies of transition in pipe flow and identification of the Reynolds number as the primary controlling parameter in the 1880s. Readers are referred elsewhere for a comprehensive account of the subject. For example, Schlichting (1979) describes measurements and theory for a range a flows. The aim here is to highlight results that have given insight into the rotating flows considered and are useful in engineering applications.

4.1 The Free Disc

Transition to turbulent flow on a free rotating disc has been subject to much research and provides a good starting point for the discussion. Measurements of moment for a free rotating disc by Theodorsen and Regier (1944) indicated transition to turbulent flow at a critical rotational Reynolds number $Re_\phi \approx 3 \times 10^5$, corresponding to a critical local Reynolds number $Re_\delta = r\sqrt{\Omega/\nu}$ (based on the laminar boundary layer length scale) of ~550. Their results are reproduced in Figure 4.1. At low Reynolds numbers, the measured moment agrees with von Kármán's (1921) laminar solution given by equation 3.1.11. At high Reynolds number, measurements for a smooth disc in quiescent surroundings agree with von Kármán's turbulent flow solution, which will be discussed in Chapter 5. Calculated disc moments assuming an inner laminar flow region and outer turbulent region with transition at a local Reynolds number $Re_\delta = 557$ match the data bridging the low and high Reynolds number smooth disc asymptotes. Earlier transition was induced through excitation of the flow by an air jet and by a rough disc surface which also gives higher moments in the turbulent regime.

Further investigations have been conducted by many authors with Gregory *et al.* (1955), Lingwood (1995, 1996) and Imayama *et al.* (2014) giving interesting discussions, stability analysis, velocity measurements and interpretation. Above certain local Reynolds numbers Re_δ (typically ~300), it has been found experimentally and confirmed analytically that the laminar boundary layer on a free disc becomes unstable to small disturbances. This leads to the formation of spiral vortices that rotate with the disc and to travelling disturbances. Sensitivity to the roughness size has also been observed. At higher Reynolds number (or higher radii), the flow becomes turbulent with $Re_\delta \approx 510$ marking the start of transition to fully turbulent flow. Lingwood associates this condition with an absolute instability leading to growth of disturbances at this position. This is consistent with the critical Reynolds number for transition being relatively insensitive to experimental conditions (compared, for example, to transition in pipe flow). The spiral vortices at lower radii are associated with convective instabilities in which a disturbance grows as it is

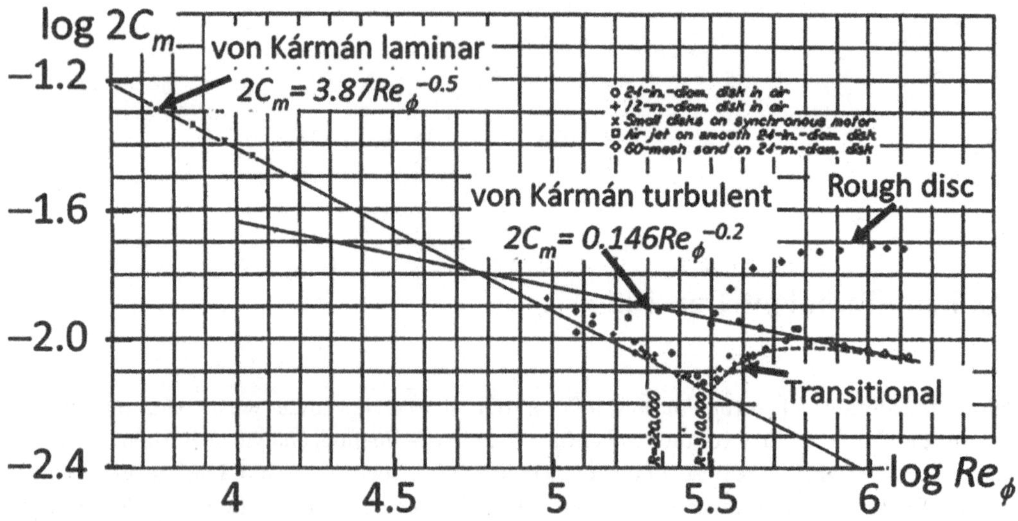

Figure 4.1 Free disc moment coefficient as a function of Reynolds number. *Source:* Adapted from Theodorsen and Regier (1944).

advected with the flow. Typically, about 30 vortices with a wave angle of about 14° to the radial direction (in the direction of rotation) have been observed in the pretransitional region. Velocity and moment measurements indicate that these have limited effect on the mean flow. A simplified interpretation of the transition from stable laminar flow, to convectively unstable conditions, and then to absolute instability and turbulent flow is shown in Figure 4.2.

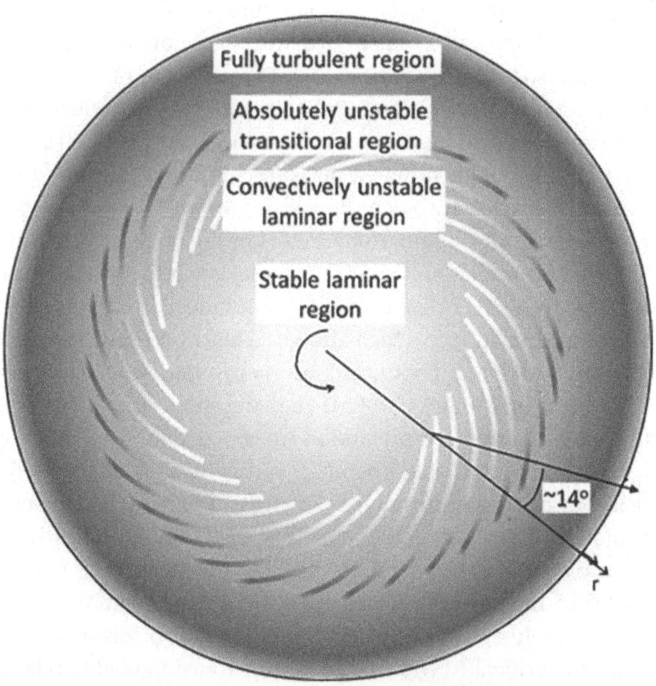

Figure 4.2 Development of spiral vortices and transition to turbulence on a smooth free rotating disc. Disc rotates anticlockwise.

4.2 BEK Flows (Bödewadt, Ekman, von Kármán)

Similar phenomena to those identified for the free disc occur for other flows in the BEK (Bödewadt, Ekman, von Kármán) family. For example, Savas (1987) observed circular and spiral waves with radial inflow on the end disc of a cylinder during spin down. Faller (1963) observed spiral vortices for Ekman layer flow above a critical Reynolds number $Re_\delta \approx 125$. Experiments on Ekman layer flows reported by Faller and Kaylor (1966) for radial inflow and Owen *et al.* (1985) for radial outflow between corotating discs indicate a critical Reynolds number for transition of $Re_\delta \approx 180$–200. Lingwood's (1997) analysis of the BEK class of flows considered both convective instabilities and absolute instabilities. For the free disc (as discussed above) and Ekman layer flow the onset of absolute instability correlates well with observations of the start of transition to full turbulence. Lingwood's absolute instability results for a range of BEK flows are summarised in Table 4.1 (following Gao and Chew, 2021b). These could be used in estimating the state of the boundary layer in engineering applications. However, it should be noted that roughness elements or other flow disturbances often cause earlier transition and that departure of the far field flow from a forced vortex may also affect transition.

The limiting cases of the free rotating disc, linear Ekman layer flow and forced vortex flow over a stationary disc correspond to $\Omega_\infty/\Omega_0 = 0$, 1 and ∞ in Table 4.1. Critical boundary layer Reynolds numbers based on the length scale $\sqrt{\nu/\Omega_0}$ or $\sqrt{\nu/\Omega_\infty}$ are given. The rotational Reynolds number Re_ϕ is also included as this is commonly used to characterise experimental or engine conditions. For $\Omega_\infty/\Omega_0 < 1$ the net flow in the disc boundary layer is radially outwards. For $\Omega_\infty/\Omega_0 > 1$ the fluid rotates faster than the disc the net flow is radially inwards. It is noticeable that transition is predicted at lower Reynolds numbers for inflow than for outflow cases. This may be associated with the inflow being a decelerating flow, while the outflow is an accelerating flow.

Table 4.1 Lingwood's (1997) predictions of absolute instability for the BEK class of flows. Ω_0 is disc angular speed. Ω_∞ is the angular speed of the 'freestream' forced vortex. *Source:* Adapted from Lingwood, (1997).

	Critical boundary layer Reynolds numbers		Critical local Re_ϕ
Ω_∞/Ω_0	$\lvert\Omega_0 r - v_{\phi,\delta}\rvert/\sqrt{\nu\Omega_0}$	$\lvert\Omega_0 r - v_{\phi,\delta}\rvert/\sqrt{\nu\Omega_\infty}$	$\rho\Omega_0 r^2/\mu$
0	507	∞	2.57×10^5
0.259	418	822	3.19×10^5
0.464	326	479	3.71×10^5
0.643	269	336	5.67×10^5
0.815	229	254	1.54×10^6
1.00±	198	198	∞
1.23	172	155	5.74×10^5
1.56	147	118	6.98×10^4
2.15	121	82.5	1.10×10^4
3.86	97.1	49.5	1.16×10^3
7.21	90.6	33.7	213
∞	∞	21.6	0

For the Bödewadt flow, with a stationary disc giving $\Omega_\infty/\Omega_0 = \infty$, transition is predicted for a boundary layer Reynolds number of just 21.6, corresponding to a radius on the disc of just a few boundary layer thicknesses. There is some experimental evidence of relatively low Reynolds number transition in Bödewadt flow, but this is obscured by difficulties in achieving this exact flow experimentally.

The critical values of the Re_ϕ approach infinity in the Ekman layer limit (as the velocity difference across the boundary layer tends to zero). For high-speed experimental rigs or engines with $Re_\phi \sim 10^6$ or higher, predominantly turbulent flow can be expected except where the flow angular velocity approaches that of the disc. Laminar flow may occur in these regions and, referring back to Section 3.3.2, the linear Ekman layer solution may apply.

4.3 Enclosed Rotor–Stator Disc Cavities

The flow in a simple closed rotor–stator disc cavity, as illustrated in Figure 4.3, was investigated by Daily and Nece (1960). For sufficiently wide cavities, separate boundary layers develop on the rotor and stator discs, with the shroud boundary layer recirculating the outflow on the rotor to supply the inflow on the stator. Away from the boundary layers the radial velocity is approximately zero, there is a weak axial flow from the stator to the rotor, and the core flow typically rotates at about 35% to 50% of disc speed depending on stator and rotor spacing and whether the flow is laminar or turbulent. Based on disc moment measurements, Daily and Nece deduced a critical value of rotational Reynolds number for transition of $\sim 1.5 \times 10^5$. This was inferred from the point at which two curves showing the variation of moment coefficient with Reynolds number for fully laminar and fully turbulent flow intersected. There was no sudden change in slope of the data, contrasting with Theodorsen and Regier's results for the free disc in Figure 4.1.

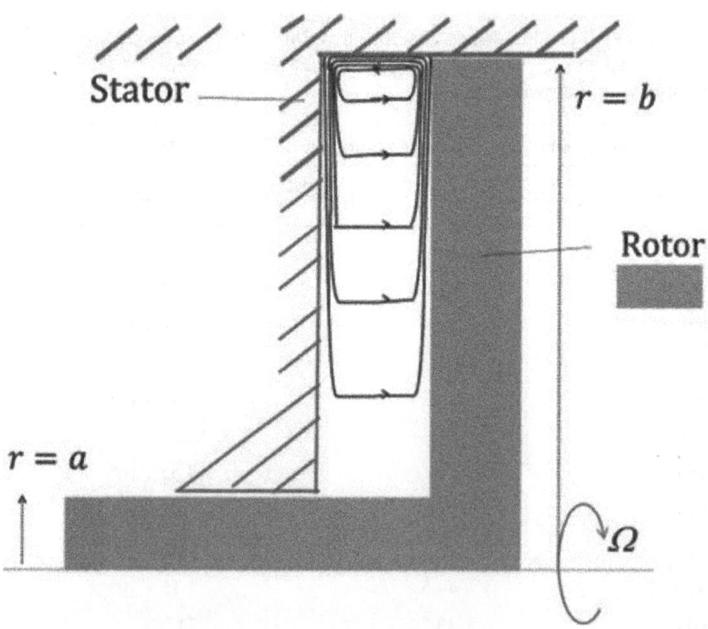

Figure 4.3 Illustration of flow in a closed rotor–stator disc cavity.

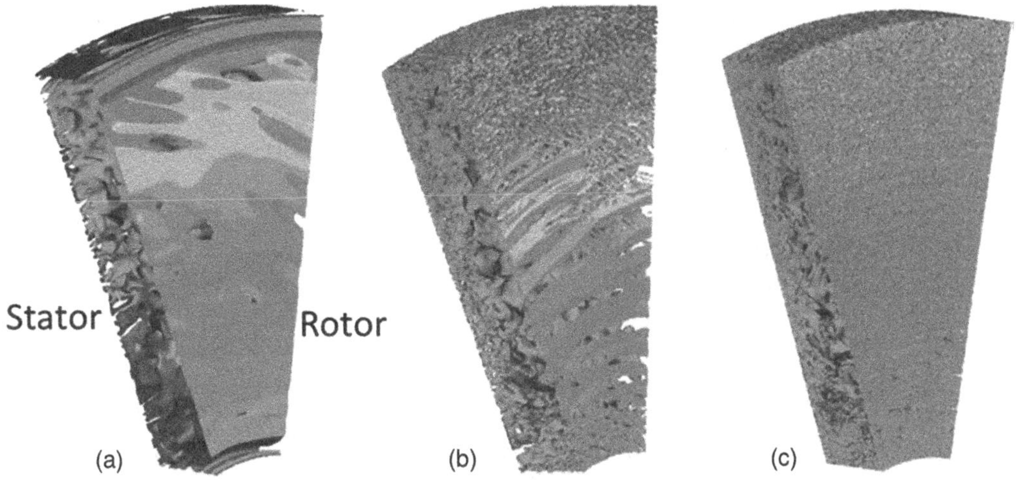

Figure 4.4 Isosurfaces of Q-criterion coloured by circumferential velocity from DNS for flow in a closed rotor–stator cavity (a) $Re_\phi = 8\times10^4$ (b) $Re_\phi = 4\times10^5$ (c) $Re_\phi = 10^6$. *Source:* Pitz *et al.* (2017a) / with permission of American Society of Mechanical Engineers (ASME).

Further experimental studies and direct numerical simulations (DNS) have largely confirmed Daily and Nece's observations. Itoh *et al.*'s (1992) measurements suggested that rotor boundary layer transition occurred for $\rho\Omega_0 r^2/\mu$ between 1.6×10^5 and 3.6×10^5. DNS, in which the full governing equations are solved numerically, has also shown initiation of transition at $\rho\Omega_0 r^2/\mu{\sim}1.6\times10^5$. Further discussion is given by, for example, Yim *et al.* (2018) and Gao and Chew (2021b). A visualisation of turbulence development from DNS is shown in Figure 4.4, following Pitz *et al.* (2017a). Instantaneous isosurfaces of Q-criterion (identifying vortices) coloured by circumferential velocity are shown for three different Reynolds numbers. The inner-to-outer radius ratio here is 0.286. Transition of the rotor boundary layer from laminar flow with some large-scale flow features at $Re_\phi = 8\times10^4$ to transitional at $Re_\phi = 4\times10^5$, and almost fully turbulent at $Re_\phi = 10^6$ is observed. For $Re_\phi = 4\times10^5$, the flow on the rotor at low radii appears to be affected by impingement of flow deflected from the radial inflow on the stator. This effect may account for the lack of an obvious small laminar region on the rotor at $Re_\phi = 10^6$, for which the condition $\rho\Omega_0 r^2/\mu \sim 1.6\times10^5$ corresponds to a radius ratio $r/b = 0.4$.

The DNS shown in Figure 4.4 and other results confirm earlier transition of the stator boundary layer, as indicated by the (ideal) Bödewadt condition in Table 4.1. Poncet *et al.* (2009) reported experiments and DNS showing that, for the stator disc boundary layer in a rotor-stator system, the first instability sets in at a boundary layer Reynolds number of 34.7 as circular rolls, and the second instability occurs at a boundary layer Reynolds number of 68. Xie *et al.* (2024) used DNS to study instability and transition on the stator disc giving further insight and showing interaction of circular and spiral waves. In their study, a rotating shroud was introduced, stabilising the flow at outer radii.

4.4 Corotating Disc Cavities with Radial Flow

For radial outflow between corotating discs, as illustrated in Figure 3.9, Owen and Pincombe's (1980) and Owen *et al.*'s (1985) core velocity measurements indicate turbulent flow in the constant mass flow boundary layers for $bC_W/2\pi r \approx 180$ and above. In this flow, the core flow rotates slower than the discs and core swirl fraction increases as radius increases. Assuming that equation 3.3.22 holds for the core velocity at this point, as supported by the experiments, this condition corresponds to $Re_\delta = |\Omega r - v_{\phi,\delta}|/\sqrt{\nu\Omega} \approx 180$. The lower core velocities at which transition was observed are well below disc speed. For these cases, the value of $Re_\delta \approx 180$ is significantly below the critical values for absolute instability from Lingwood's (1997) analysis given in Table 4.1. The assumption of forced vortex core flow in the stability analysis and/or possible effects of additional disturbances in the experiment may account for this difference.

Viaud *et al.* (2008) and Abedi *et al.* (2025) have considered transition in radial outflow using DNS. Both studies indicate transition close to Lingwood's critical Re_δ (despite the assumed forced vortex in the analysis). Abedi *et al.* made direct comparisons with Owen *et al.*'s core velocity measurements at one condition. Their plane disc results considerably underestimated the measured velocities, indicating delayed transition and weaker turbulence compared to the experiment. They also found that placing an axisymmetric groove with width and depth equal to the Ekman layer thickness upstream of expected transition moved the predictions closer to the experimental measurements. This suggests that other disturbances in the experiments, such as those induced by the outlet holes in the cylindrical shroud, may account for differences with the DNS. Further investigations of both sensitivity to conditions and numerical accuracy are needed to establish a firm understanding.

Radial inflow between corotating discs was also considered by Owen *et al.* $bC_W/2\pi r \approx 180$ was again shown to give reasonable estimates of transition to turbulence in constant mass flow boundary layers. In this case, the core flow rotates faster than the discs and core swirl fraction increases as radius decreases, with opposite trends to those for outflow. Compared to the BEK flows, which assume a forced vortex core, the difference in acceleration or deceleration between the outflow and inflow cases is considerably reduced. This may be a factor in the trends in Table 4.1 not being reproduced.

4.5 Rotating Cones

As discussed in Section 3.2, flows associated with broad cones (with high enough cone angle) have strong similarities with rotating disc flows. Tien and Campbell's (1963) experiments for rotating cones in a quiescent environment indicated similarity of disc and cone flows for half angles of 45° and above. With a cone of half angle 30° transition occurred earlier than for larger cone angles. Kreith *et al.*'s (1963) experiments indicated similarities of disc and cone flow for half angles ≥26.5°, but with the transitional Reynolds number decreasing with cone angle. Yamada and Ito's (1975) experimental study of enclosed rotating cones confirms similarity of broad cone cases with disc flows. For cone half angles ≥45°, plots of disc moment measurements against Reynolds number for conditions with separate boundary layers show little effect of cone angle. As observed by Daily and Nece (1960) for enclosed rotating discs, these plots indicate a gradual transition from fully laminar to turbulent conditions with similar transitional Reynolds number to the enclosed rotating disc.

More recent studies of transition on a rotating cone in a quiescent environment confirm and extend the earlier work mentioned above. For example, Kobayashi and Izumi's (1983) flow visualisation for transitional flows at two different cone angles is reproduced in Figure 4.5. With a cone half angle of 60°, similarities to the free disc flow illustrated in Figure 4.1 are clear (noting that the direction of rotation differs in the two figures). For the 15° half cone angle, the vortices preceding transition are cylindrical rather than spiralling. Transition occurs earlier for slender cones.

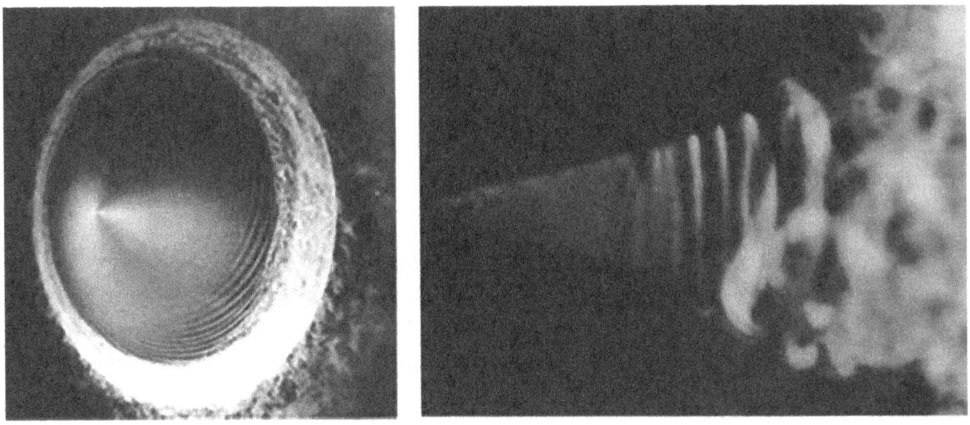

Figure 4.5 Flow visualisation of transitional flows on rotating cones with half angles of 60° and 15°, clockwise rotation. *Source:* Reproduced from Kobayashi and Izumi (1983) / Cambridge University Press.

Noting the dependency of the boundary layer length scale on cone half angle, the boundary layer Reynolds number for this case is $\sqrt{\Omega/\nu \sin\varepsilon}$. For half angles $\varepsilon \geq 60°$, the critical Reynolds number obtained by Kobayashi and Izumi from velocity measurements is fairly insensitive to cone angle, matching their free disc result. For $\varepsilon = 45°$, the critical Reynolds number is about 80% of the free disc value and reduces further at lower angles. Stability theory for onset of the spiral vortices captures this trend. Garrett and Peake's (2007) extension of Lingwood's absolute instability, free disc analysis to free cones showed reasonable agreement with the Kobayashi and Izumi's measurements of transitional Reynolds number for $\varepsilon \geq 45°$, but does not capture the steep reduction in critical Reynolds number at lower angles. This is discussed further by, amongst others, Hussain *et al.* (2014) who note the different transition mechanisms for broad and slender cones. Supported by further flow visualisation and measurements from Kobayashi and Izumi, Hussain *et al.* associate transition on slender rotating cones with counter-rotating, Görtler-type vortices arising from convective instability. As described by Saric (1994), Görtler and Taylor instabilities (that occur between rotating cylinders) share the same physical mechanism described by Rayleigh and discussed above in Section 3.1.1.

4.6 Rotating Cylinders

In contrast to their results for a free rotating disc, Theodorsen and Regier's (1944) measurements of moment for a free rotating cylinder do not show a sharp transition from laminar to turbulent flow. The free cylinder results are shown in Figure 4.6 and may be compared with the free disc results in Figure 4.1. Equations for the cylinder moment in laminar and turbulent conditions were given by Theodorsen and Regier and are included on the figure (in the present notation). The equation for laminar flow corresponds to the analytical flow solution 3.1.1 (with $A = 0$, $B = \Omega b^2$, and shear stress given by equation 2.1.29).

As discussed in Section 3.1, stability of flow between rotating cylinders has been subject to considerable research. Rayleigh's criterion for inviscid flow indicates instability of the laminar flow solution for the free cylinder and so departure of experimental data from this (for Re_ϕ above ~10) in Figure 4.6 is not surprising. Andereck *et al.*'s (1986) and many other studies of flow between rotating cylinders show a wide variety of flow types including wavy vortices, turbulent

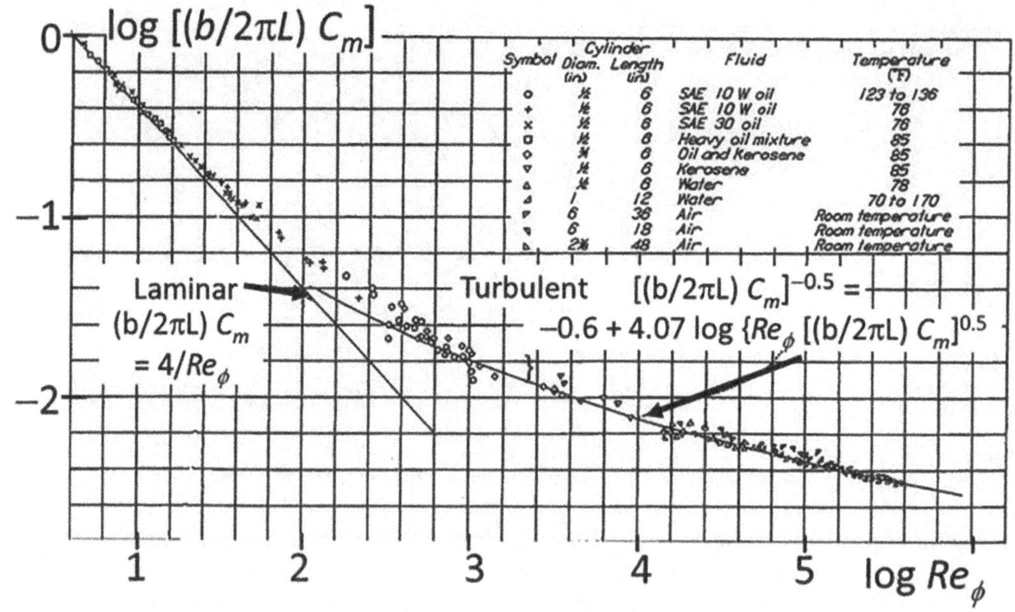

Figure 4.6 Free cylinder moment coefficient as a function of Reynolds number. *Source:* Adapted from Theodorsen and Regier (1944).

Taylor vortices and featureless turbulence, depending on the rotation rates of the two cylinders, radius ratio of the cylinders and Reynolds number. The review by Grossmann *et al.* (2016) includes further data and identifies conditions where transition in the boundary layers is delayed compared to transition in the interior flow.

For rotating cylinders enclosed by an outer stationary cylinder with a radius ratio of 0.7246 (Lathrop *et al.*'s, 1992) experiments indicate transition to shear-driven turbulence at Reynolds number based on radial spacing between the cylinders of 1.3×10^4, corresponding to a rotational Reynolds number for the inner cylinder of $Re_\phi \sim 3 \times 10^4$. Poncet *et al.* (2013) reported DNS for an inner rotating cylinder and outer stationary cylinder configuration with a radius ratio of 0.35. They state that transition to featureless turbulence occurs at Reynolds numbers below Lathrop *et al.*'s critical value of 1.3×10^4, with flow visualisation at a Reynolds number based on gap width of 10^4 (corresponding to $Re_\phi \sim 0.54 \times 10^4$) indicating featureless turbulent flow. This is consistent with the expected change in balance between shear flow effects and centrifugal effects as radius ratio changes.

4.7 Centrifugal Free Convection

4.7.1 Radial Convection

As discussed in Section 2.5.2, positive radial temperature gradients in a rotating flow can give rise to centrifugal Rayleigh–Bénard convection (for which analogies can be drawn with Taylor-Couette flow between rotating cylinders). Rayleigh–Bénard convection under gravity has been the subject of numerous investigations and continues to the present day. For flow between two horizontal plates with the lower plate hotter than the upper plate, convection occurs for Rayleigh numbers above a critical value. As discussed for example by Busse (1978), for Prandtl numbers of order

1 and less, steady convective rolls first occur, becoming unsteady and then turbulent as Rayleigh number rises. For $Pr = 0.7$, the flow is turbulent for Rayleigh numbers greater than $\sim 5 \times 10^3$. Lloyd and Moran (1974) suggested a considerably higher transition Rayleigh number based on their experiments on convection of flat plates immersed in a tank of liquid. This raises question as to how turbulence is defined and relates to larger scale flow structures.

Onset of centrifugal Rayleigh–Bénard convection between corotating cylinders, with the outer cylinder hotter than the inner cylinder was considered by Pitz *et al.* (2017b). This study presented numerical solutions of the linearised and nonlinear flow equations assuming the Boussinesq approximation for density variation and a value of $\beta \Delta T = 0.1$. Pitz *et al.* used a centrifugal Rayleigh number defined as $Pr\beta\Delta TRe_\phi^2(b-a)^3(b+a)/2b^4$ which equals $Ra(b-a)^3(b+a)/2b^4$ in terms of the rotational Rayleigh number used in this book. In the limit of radius ratio $a/b \rightarrow 1$, Pitz *et al.*'s definition corresponds to that commonly used for gravitational convection (with $\Omega^2(a+b)/2$ replacing gravity). Pitz *et al.*'s critical Rayleigh number for onset of convection was found to rise from the value for gravitational convection (1708) in the limit $a/b \rightarrow 1$ to $\sim$5600 at the lowest a/b considered of 0.2. The initial perturbation from the conductive state involves well-defined pairs of cells distributed circumferentially around the annulus. Variations in the axial direction only occur close to the adiabatic discs that form the end walls of the annulus. This shows qualitative agreement with the analysis for inertial waves in Section 3.3.4.

For the case $a/b = 0.52$, Pitz *et al.* presented nonlinear flow solutions for their Rayleigh numbers of 10^5 and 10^8 (corresponding to $Ra = Pr\beta\Delta TRe_\phi^2 = 1.19 \times 10^6$ and 1.19×10^9). Temperature contours on a mid-axial plane through the annulus are shown in Figure 4.7 at times up to 27.7 and 60 revolutions from start-up for the lower and higher Rayleigh numbers, respectively. A distinct flow pattern with six pairs of cells is seen at the lower Rayleigh number. Development of turbulent flow is apparent for the higher Rayleigh number, but with some regularity still discernible in the circumferential variation. A full systematic investigation of the onset of turbulence has not been conducted but DNS for this geometry by Saini and Sandberg (2020) has confirmed turbulence for $Ra = 1.19 \times 10^7$ and higher. Simulations by Gao and Chew (2021b) for $Ra = 1.19 \times 10^{11}$ still show some regularity in formation of turbulent plumes on the cylindrical surfaces.

As indicated in Sections 2.2.3 and 2.5.2, when the Eckert number ($Ec = \Omega^2 b^2/C_p \Delta T_{ref}$) is not negligible the rotary stagnation temperature T^* replaces the static temperature as a conserved property for isentropic flow in a rotating system. If an element of fluid in a forced vortex with $Ec = 0$ and $dT/dr > 0$ moves radially outwards in an isentropic process, its static temperature is unchanged and its density will be greater than that of the surrounding fluid. Centrifugal force will then accelerate the radial flow (as this is an unstable condition). For $Ec > 0$, the work done on the element, as its angular momentum is increased to that of the surrounding fluid, is significant, resulting in constant T^* and a rise in static temperature. Hence, the condition for instability becomes $dT^*/dr > 0$. Viscosity and thermal conductivity may modify this result. Note that the static temperature distribution in the isentropic forced vortex corresponds to radially inwards thermal conduction which may disturb the equilibrium. Nevertheless, as conduction is generally weak compared to convection, dT^*/dr is expected to have a strong effect on stability and development of turbulence.

4.7.2 Axial Convection

Free convection in a rotating cavity with a cylindrical shroud and discs at different (but uniform) temperatures was described in Section 3.3.3 and was studied experimentally by Hudson *et al.* (1978). For sufficiently high rotational Reynolds number and spacing between the discs,

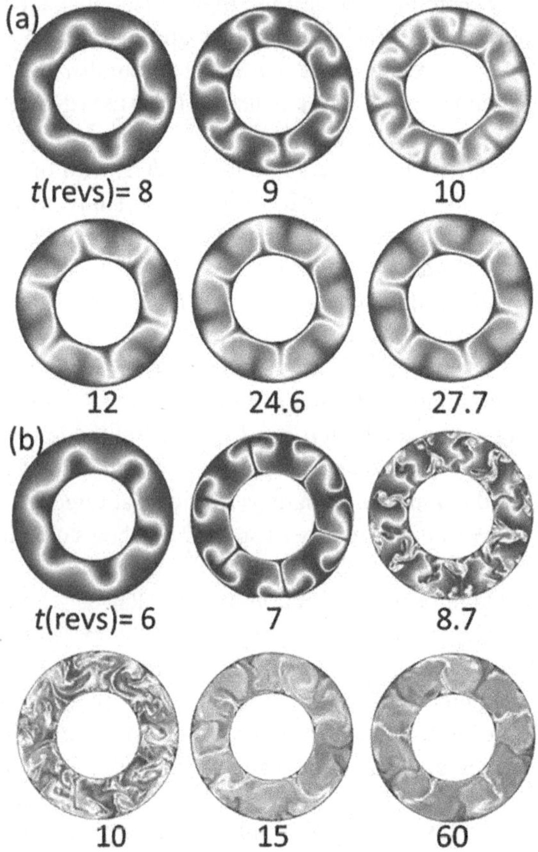

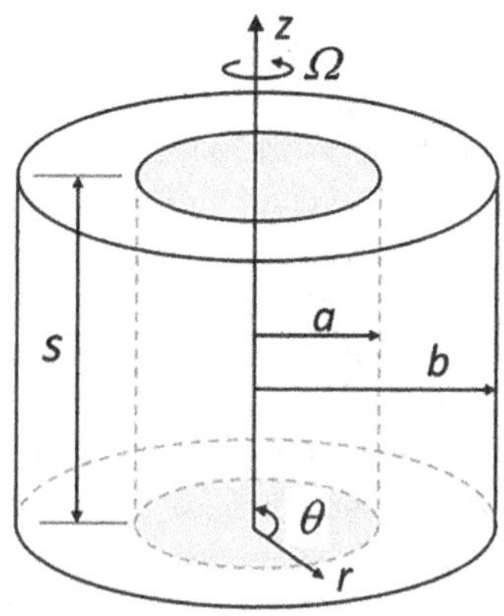

Figure 4.7 Instantaneous temperature contours in centrifugal free convection in a rotating annular cavity (a) $Ra = 1.19 \times 10^6$ (b) $Ra = 1.19 \times 10^9$. *Source:* Pitz *et al.* (2017b) / with permission of Cambridge University Press.

heat transfer measurements were consistent with laminar Ekman layers forming on the disc. Wang *et al.* (2024) considered axial convection in a cavity with higher aspect ratio $s/b = 0.5$ and radius ratio $a/b = 0.52$. Their simulations showed some turbulence in the core flow at $Ra = 1.2 \times 10^{10}$, but the Ekman layers remained laminar. Transition in the Ekman layers is expected at very high Reynolds numbers and may be preceded by transition in the shroud boundary layer.

A more common situation in turbomachinery is that the cylindrical shroud is heated and the discs have positive radial temperature gradients. In these conditions, mixed axial/radial convection may occur, and the core flow is expected to become non-axisymmetric, unsteady and turbulent at moderate Rayleigh numbers, as discussed for radial convection. However, as shown by Pitz *et al.*'s (2019) DNS, laminar Ekman layers may still form on the discs, driven by the unsteady interior flow structure. The criterion for transition given in Table 4.1 indicates that laminar Ekman layers can persist up to very high Rayleigh number. Consideration of scaling in radial convective flows by Gao and Chew (2021a) and simulations at high Reynolds numbers by Wang *et al.* (2025) indicate that the disc Ekman layers become intermittently or fully turbulent at some engine conditions. This is further discussed in Chapter 8.

4.8 Concluding Remark

While prediction of transition in engineering flows will be subject to uncertainties regarding exact boundary conditions and vibration levels, the results presented above give a useful indication of whether the flow will be laminar or turbulent for many applications. However, the discussion is far from exhaustive, and further information is available for other cases, such as closely spaced rotating and stationary surfaces where the boundary layers on the two surfaces merge. The material selected illustrates two key physical phenomena: centrifugal instability on rotating cylinders and narrow cones, and Coriolis force effects on rotating discs and wide cones. Both effects lead to formation of three dimensional flow structures prior to transition to full turbulence, and there is much agreement between experimental, analytical and DNS studies. Considerable insight has been achieved since Theodorsen and Regier's seminal experiments were published in 1944. The relatively recent emergence of DNS as a research tool promises further advances.

In gas turbine engines, most disc cavity flows will be fully turbulent at normal operating conditions. Much experimental research has been conducted at rotational Reynolds numbers of around 10^6 which is an order of magnitude lower than some engine conditions. Thus, transitional effects are more likely to be encountered in research data than in engine tests. Exceptions occur when the flow rotates with a similar velocity to the adjacent surface. A notable example occurs in cavities where centrifugal buoyancy dominates and the disc Ekman layers may remain laminar at quite extreme conditions and despite turbulence associated with centrifugal Rayleigh–Bénard convection at the cylindrical shroud. At some engine operating conditions in a flight cycle stabilising radial temperature gradients may occur, producing relatively quiescent rotating flow in some cavities.

References

Abedi, H., Pitz, B., Marxen, O., and Chew, J., 2025. Numerical investigation of Ekman layer transition in rotating flows. ASME Turbo Expo, Paper GT2025-152476. https://doi.org/10.1115/GT2025-152476.

Andereck, C. D., Liu, S. S., and Swinney, H. L., 1986. Flow regimes in a circular Couette system with independently rotating cylinders. J. Fluid Mech., 164, 155–183. https://doi.org/10.1017/S0022112086002513.

Busse, F. H., 1978. Non-linear properties of thermal convection. Rep. Prog. Phys., 41, 1929. https://doi.org/10.1088/0034-4885/41/12/003.

Daily, J. W. and Nece, R. E., 1960. Chamber dimension effects on induced flow and frictional resistance of enclosed rotating disks. ASME J. Basic Eng., 82(1), 217–230. https://doi.org/10.1115/1.3662532.

Faller, A., 1963. An experimental study of the instability of the laminar Ekman boundary layer. J. Fluid Mech., 15, 560–576. https://doi.org/10.1017/S0022112063000458.

Faller, A. J. and Kaylor, R. E., 1966. A numerical study of the instability of the Laminar Ekman boundary layer. J. Atmos. Sci., 23, 466–480. https://doi.org/10.1175/1520-0469(1966)023<0466:ANSOTI>2.0.CO;2.

Gao, F. and Chew, J. W., 2021a. Ekman layer scrubbing and shroud heat transfer in centrifugal buoyancy-driven free convection. ASME J. Eng. Gas Turbines Power, 143, 071010. https://doi.org/10.1115/1.4050366.

Gao, F. and Chew, J. W., 2021b. Evaluation and application of advanced CFD models for rotating disc flows. J. Mech. Eng. Sci. (Part C, Proc. IMechE), 235, 6847–6864. https://doi.org/10.1177/09544062211013850.

Garrett, S. J. and Peake, N., 2007. The absolute instability of the boundary layer on a rotating cone. Eur. J. Mech. B/Fluids, 26, 344–353. https://doi.org/10.1016/j.euromechflu.2006.08.002.

Gregory, N., Stuart, J. T., and Walker, W. S., 1955. On the stability of three-dimensional boundary layers with application to the flow due to a rotating disk. Philos. Trans. R. Soc. London, Ser. A, 248, 155–199. https://doi.org/10.1098/rsta.1955.0013.

Grossmann, S, Lohse, D. and Sun, C., 2016. High–Reynolds number Taylor–Couette turbulence. Annu. Rev. Fluid Mech., 48, 53–80. https://doi.org/10.1146/annurev-fluid-122414-034353.

Hudson, J., Tang, D., and Abell, S., 1978. Experiments on centrifugally driven thermal convection in a rotating cylinder. J. Fluid Mech., 86(1), 147–159. https://doi.org/10.1017/S0022112078001044.

Hussain, Z., Garrett, S., and Stephen, S., 2014. The centrifugal instability of the boundary-layer flow over slender rotating cones. J. Fluid Mech., 755, 274–293. https://doi.org/10.1017/jfm.2014.417.

Imayama, S., Alfredsson, P. H., and Lingwood, R. J., 2014. On the laminar–turbulent transition of the rotating-disk flow: The role of absolute instability. J. Fluid Mech., 745, 132–163. https://doi.org/10.1017/jfm.2014.80.

Itoh, M., Yamada, Y., Imao, S., and Gonda, M., 1992. Experiments on turbulent flow due to an enclosed rotating disk. Exp. Therm. Fluid Sci., 5, 359–368. https://doi.org/10.1016/0894-1777(92)90081-F.

Kobayashi, R. and Izumi, H., 1983. Boundary-layer transition on a rotating cone in still fluid. J. Fluid Mech., 127, 353–364. https://doi.org/10.1017/S0022112083002761.

Kreith, F., Ellis, D., and Giesing, J., 1963. An experimental investigation of the flow engendered by a rotating cone. Appl. Sci. Res., 11, 430–440. https://doi.org/10.1007/BF03184627.

Lathrop, D. P., Fineberg, J. and Swinney, H. L., 1992. Turbulent flow between concentric rotating cylinders at large Reynolds number. Phys. Rev. Lett. 68, 1515. https://doi.org/10.1103/PhysRevLett.68.1515.

Lingwood, R. J., 1995. Absolute instability of the boundary layer on a rotating disk. J. Fluid Mech., 299, 17–33. https://doi.org/10.1017/S0022112095003405.

Lingwood, R. J., 1996. An experimental study of absolute instability of the rotating-disk boundary-layer flow. J. Fluid Mech., 314, 373–405. https://doi.org/10.1017/S0022112096000365.

Lingwood, R., 1997. Absolute instability of the Ekman layer and related rotating flows. J. Fluid Mech., 331, 405–428. https://doi.org/10.1017/S0022112096004144.

Lloyd, J. R. and Moran, W. R., 1974. Natural convection adjacent to horizontal surface of various planforms. ASME J. Heat Transfer, 96, 443–447. https://doi.org/10.1115/1.3450224.

Owen, J. M. and Pincombe, J. R., 1980. Velocity measurements inside a rotating cylindrical cavity with a radial outflow of fluid. J. Fluid Mech., 99, 111–127. https://doi.org/10.1017/S0022112080000547.

Owen, J. M., Pincombe, J. R., and Rogers, R. H., 1985. Source–sink flow inside a rotating cylindrical cavity. J. Fluid Mech., 155, 233–265. https://doi.org/10.1017/S0022112085001793.

Pitz, D. B., Chew, J. W., Marxen, O., and Hills, N. J., 2017a. Direct numerical simulation of rotating cavity flows using a spectral element-Fourier method. ASME J. Eng. Gas Turbines Power, 139, 072602. https://doi.org/10.1115/1.4035593.

Pitz, D. B., Marxen, O., and Chew, J. W., 2017b. Onset of convection induced by centrifugal buoyancy in a rotating cavity. J. Fluid Mech., 826, 484–502. https://doi.org/10.1017/jfm.2017.451.

Pitz, D. B., Chew, J. W., and Marxen, O., 2019. Large-eddy simulation of buoyancy-induced flow in a sealed rotating cavity. ASME J. Eng. Gas Turbines Power, 141, 21020. https://doi.org/10.1115/1.4041113.

Poncet, S., Da Soghe, R., Bianchini, C., Viazzo, S., and Aubert, A., 2013. Turbulent Couette–Taylor flows with endwall effects: A numerical benchmark. Int. J. Heat Fluid Flow, 44, 229–238. https://doi.org/10.1016/j.ijheatfluidflow.2013.05.018.

Poncet, S., Serre, E., and Le Gal, P., 2009. Revisiting the two first instabilities of the flow in an annular rotor-stator cavity. Phys. Fluids, 21, 064106. https://doi.org/10.1063/1.3156859.

Saini, D. and Sandberg, R. D., 2020. Simulations of compressibility effects in centrifugal buoyancy-induced flow in a closed rotating cavity. Int. J. Heat Fluid Flow, 85, 108656. https://doi.org/10.1016/j.ijheatfluidflow.2020.108656.

Saric, W. S., 1994. Görtler votices. Annu. Rev. Fluid Mech., 26, 370–409. https://doi.org/10.1146/annurev.fl.26.010194.002115.

Savaş, Ö., 1987. Stability of Bödewadt flow. J. Fluid Mech., 183, 77–94. https://doi.org/10.1017/S0022112087002532.

Schlichting, H., 1979. Boundary-layer theory, Seventh edition, McGraw-Hill.

Theodorsen, T. and Regier, A., 1944. Experiments on drag of revolving disks, cylinders, and streamline rods at high speeds. NACA Report 793. https://ntrs.nasa.gov/api/citations/20050241738/downloads/20050241738.pdf.

Tien, C. and Campbell, D., 1963. Heat and mass transfer from rotating cones. J. Fluid Mech., 17(1), 105–112. https://doi.org/10.1017/S0022112063001142.

Viaud, B., Serre, E., and Chomaz, J. M., 2008. The elephant mode between two rotating disks. J. Fluid Mech., 598, 451–464. https://doi.org/10.1017/S0022112007009962.

von Kármán, T., 1921. Über laminare und turbulente Reibung. Z. Angew. Math. Mech., 1, 233–252. https://doi.org/10.1002/zamm.19210010401.

Wang, R., Gao, F., Chew, J. W., and Marxen, O., 2024. Large-eddy simulation of axial, radial and mixed centrifugal convection in a closed rotating cavity. Int. J. Heat Mass Transfer, 227, 125559. https://doi.org/10.1016/j.ijheatmasstransfer.2024.125559.

Wang, R., Gao, F., Chew, J. W., and Marxen, O., 2025. Flow and heat transfer in a rotating disc cavity with axial throughflow at high speed conditions. ASME J. Turbomach., 147, 091005. https://doi.org/10.1115/1.4067437.

Xie, Y., Du, Q., Xie, L., Wang, Z., and Li, S., 2024. Onset of turbulence in rotor–stator cavity flows. J. Fluid Mech., 991, A3. https://doi.org/10.1017/jfm.2024.536.

Yamada, Y. and Ito, M., 1975. On the frictional resistance of enclosed rotating cones. Bull. JSME, 18(123), 1026–1034. https://doi.org/10.1299/jsme1958.18.1026.

Yim, E., Chomaz, J. M., Martinand, D., and Serre, E., 2018. Transition to turbulence in the rotating disk boundary layer of a rotor–stator cavity. J. Fluid Mech., 848, 631–647. https://doi.org/10.1017/jfm.2018.239.

5

Integral Boundary-Layer Methods for Turbulent Flow

Although integral boundary-layer methods may be applied to laminar or turbulent flows, the focus here is on turbulent flows, which are beyond the analytical methods described in Chapter 2 and are most common in engineering. Such methods were widely applied in fluid dynamics research before computational methods became prevalent and are still used in industry. The developments for rotating disc flows date back to von Kármán's (1921) analysis of the free rotating disc and continued to at least the 1990s. Their popularity is due to their success in providing quick and convenient estimates in engineering calculations. A drawback of the method is that it is sensitive to assumptions made regarding the velocity and temperature profiles across the boundary-layer and relations for surface shear stress and heat transfer.

More general introductions to integral methods are given in many books on boundary-layer theory, for example, by Schlichting (1979). Treatments for rotating disc flows are described in books by Dorfman (1963), Owen and Rogers (1989, 1995), Shevchuk (2009) and Childs (2011). Treatments and assumptions vary, but all are based on the same basic principles. The approaches discussed below are representative in terms of assumptions made and capability. Many of the results and methods described have been used in industrial design.

5.1 Axisymmetric Integral Boundary-Layer Equations

5.1.1 Mass and Momentum Equations

Steady, axisymmetric flow is assumed and the (s, n) coordinate system of Figure 5.1 is adopted to consider boundary-layer flow on a rotating, axisymmetric body. With boundary-layer assumptions, $d(\sin\varepsilon)/dr \ll 1$, and the change of coordinate system, equations 2.1.30–2.1.33 give the following equations:

$$\frac{\partial(\rho r v_s)}{\partial s} + \frac{\partial(\rho r v_n)}{\partial n} = 0 \tag{5.1.1}$$

$$\frac{1}{r}\left\{\frac{\partial\left(\rho r v_s^2\right)}{\partial s} + \frac{\partial\left(\rho r v_n v_s\right)}{\partial n} - \sin(\varepsilon)\rho v_\phi^2\right\} = -\frac{\partial p}{\partial s} + \frac{\partial \tau_{sn}}{\partial n} \tag{5.1.2}$$

$$\frac{1}{r^2}\left\{\frac{\partial\left(\rho r^2 v_s v_\phi\right)}{\partial s} + \frac{\partial\left(\rho r^2 v_n v_\phi\right)}{\partial n}\right\} = \frac{\partial \tau_{n\phi}}{\partial n} \tag{5.1.3}$$

$$\frac{\partial p}{\partial n} = 0. \tag{5.1.4}$$

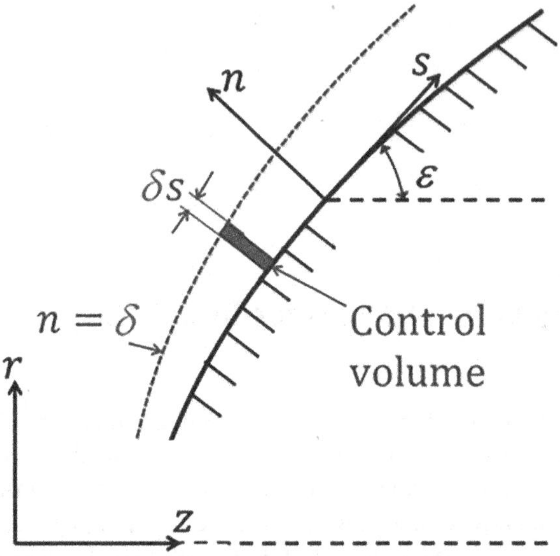

Figure 5.1 Boundary-layer coordinates for an axisymmetric body and control volume for integral equations.

At the boundary-layer edge $n = \delta$, it is assumed that $v_{s,\delta} \approx 0$, $v_\phi = v_{\phi,\delta}$ and $\tau_{sn} = \tau_{n\phi} = 0$. Assuming steady, inviscid, axisymmetric flow outside the boundary layer, equation 3.3.7 gives

$$\frac{\partial p}{\partial s} = \sin(\varepsilon)\frac{\rho v_{\phi,\delta}^2}{r}. \tag{5.1.5}$$

Note that equation 3.3.7 allows non-zero $v_{s,\delta}$ for flow on a conical body with $\sin\varepsilon \neq 1$. This remains compatible with $v_{s,\delta} \approx 0$ in the boundary-layer treatment since $v_{s,\delta}$ is of the same order as $v_{n,\delta}$ and $v_{n,\delta} \ll v_s$ in the boundary layer. Combining equations (5.1.2) and (5.1.5)

$$\frac{1}{r}\left\{\frac{\partial\left(\rho r v_s^2\right)}{\partial s} + \frac{\partial\left(\rho r v_n v_s\right)}{\partial n} - \sin(\varepsilon)\rho\left[v_\phi^2 - v_{\phi,\delta}^2\right]\right\} = \frac{\partial \tau_{sn}}{\partial n}. \tag{5.1.6}$$

The integral equations are obtained from the integration of equations (5.1.1), (5.1.6) and (5.1.3) across the boundary layer or, equivalently, from mass and momentum balances for the control volume in Figure 5.1 in the limit $\delta s \to 0$. First it is assumed that the shape of the velocity profile across the boundary layer is independent of s, and v_s and v_ϕ have the following form:

$$v_s = v_{s0}\eta^{1/n}(1-\eta), \quad v_\phi = v_{\phi,0} - (v_{\phi,0} - v_{\phi,\delta})\eta^{1/n} \quad \text{where} \quad \eta = n/\delta \tag{5.1.7a}$$

or

$$v_s = v_{s0}f(\eta), \quad v_\phi = v_{\phi,0} - (v_{\phi,0} - v_{\phi,\delta})g(\eta) \text{ where } f(\eta) = \eta^{1/n}(1-\eta), \quad g(\eta) = \eta^{1/n}. \tag{5.1.7b}$$

The boundary conditions for f and g are $f(0) = f(1) = g(0) = 0$ and $g(1) = 1$. Near the surface, the assumed profiles follow a $1/n$ power law as frequently used to approximate turbulent boundary layers and described by Schlichting (1979). The expression used for v_s allows for the crossflow generated by the rotation and generalises the form introduced by von Kármán for $n = 7$. Integration from $\eta = 0$ to 1, allowing for variation of density with the streamwise coordinate s but neglecting any variation across the boundary layer, then gives the following ordinary differential equations:

$$v_{n,\delta} = -\frac{I_1}{\rho r}\frac{d(\rho r v_{s0}\delta)}{ds} \tag{5.1.8}$$

Table 5.1 Summary of coefficients for the integral equations and indicative Reynolds number range.

n	K_n^2	I_1	I_2	I_3	I_4	I_5	$\sim\delta\lvert v_{\phi,0}-v_{\phi,\delta}\rvert/\nu$ Newman (1983)	Corresponding Re_ϕ for Free Disc
5	0.0384	0.3788	0.1751	0.2976	0.1667	0.2857		
6	0.029	0.3956	0.1929	0.3214	0.1429	0.25		
7	0.0225	0.4083	0.2071	0.3403	0.125	0.2222	0.6×10^5	2×10^6
8	0.0179	0.4183	0.2188	0.3556	0.1111	0.2	3×10^5	1.1×10^7
9	0.0145	0.4263	0.2285	0.3682	0.1	0.1818	7×10^5	2.5×10^7
10	0.0118	0.4329	0.2367	0.3788	0.09091	0.1667	1.7×10^6	5.7×10^7

$$I_2\frac{d\left(\rho r v_{s0}^2\delta\right)}{ds}+\sin(\varepsilon)\rho\delta\left[(v_{\phi,0}-v_{\phi,\delta})^2I_5-2v_{\phi,0}(v_{\phi,0}-v_{\phi,\delta})I_4\right]=-r\tau_{s,0} \tag{5.1.9}$$

$$-\frac{d(\rho r^2v_{s0}[v_{\phi,0}-v_{\phi,\delta}]\delta)}{r\,ds}I_3+\rho r v_{s,0}\delta\frac{d(rv_{\phi,0})}{r\,ds}I_1+\frac{d(\rho r v_{s0}\delta)}{ds}(v_{\phi,0}-v_{\phi,\delta})I_1=-r\tau_{\phi,0} \tag{5.1.10}$$

where (5.1.8) has been used in deriving (5.1.10) and the integrals I are defined, following Chew and Rogers (1988), below:

$$I_1=\int_0^1 f(\eta)d\eta,\quad I_2=\int_0^1 f^2(\eta)d\eta,\quad I_3=\int_0^1 f(\eta)g(\eta)d\eta,$$
$$I_4=1-\int_0^1 g(\eta)d\eta,\quad I_5=1-\int_0^1 g^2(\eta)d\eta. \tag{5.1.11}$$

Numerical values of these integrals are tabulated in Table 5.1.

To complete the integral momentum equations, the following expressions are assumed for the surface shear stress:

$$\tau_{\phi,0}=-\rho K_n^2\left(\frac{\mu}{\rho\delta}\right)^{2/(n+1)}(v_{\phi,0}-v_{\phi,\delta})\left[v_{s0}^2+(v_{\phi,0}-v_{\phi,\delta})^2\right]^{(n-1)/2(n+1)} \tag{5.1.12}$$
$$\tau_{s,0}=\frac{-v_{s,0}}{(v_{\phi,0}-v_{\phi,\delta})}\tau_{\phi,0}.$$

As described by Schlichting, values of K_n^2 may be deduced from data for turbulent pipe flows. Newman (1983) and Owen *et al.* (1985) give values for K_n^2; Newman for $n=7$–10 and Owen *et al.* for $n=5$–9. For $n=7$–9, the two sets of values agree to within 2%. Table 5.1 gives Owen and Rogers's values where they are available and Newman's for $n=10$. Some workers have adjusted K_n^2 to achieve a better match with experimental data. For example, Wormley (1969) used a value of 0.021 rather than 0.0225 with $n=7$ for his study of flow in a vortex chamber.

Indicative Reynolds numbers (based on boundary-layer thickness) where different values of n are appropriate are given by Newman and are included in Table 5.1. These were based on the consideration of values given by Schlichting for pipe flows. Based on experimental data for free discs with Re_ϕ up to 2×10^6, Newman recommended using the $1/7th$ power law for lower Reynolds numbers with departure of the calculated entrainment flow rate from experiment for Re_ϕ below $\sim8\times10^5$ attributed to laminar flow regions on the disc. Further examination of friction factor results for pipe flow suggests that predictions for $n=7$–10 converge to values within 5% at $\delta|v_{\phi,0}-v_{\phi,\delta}|/\nu\approx0.5\times10^5$, and that a reasonable choice would be $n=7$ and $n=10$ below and above this value up to boundary-layer Reynolds numbers around 2×10^6. However, pipe flow results may not read across directly to rotating discs. Based on comparison with experimental data, $n=7$ has been widely used for rotating disc flows. The choice of n and validity of the free disc solutions are discussed further in section 5.2.

Nondimensional variables and scaling of the boundary-layer thickness with rotational Reynolds number ($Re_\phi = \rho_{ref}\Omega b^2/\mu_{ref}$) and cone half angle ε, where b is a representative radius, may be introduced as follows:

$$\check{r} = r/b, \qquad \check{v}_i = v_i/\Omega b, \quad \delta^* = Re_\phi^{2/(n+3)} \sin(\varepsilon)^{(n+1)/(n+3)} \check{\rho}\delta/b,$$
$$\check{\rho} = \rho/\rho_{ref}, \quad \check{\mu} = \frac{\mu}{\mu_{ref}}. \tag{5.1.13}$$

The integral equations (5.1.9) and (5.1.10), with use of (5.1.12) and $ds = dr/\sin\varepsilon$, then have the nondimensional form

$$I_2\frac{d\left(\check{r}\check{v}_{s0}^2\delta^*\right)}{d\check{r}} + \delta^*\left[(\check{v}_{\phi,0} - \check{v}_{\phi,\delta})^2 I_5 - 2\check{v}_{\phi,0}(\check{v}_{\phi,0} - \check{v}_{\phi,\delta})I_4\right]$$
$$= -\check{r}\check{\rho}K_n^2\left(\frac{\check{\mu}}{\delta^*}\right)^{2/(n+1)} \check{v}_{s0}\left[\check{v}_{s0}^2 + (\check{v}_{\phi,0} - \check{v}_{\phi,\delta})^2\right]^{(n-1)/2(n+1)},$$
$$-\frac{d(\check{r}^2\check{v}_{s,0}[\check{v}_{\phi,0} - \check{v}_{\phi,\delta}]\delta^*)}{\check{r}d\check{r}} I_3 + \check{r}\check{v}_{s,0}\delta^*\frac{d(\check{r}\check{v}_{\phi,0})}{\check{r}d\check{r}} I_1 + \frac{d(\check{r}v_{s0}\delta^*)}{d\check{r}}(\check{v}_{\phi,0} - \check{v}_{\phi,\delta})\, I_1$$
$$= \check{r}\check{\rho}K_n^2\left(\frac{\check{\mu}}{\delta^*}\right)^{2/(n+1)} (\check{v}_{\phi,0} - \check{v}_{\phi,\delta})\left[\check{v}_{s0}^2 + (\check{v}_{\phi,0} - \check{v}_{\phi,\delta})^2\right]^{(n-1)/2(n+1)}. \tag{5.1.14}$$

These equations are independent of Re_ϕ and ε, which indicates the following dependencies on these parameters:

Boundary-layer thickness	$\delta/b \propto Re_\phi^{-2/(n+3)} \sin(\varepsilon)^{-(n+1)/(n+3)}$
Boundary-layer mass flow	$\dot{m}/\rho_{ref}\Omega b^3 \propto Re_\phi^{-2/(n+3)} \sin(\varepsilon)^{-(n+1)/(n+3)}$
Surface shear stress	$\tau_0/\rho_{ref}\Omega^2 b^2 \propto Re_\phi^{-2/(n+3)} \sin(\varepsilon)^{2/(n+3)}$
Moment	$\dfrac{M}{\rho_{ref}\Omega^2 b^5} \propto Re_\phi^{-2/(n+3)} \sin(\varepsilon)^{-(n+1)/(n+3)}$. (5.1.15)

5.1.2 The Energy Equation

For steady, axisymmetric flow in the (s, n) coordinate system, the boundary-layer energy equation obtained from 2.1.34 is as follows:

$$\frac{\partial(\rho r v_s h_t)}{r\partial s} + \frac{\partial(\rho r v_n h_t)}{r\partial n} = -\frac{\partial}{\partial n}(q_n - \tau_{n\phi}v_\phi - \tau_{sn}v_s). \tag{5.1.16}$$

Motivated by the Reynolds analogy, the variation of enthalpy across the boundary layer is assumed to be similar to that of the circumferential velocity, so that

$$h_t = h_{t,0} - (h_{t,0} - h_{t,\delta})g(\eta) \quad \text{where} \quad g(\eta) = \eta^{1/n}. \tag{5.1.17}$$

As discussed by Chew and Rogers (1988), other studies have introduced modifications to the enthalpy profile, which allow it to vary with streamwise position and aim to account for departures from conditions in which the Reynolds analogy is exact. Here, the simplest method given by Chew and Rogers is used, noting that there are additional uncertainties in the treatment of the integral energy equation to those in the momentum equations and that there is some evidence that a more complex treatment gives better agreement with experimental data.

From equations 2.5.28 and 2.5.29, the heat flux to the surface is given by

$$\frac{q_0}{\left(h_\delta - h_0 + R_f(v_{\phi,0} - v_{\phi,\delta})^2/2\right)} = \frac{\psi(Pr)}{Pr}\left[\frac{\tau_{\phi,0}}{(v_{\phi,\delta} - v_{\phi,0})}\right] \tag{5.1.18}$$

with R_f taken as $Pr^{1/3}$ and $\psi(Pr) = Pr^{0.6}$, as discussed in Section 2.5. Integrating the energy equation across the boundary layer with the use of equation (5.1.8) gives the following equation:

$$-\frac{d(\rho r v_{s,0}[h_{t,0} - h_{t,\delta}]\delta)}{r\,ds} I_3 + \frac{d(\rho r v_{s0} h_{t,0}\delta)}{r\,ds} I_1 - \frac{d(\rho r v_{s0}\delta)}{r\,ds} h_{t,\delta} I_1 = -q_0 - v_{\phi,0}\tau_{\phi,0}. \tag{5.1.19}$$

Defining $\check{h} = h_t/\Delta h_{ref}$ where Δh_{ref} is a representative enthalpy difference, the nondimensional integral energy equation may be written as

$$\begin{aligned}
&-\frac{d(\check{r}\check{v}_{s0}[\check{h}_{t,0} - \check{h}_{t,\delta}]\delta^*)}{\check{r}d\check{r}} I_3 + \frac{d(\check{r}\check{v}_{s0}\check{h}_{t,0}\delta^*)}{\check{r}\,d\check{r}} I_1 - \frac{d(\check{r}\check{v}_{s0}\delta^*)}{\check{r}\,d\check{r}}\check{h}_{t,\delta} I_1 \\
&= -\check{\rho}K_n^2\left(\frac{\check{\mu}}{\delta^*}\right)^{2/(n+1)} \left[\check{v}_{s0}^2 + (\check{v}_{\phi,0} - \check{v}_{\phi,\delta})^2\right]^{(n-1)/2(n+1)} \\
&\times\left\{\frac{\psi(Pr)}{Pr}(\check{h}_{t,\delta} - \check{h}_{t,0}) + \left(1 - \frac{\psi(Pr)}{Pr}\right)\left(\frac{\Omega^2 b^2}{\Delta h_{ref}}\check{v}_{\phi,0}(\check{v}_{\phi,\delta} - \check{v}_{\phi,0})\right)\right. \\
&\left.\quad -(1 - R_f)\frac{\psi(Pr)}{Pr}\frac{\Omega^2 b^2}{2\Delta h_{ref}}(\check{v}_{\phi,0} - \check{v}_{\phi,\delta})^2\right\}.
\end{aligned} \tag{5.1.20}$$

It may be noted that the heat and work transfer terms on the right-hand side of this equation are simplified considerably when either the Prandtl number is unity or the Eckert number $\Omega^2 b^2/\Delta h_{ref}$ approaches zero. Simplifications might also be achieved using rothalpy in place of total enthalpy.

As for the momentum equations, the nondimensional integral energy equation is independent of Re_ϕ and ε. This indicates the following dependency for heat transfer:

$$\text{Surface heat transfer} \qquad q_0 b/\kappa\Delta T_{ref} \propto Re_\phi^{(n+1)/(n+3)}(\sin\varepsilon)^{2/(n+3)}. \tag{5.1.21}$$

5.2 Free and Enclosed Rotating Disc Flows

5.2.1 A Disc Rotating in a Quiescent Environment

A disc rotating in a quiescent environment is a relatively well-understood and well-investigated test case and provides a reference for other configurations. Hence, the integral solution and comparison with experimental data is considered in some detail. As described in the next section, the free disc is a special case of a similarity solution of the integral momentum equations for forced vortex flow above a rotating disc or cone. The solution for the free disc with various values of n is summarised as follows and in Table 5.2:

$$\begin{aligned}
&\frac{\delta}{r} = B_\delta\left(\frac{\rho\Omega r^2}{\mu}\right)^{-2/(n+3)}, \qquad \frac{\tau_{\phi,0}}{\rho\Omega^2 r^2} = -B_\tau\left(\frac{\rho\Omega r^2}{\mu}\right)^{-2/(n+3)}, \\
&\frac{\dot{m}}{\rho\Omega r^3} = B_m\left(\frac{\rho\Omega r^2}{\mu}\right)^{-2/(n+3)}, \qquad \frac{2|M|}{\rho\Omega^2 b^5} = B_M\left(\frac{\rho\Omega b^2}{\mu}\right)^{-2/(n+3)}.
\end{aligned} \tag{5.2.1}$$

Comparison of von Kármán's velocity profiles for the above solution with measurements is shown in Figure 5.2. The measurements are as plotted by Itoh and Hasegawa (1994) and include

Table 5.2 Coefficients for the free disc integral solution.

n	$v_{s0}/\Omega r$	B_δ	B_τ	B_m	B_M
7	0.162	0.526	0.0267	0.219	0.0729
8	0.143	0.505	0.0210	0.190	0.0569
9	0.128	0.486	0.0169	0.167	0.0454
10	0.116	0.463	0.0136	0.146	0.0365

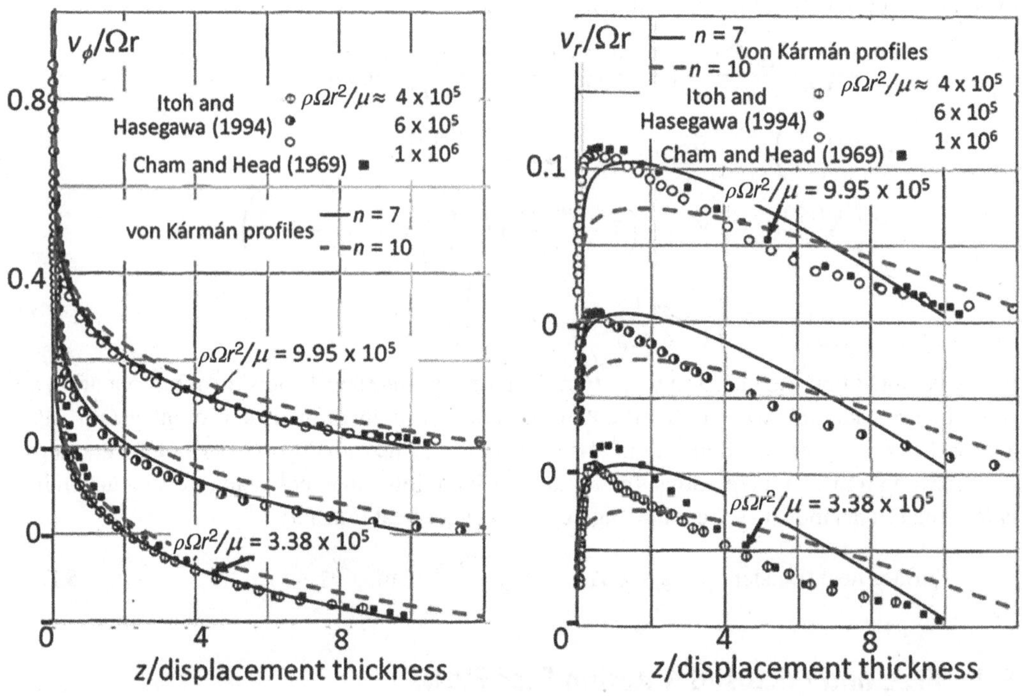

Figure 5.2 Comparison of von Kármán profiles with measurements for the free disc. *Source:* Adapted from Itoh and Hasegawa (1994).

those of Cham and Head (1969), covering the Re_ϕ range of ~4×10^5 to 10^6. The integral solution profiles for $n = 7$ and $n = 10$ are given to indicate trends, but $n = 7$ is expected to be the most appropriate at these Reynolds numbers. This is confirmed by comparisons with measurements that also show that the assumed form of the radial velocity is a poorer match to the experiments than is the case for the circumferential velocity. As reported by Newman (1983), the integral solution gives good agreement with Cham and Head's measured flow entrainment for $Re_\phi = 10^6$ to 2×10^6. At higher Re_ϕ, any improvements expected by the use of higher n may be offset by changes in accuracy of the assumed radial velocity profile.

Although the $n = 7$ solution has proven effective for Re_ϕ values typical of those in laboratory studies, the scaling at higher Reynolds numbers is very relevant to modern, high-pressure ratio, gas turbine engines and, for example, to supercritical CO_2 turbines, where values well above 10^7 occur. Dorfman's (1963) comparison of von Kármán's solution with Theodorsen and Regier's (1944) disc moment measurements shows some divergence of the solution and experiment at higher

Re_ϕ. For the highest value $Re_\phi \sim 7 \times 10^6$, the discrepancy is about 10%. Dorfman derived an alternative formula for the moment coefficient with logarithmic scaling that gives a good match to Theodorsen and Regier's data and agrees with the Reynolds number scaling for boundary-layer drag on a smooth flat plate. Dorfman also gives a revised expression for disc boundary-layer thickness, but this does not agree well with the von Kármán solution. If it is supposed that similar velocity profiles apply at all Reynolds numbers, then the logarithmic scaling for mass flow entrainment and boundary-layer thickness will follow that for the moment coefficient. Matching this scaling with the von Kármán solution at $Re_\phi = 2 \times 10^6$, and differentiating the moment coefficient with respect to the radius to find the local shear stress gives a revised solution for the free disc. This revised solution is compared with von Kármán's as follows:

Von Kármán free disc solution ($n = 7$)

$$\frac{\delta}{r} = 0.526\left(\frac{\rho\Omega r^2}{\mu}\right)^{-0.2}, \quad \frac{\tau_{\phi,0}}{\rho\Omega^2 r^2} = -0.0267\left(\frac{\rho\Omega r^2}{\mu}\right)^{-0.2},$$

$$C_{W,fd} = 0.219\, Re_\varphi^{0.8}, \qquad C_{M,fd} = 0.0729\, Re_\varphi^{-0.2}. \tag{5.2.2}$$

Free disc solution with revised (logarithmic) Re_ϕ scaling

$$\frac{\delta}{r} = 3.34\left[\log\left(\frac{\rho\Omega r^2}{\mu}\right)\right]^{-2.58}, \quad \frac{\tau_{\phi,0}}{\rho\Omega^2 r^2} = -0.118\left[\log\left(\frac{\rho\Omega r^2}{\mu}\right)\right]^{-2.4},$$

$$C_{W,fd} = 1.39\, Re_\varphi[\log(Re_\varphi)]^{-2.58}, \quad C_{M,fd} = 0.491[\log(Re_\varphi)]^{-2.58}. \tag{5.2.3}$$

For convenience, the expression given for $\tau_{\phi,0}$ in (5.2.3) is a curve fit rather than the exact expression derived by differentiation of C_M.

Figure 5.3 compares the above solutions and also includes the integral solution for $n = 10$ and the laminar solution given in equation 3.1.11 (with $5\sqrt{\nu/\Omega}$ taken as indicative of the laminar boundary thickness). As noted in the previous chapter, transition to turbulence has been observed in the region of $Re_\phi = 3 \times 10^5$. Dorfman's logarithmic scaling result for moment coefficient gives significantly higher values than von Kármán's solution at high Reynolds numbers with 20% difference at $Re_\phi = 10^8$. For lower Re_ϕ turbulent flow, the log scaling result is close to the $n = 7$ solution, but it better matches the $n = 10$ solution at high Re_ϕ, as expected. The disc flow entrainment and boundary-layer thickness given by the integral solutions for $n = 7$ and 10 show large differences. This is consistent with the velocity profile result in Figure 5.3 and suggests caution in using results unsupported by experimental data for values of n other than 7. Extrapolation to higher Reynolds numbers inevitably involves some uncertainty. Equation (5.2.3) might be preferred over the von Kármán solution for high Reynolds number applications as the logarithmic scaling better matches the highest Re_ϕ data available and is in line with the experience for other flows.

Heat transfer is related to shear stress on the disc through the Reynolds analogy, as described in Section 2.5.3. The integral energy and angular momentum equations (5.1.19) and (5.1.10) can be shown to have similar form when $Pr = 1$. For the case of a disc rotating in quiescent surroundings at temperature T_{ref} and with a surface temperature distribution of the form $T_0 - T_{ref} \propto r^2$, the energy equation solution is directly related to the solution for angular momentum. Considering the effect of the Prandtl number and temperature distributions of the form $T_0 - T_{ref} \propto r^p$, Dorfman obtained the following expressions for disc heat transfer, based on von Kármán's momentum solution:

$$\frac{rq_0}{\kappa(T_{ref} - T_0)} = 0.0197(p + 2.6)^{0.2} Pr^{0.6}\left(\frac{\rho\Omega r^2}{\mu}\right)^{0.8}$$

$$Nu_{av} = \frac{bq_{0,av}}{\kappa(T_{ref} - T_{0,av})} = 0.0197\frac{(p+2)}{(p+2.6)^{0.8}} Pr^{0.6} Re_\phi^{0.8}. \tag{5.2.4}$$

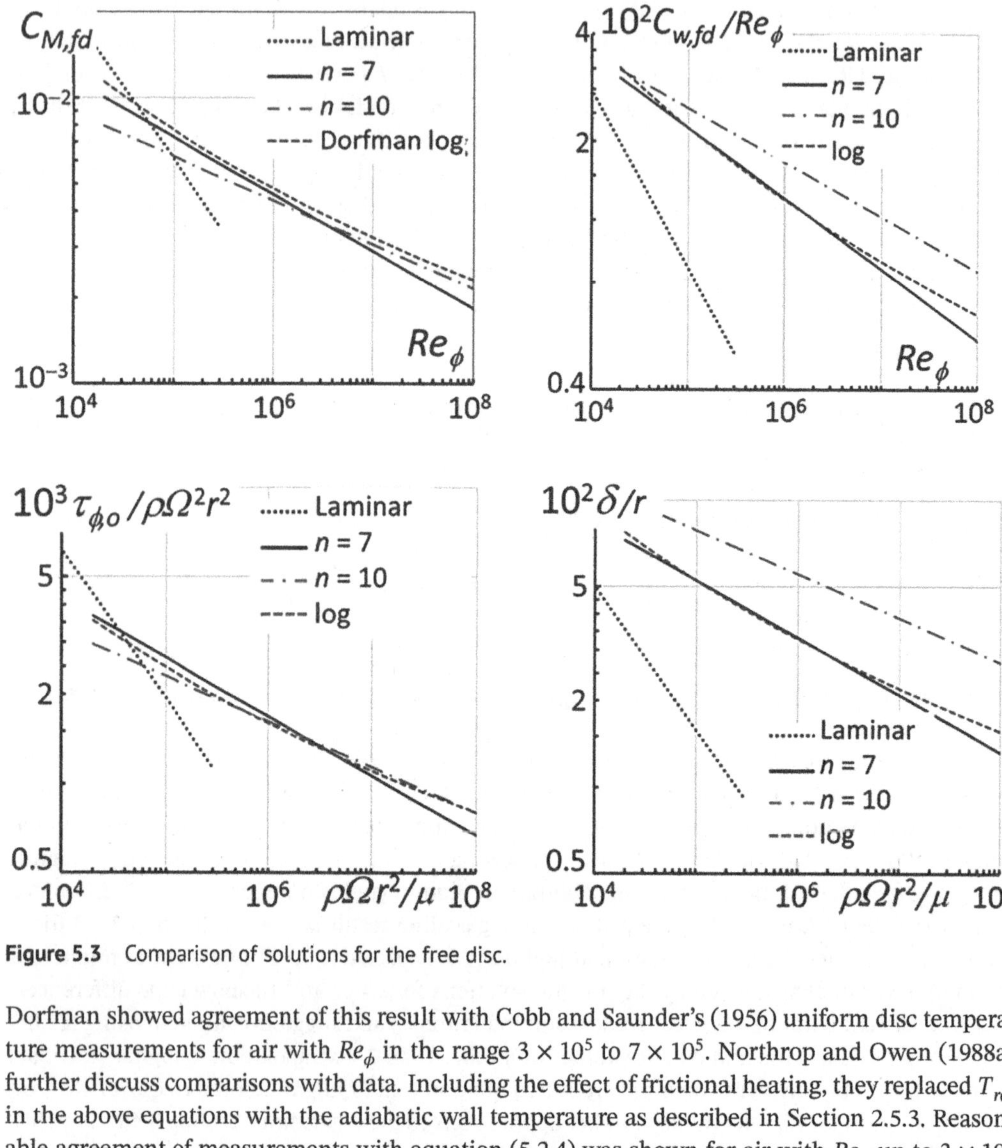

Figure 5.3 Comparison of solutions for the free disc.

Dorfman showed agreement of this result with Cobb and Saunder's (1956) uniform disc temperature measurements for air with Re_ϕ in the range 3×10^5 to 7×10^5. Northrop and Owen (1988a) further discuss comparisons with data. Including the effect of frictional heating, they replaced T_{ref} in the above equations with the adiabatic wall temperature as described in Section 2.5.3. Reasonable agreement of measurements with equation (5.2.4) was shown for air with Re_ϕ up to 3×10^6 and disc temperature variations corresponding approximately to $p = -0.2$, 0.1, 0.4 and 0.6.

As discussed above and shown in Figure 5.3, the disc shear stress may be underestimated by von Kármán's solution at higher Re_ϕ. Using the logarithmic scaling in equation (5.2.3) rather than the $n = 7$ solution, and including frictional effects, Dorfman's expression for the local Nusselt number is modified to

$$\frac{rq_0}{\kappa(T_{ref} + R_f \Omega^2 r^2/2C_p - T_0)} = 0.0871(p + 2.6)^{0.2} Pr^{0.6} \left(\frac{\rho\Omega r^2}{\mu}\right) \left[\log\left(\frac{\rho\Omega r^2}{\mu}\right)\right]^{-2.4}. \quad (5.2.5)$$

Noting the lack of experimental data, at higher Re_ϕ, the use of this expression would be reasonably consistent with the use of equation (5.2.3) to extrapolate other flow properties to high Reynolds numbers. No attempt has been made to modify Dorfman's results for the effect of disc temperature distribution to account for logarithmic scaling or frictional heating.

The effect of cone angle can be incorporated in the above results using the scaling given in equations (5.1.15) and (5.1.21). As discussed in Chapters 3 and 4, this relies on the boundary-layer assumptions that will break down for small cone angles. Tien and Campbell's (1963) experiments for rotating cones in a quiescent environment indicated similarity of disc and cone turbulent flows for half angles of 30° and above for Re_ϕ up to $\sim 7 \times 10^5$. Incorporating the effect of cone angle in the logarithmically scaled formulae for disc flow and heat transfer gives

$$\frac{\delta}{r} = \frac{3.34}{\sin\varepsilon}\left[\log\left(\frac{\rho\Omega r^2}{\mu\,\sin\varepsilon}\right)\right]^{-2.58}, \qquad \frac{\tau_{\phi,0}}{\rho\Omega^2 r^2} = -0.118\left[\log\left(\frac{\rho\Omega r^2}{\mu\,\sin\varepsilon}\right)\right]^{-2.4},$$

$$C_{W\,fd} = \frac{1.39}{\sin\varepsilon} Re_\varphi\left[\log\left(\frac{Re_\varphi}{\sin\varepsilon}\right)\right]^{-2.58}, \qquad C_{M\,fd} = \frac{0.491}{\sin\varepsilon}\left[\log\left(\frac{Re_\varphi}{\sin\varepsilon}\right)\right]^{-2.58},$$

$$\frac{rq_0}{\kappa(T_{ref} + R_f\Omega^2 r^2/2C_p - T_0)} = 0.0871(p+2.6)^{0.2} Pr^{0.6}\left(\frac{\rho\Omega r^2}{\mu}\right)\left[\log\left(\frac{\rho\Omega r^2}{\mu\,\sin\varepsilon}\right)\right]^{-2.4}. \tag{5.2.6}$$

The free disc solution may be modified to account for the effect of a non-zero hub radius (a) using the following adjustments to the solution for $a = 0$:

$$\frac{C_W}{C_{W,a=0}} = \frac{C_M}{C_{M,a=0}} = 1 - \left(\frac{a}{b}\right)^5$$

$$\frac{\tau_{\phi,0}}{\tau_{\phi,0,a=0}} = \frac{q_0}{q_{0,a=0}} = 1 + 0.087\left(\frac{a}{r}\right)^5. \tag{5.2.7}$$

These equations are based on the numerical solution of the integral equations, as reported by Chew (1988), with shear stress obtained from differentiation of the disc moment and heat flux estimated from the Reynolds analogy.

5.2.2 Forced Vortex Flow Above a Rotating Disc or Cone with $0 \leq k_\delta < 1$

As described by Newman (1983), von Kármán's similarity solution of the integral momentum equations for the free disc can be extended to the case of forced vortex flow above a rotating disc. The solution for a rotating cone may be obtained from that of a rotating disc, as shown by the nondimensional equations above. It is assumed that $v_{\phi,0} = \Omega r$ and $v_{\phi,\delta} = k_\delta \Omega r$ with constant k_δ. Assuming constant density, that radial and circumferential velocity components v_s and v_ϕ are proportional to r and that the boundary-layer thickness δ is proportional to r^m where $m = (n-1)/(n+3)$, differential equations (5.1.9) and (5.1.10) may be reduced to algebraic equations. Using the relationship between the surface shear stress components in the second part of equation (5.1.12) then allows δ to be eliminated, resulting in the following relation between $v_{s0}/\Omega r$ and k_δ:

$$\left(\frac{v_{s0}}{\Omega r(1-k_\delta)}\right)^2 [A1 + A2/(1-k_\delta)] = A3/(1-k_\delta) + A4 \tag{5.2.8}$$

where $A1$ to $A4$ are functions of n only and are tabulated in Table 5.3. The positive square root should be taken in calculating v_{s0}. The corresponding results for δ/r and $\tau_\phi/\rho\Omega^2 r^2$ as functions of the freestream swirl fraction k_δ and n can be obtained by substitution for v_{s0} in equations (5.1.9), (5.1.10) and (5.1.12). The Reynolds number and cone half-angle dependency follows that deduced in Section 5.1. Integration across the boundary layer and over the cone area gives the flow entrainment at radius r and disc moment for a body of radius b.

Table 5.3 Coefficients for the integral solution with $0 \le k_\delta \le 1$.

n	*m*	*A1*	*A2*	*A3*	*A4*
7	0.6	0.2420	0.8167	0.25	−0.2222
8	0.6364	0.2500	0.8366	0.2222	−0.2
9	0.6667	0.2566	0.8526	0.2	−0.1818
10	0.6923	0.2622	0.8658	0.1818	−0.1667

Focussing on conditions with flow rotating at lower speeds than the disc (with $0 \le k_\delta \le 1$), and defining a local rotational Reynolds number $Re_k = \rho|1 - k_\delta|\Omega r^2/\mu$, the solution may be related to the free disc case $k = 0$ at the same rotational Reynolds number Re_ϕ, as follows:

$$\frac{\delta}{\delta_{fd}} = \frac{B_\delta}{B_{\delta fd}}(1-k_\delta)^{(n-1)/(n+3)}, \qquad \frac{\tau_{\phi,0}}{\tau_{\phi,0,fd}} = -\frac{B_\tau}{B_{\tau fd}}(1-k_\delta)^{2(n+1)/(n+3)},$$

$$\frac{\dot{m}}{\dot{m}_{fd}} = \frac{B_m}{B_{m,fd}}(1-k_\delta)^{2(n+1)/(n+3)}, \qquad \frac{M}{M_{fd}} = \frac{B_M}{B_{M,fd}}(1-k_\delta)^{2(n+1)/(n+3)} = \frac{\tau_{\phi,0}}{\tau_{\phi,0,fd}}. \qquad (5.2.9)$$

Here, the subscript *fd* denotes the free disc solution or, if appropriate, the free cone solution. The coefficients B_δ, B_τ, B_m and B_M depend on n and k_δ only and are defined below. In deriving the moment, it is assumed that the angle ε is constant, although the result given still applies for $k_\delta = 0$ for bodies with (slowly) variable angle, in which case the value of ε at $r = b$ should be used in the equations.

$$B_\delta = [1+\alpha^2]^{(n-1)/2(n+3)}\left(\frac{K_n{}^2\alpha}{2I_4 - [(m+3)I_2\alpha^2 + I_5](1-k_\delta)}\right)^{(n+1)/(n+3)},$$

$$B_\tau = B_\delta\alpha\,[\{(2+m)I_1 - (4+m)I_3\}\,(1-k_\delta) + 2I_1]. \qquad (5.2.10)$$

$$B_m = 2\pi\, I_1\alpha B_\delta, \quad B_M = \frac{4\pi\, B_\tau}{5 - 4/(n+3)},$$

with

$$\alpha = \frac{v_{s0}}{\Omega r(1-k_\delta)}.$$

Values of the above parameters for the case of a disc rotating in quiescent surroundings ($k_\delta = 0$) were given in Table 5.2 for n = 7–10. The solutions relative to those for the free disc are shown graphically in Figure 5.4, noting that the moment coefficient ratio equals the shear stress ratio. Similar trends are shown in all cases. The linear solution (with n = 7) in Figure 5.4 is derived assuming the flow to be a small perturbation from a forced vortex with $k_\delta = 1$. This is discussed in Section 5.3.1.

For an enclosed rotating disc with separate turbulent boundary layers on the rotor and stator, Daily and Nece (1960) found a core flow rotating at around 40% to 46% of disc speed and correlated their torque measurements as $C_M = 0.051Re_\phi^{-0.2}(s/b)^{0.1}$ for the gap to radius ratio s/b in the range 0.0636–0.217. Thus, the experimental range of $C_M Re_\phi^{0.2}$ was around 0.039–0.044. Values of $C_M Re_\phi^{0.2}$ from equation (5.2.9) for $n = 7$ and $k_\delta = 0.46$ and 0.4 are 0.037 and 0.0425, comparing well to the experimental data.

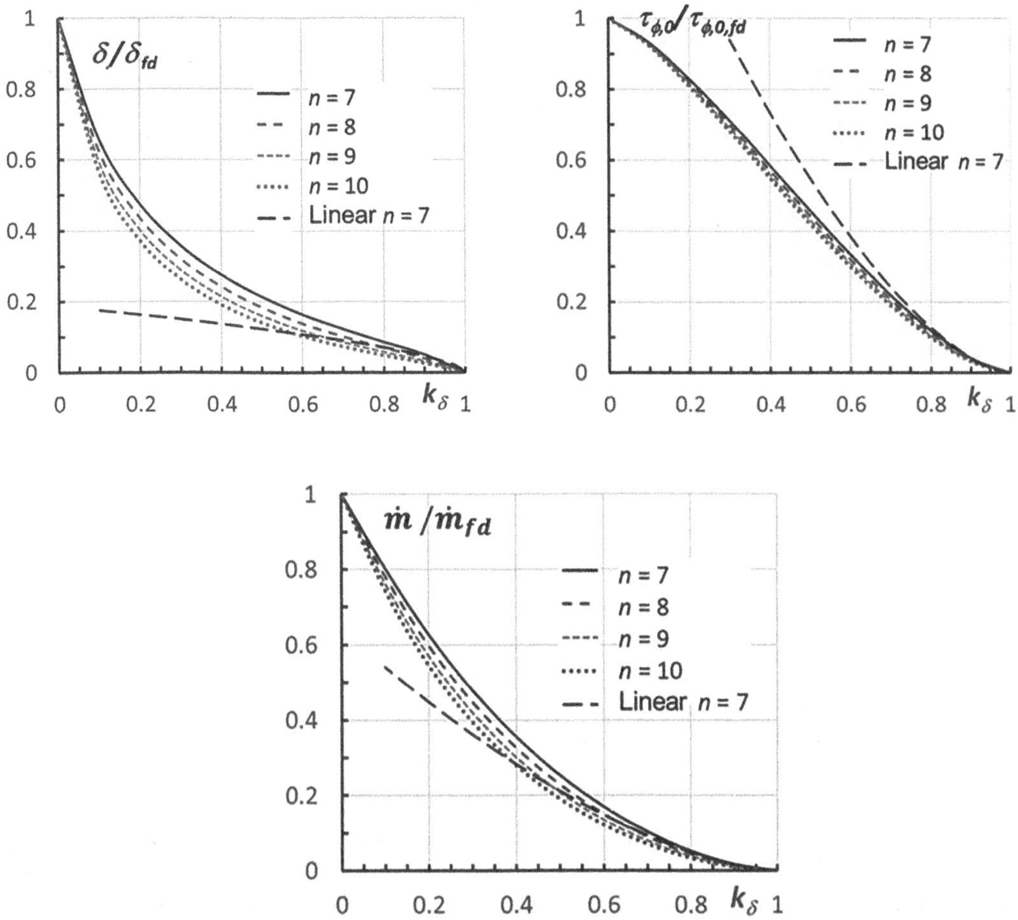

Figure 5.4 Newman's solution for forced vortex flow above a rotating disc normalised using the free disc solutions at the same Re_ϕ.

5.2.3 Forced Vortex Flow Above a Rotating Disc or Cone with $1 < k_\delta \leq \infty$

Flow rotating as a forced vortex over a stationary disc corresponds to the limit $k_\delta \to \infty$ in the case considered above. However, examination of equation (5.2.8) shows that there is no real solution for v_{s0} at high values k_δ. For $n = 7$, the highest value of k_δ with a real solution is 4.37. This may be associated with differences between the velocity profiles for rotating and stationary discs. As shown by Daily and Nece's (1960) velocity measurements for turbulent flow in a rotor-stator disc cavity and illustrated for laminar flow in Figures 3.2 and 3.3, the stator boundary layer shows a relatively strong oscillatory behaviour. The velocity profiles assumed above do not take this into account.

Daily and Nece's experiments indicated that in a closed rotor-stator cavity of sufficient width, boundary layers on the rotating and stationary discs were separated by a core flow rotating at around 46% of the disc speed. This configuration was considered by Chew (1991) in modifying the integral treatment (with $n = 7$) for the stator. CFD solutions using a mixing length model of turbulence showed good overall agreement with Daily and Nece's data, and the CFD results were used to guide modifications to the integral method. Equation (5.1.10) for the circumferential velocity was retained, but the coefficient I_3 was modified and the radial momentum equation was replaced with the assumption that $v_{s0}/(v_{\phi,0} - v_{\phi,\delta}) = 0.364$. This value corresponds to a limiting flow angle to the circumferential direction at the stator surface of 20°, approximating observations from the CFD solutions. I_3 was adjusted from the von Kármán profile value of 0.3403 to 0.3743, modifying the term representing radial advection of angular momentum. With this adjustment, the radial mass flows in stator and rotor boundary layers were equal but in the opposite direction for a core circumferential velocity of $0.42\Omega r$, as in the CFD.

With the revised value of I_3, the fixed value of limiting flow angle and $n = 7$, the solution of the integral angular momentum equation for forced vortex flow over a rotating disc gives

$$\frac{\delta}{r} = \frac{0.132|1 - k_\delta|^{0.8}}{[1 - 0.8083(1 - k_\delta)]^{0.8}} Re_k^{-0.2} (\sin\varepsilon)^{-0.8},$$

$$\frac{\tau_{\phi,0}}{\rho\Omega^2 r^2} = -0.0391(1 - k_\delta)[1 - 0.8083(1 - k_\delta)]^{0.2}|1 - k_\delta|^{0.8} Re_k^{-0.2} (\sin\varepsilon)^{0.2},$$

$$\frac{\dot{m}}{\rho\Omega r^3} = \frac{0.123(1 - k_\delta)|1 - k_\delta|^{0.8}}{[1 - 0.8083(1 - k_\delta)]^{0.8}} Re_k^{-0.2} (\sin\varepsilon)^{-0.8},$$

$$\frac{2|M|}{\rho\Omega^2 b^5} = 0.107\,|1 - k_\delta|^{1.8}[1 - 0.8083(1 - k_\delta)]^{0.2} Re_k^{-0.2} (\sin\varepsilon)^{-0.8}. \tag{5.2.11}$$

The surface heat transfer can be estimated using equation 2.5.29, based on the Reynolds analogy. In terms of a local Nusselt number, this gives

$$Nu = \frac{rq_0}{\kappa(T_{0,ad} - T_0)} = \frac{Pr^{0.6}}{(1 - k_\delta)^2} Re_k \frac{\tau_{\phi,0}}{\rho\Omega^2 r^2}. \tag{5.2.12}$$

This equation applies to both the solutions (5.2.9) and 5.2.11.

The Newman and modified solutions for boundary-layer thickness, mass flow, shear stress and heat transfer (with $n = 7$) are shown in Figure 5.5. Shear stress and heat transfer are directly related through the Reynolds analogy and two different scalings are used for this to show the limiting behaviour at $k_\delta = 1$ and as $k_\delta \to \infty$. The modified method greatly overestimates boundary-layer mass flow at low values of k_δ but shows good agreement with Newman's solution in the region $0.6 \le k_\delta \le 1.2$, where the solutions agree to within 1% of the free disc mass flow entrainment. At higher values of k_δ, Newman's solution becomes implausible as discussed above. Similar trends are seen in other parameters, with the modified integral method providing plausible results for $k_\delta \ge 0.6$.

For a forced vortex with angular velocity Ω_f above a stationary disc, equations (5.2.11) and (5.2.12) give

$$\frac{\delta}{r} = \frac{0.156}{\sin\varepsilon}\left(\frac{Re_k}{\sin\varepsilon}\right)^{-0.2}, \quad \frac{\tau_{\phi,0}}{\rho\Omega_f^2 r^2} = \frac{Nu}{Pr^{0.6} Re_k} = -0.0375\left(\frac{Re_k}{\sin\varepsilon}\right)^{-0.2}, \tag{5.2.13}$$

with $Re_k = \rho\Omega_f r^2/\mu$, and using the fluid angular velocity in Re_ϕ

$$C_{W,sd} = 0.146\left(\frac{Re_\phi}{\sin\varepsilon}\right)^{0.8}, \quad C_{M,sd} = \frac{0.102}{\sin\varepsilon}\left(\frac{Re_\phi}{\sin\varepsilon}\right)^{-0.2}.$$

5.2.4 Free Vortex Flow Above a Rotating Disc or Cone

A further, relatively simple, case of interest is free vortex flow, in which the radial velocity component is small compared to the circumferential component, above a rotating disc or cone. In a free vortex, rv_ϕ is constant and it is convenient to define the radius at which the disc speed

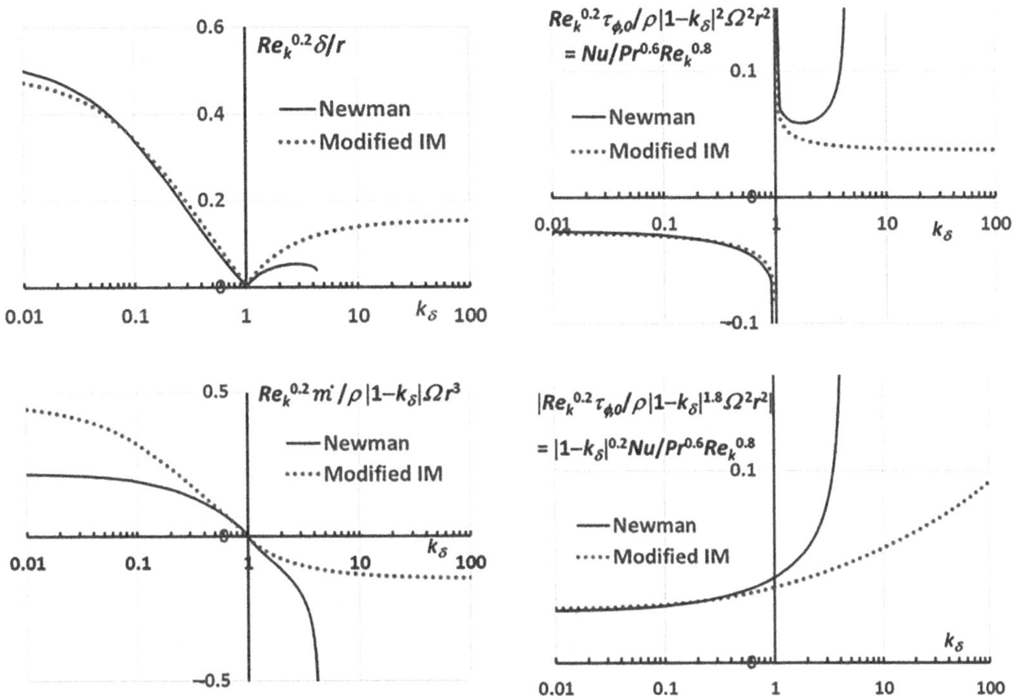

Figure 5.5 Comparison of solutions for forced vortex flow above a rotating disc.

equals the vortex speed as r_s so that $rv_\phi = \Omega r_s^2$. At this radius, there will be a stagnation line on the surface marking the start of the radial outflow on the rotor for $r > r_s$ and radial inflow for $r < r_s$. Equations (5.1.9) and (5.1.10) can be integrated numerically starting at $r = r_s$ and advancing radially outwards or inwards to obtain solutions for the two regions. Solutions with $n = 7$ and assuming uniform fluid properties are given in Figure 5.6. The scaling allows solutions to be obtained for any Reynolds number. Note that different forms are used for the two solutions. The outflow solution is scaled with r and Ωr to show the convergence to the free disc solution at high radii. Scaling for the inflow solution is chosen to avoid extreme values at lower radii.

To a good approximation, the numerical solutions for $r > r_s$, with radial outflow, are represented by the following curve fits:

$$\frac{\delta}{r} = 0.525\left(1 - e^{-0.8\left(\frac{r}{r_s}-1\right)}\right)^{0.65}\left(\frac{\rho\Omega r^2}{\mu}\right)^{-0.2}(\sin\varepsilon)^{-0.8},$$
$$\frac{\tau_{\phi,0}}{\rho\Omega^2 r^2} = -0.0267\left(1 - e^{-1.15\left(\frac{r}{r_s}-1\right)}\right)\left(\frac{\rho\Omega r^2}{\mu}\right)^{-0.2}(\sin\varepsilon)^{0.2},$$
$$\frac{\dot{m}}{\rho\Omega r^3} = 0.219\left(1 - e^{-1.15\left(\frac{r}{r_s}-1\right)}\right)\left(\frac{\rho\Omega r^2}{\mu}\right)^{-0.2}(\sin\varepsilon)^{-0.8}. \tag{5.2.14}$$

The expected scaling with cone half angle has been included in these equations. An angular momentum balance, equating the increase in angular momentum of the fluid to the surface moment, yields the following expression for the moment on the disc in the region r_s to r:

$$\frac{2|M|}{\rho\Omega^2 r^5} = \frac{|\dot{m}|}{\rho\Omega r^3}0.333\left[1 - \left(\frac{r_s}{r}\right)^2\right]. \tag{5.2.15}$$

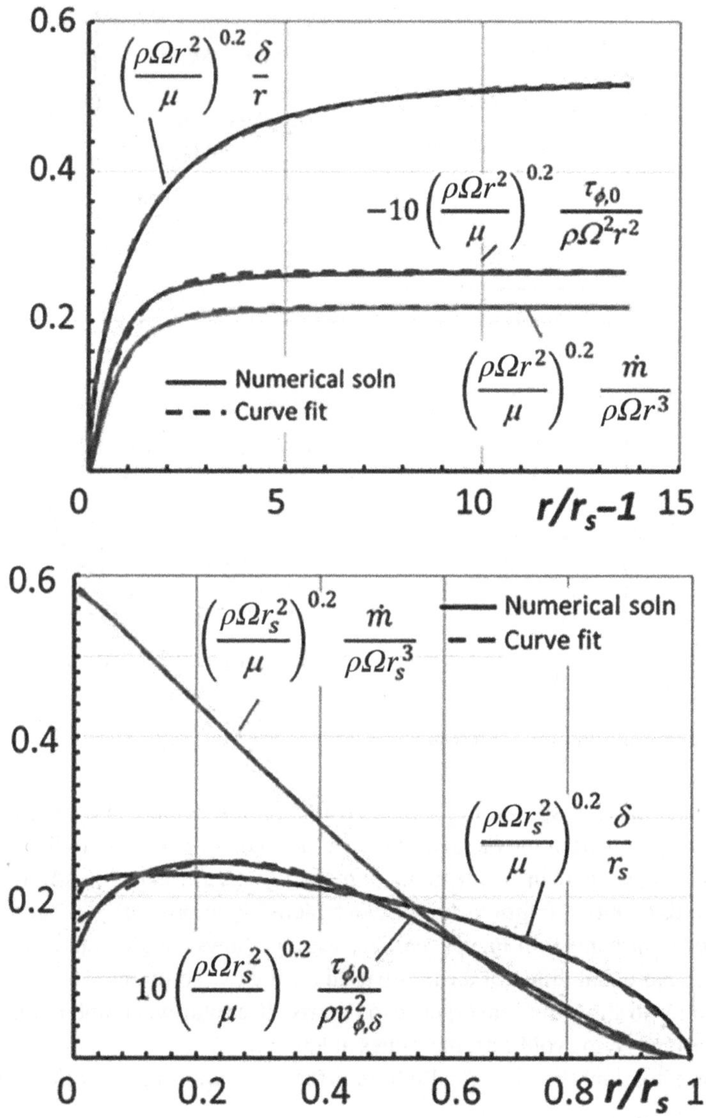

Figure 5.6 Solutions for free vortex flow above a rotating disc with outflow ($r > r_s$ top) and inflow ($r < r_s$ bottom). Abscissa gives values of the labelled quantity for each curve.

The swirl of the freestream flow reduces the disc pumping effect significantly beyond that expected from equation (5.2.7), with $a = r_s$ due to a non-zero disc radius alone.

The inflow solution, for $r < r_s$, in Figure 5.6 is approximated by the following equations:

$$\frac{\delta}{r_s} = \left[0.2126 + 0.2494\frac{r}{r_s} - 1.096\left(\frac{r}{r_s}\right)^2 + 1.484\left(\frac{r}{r_s}\right)^3 - 0.836\left(\frac{r}{r_s}\right)^4\right] \times \left(\frac{\rho\Omega r_s^2}{\mu}\right)^{-0.2} (\sin\varepsilon)^{-0.8},$$

$$\frac{\tau_{\phi,0}}{\rho v_{\phi,\delta}^2} = \left[0.0164 + 0.0686\frac{r}{r_s} - 0.167\left(\frac{r}{r_s}\right)^2 + 0.0813\left(\frac{r}{r_s}\right)^3\right]\left(\frac{\rho\Omega r_s^2}{\mu}\right)^{-0.2} (\sin\varepsilon)^{0.2},$$

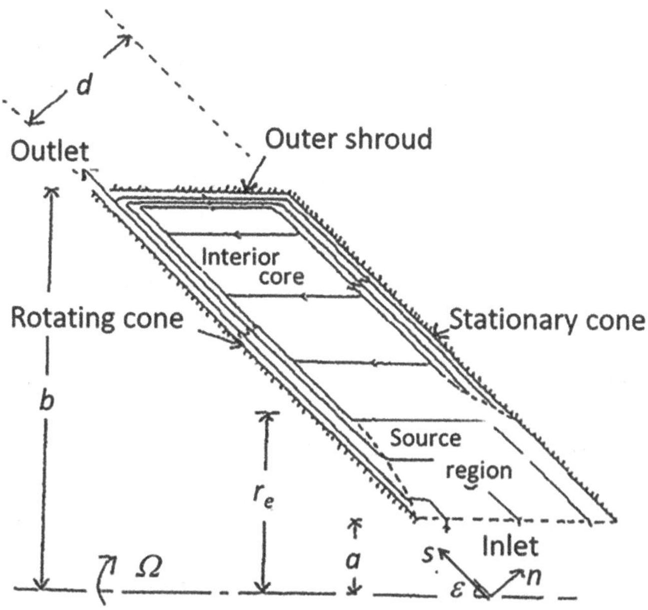

Figure 5.7 Postulated flow pattern in rotor-stator cone system. *Source:* May *et al.* (1994) / with permission of American Society of Mechanical Engineers (ASME).

$$\frac{\dot{m}}{\rho\Omega r_s{}^3} = \left[0.59 - 0.714\frac{r}{r_s} - 0.2023\left(\frac{r}{r_s}\right)^2 + 0.3242\left(\frac{r}{r_s}\right)^3\right]\left(\frac{\rho\Omega r_s^2}{\mu}\right)^{-0.2}(\sin\varepsilon)^{-0.8},$$

$$\frac{2|M|}{\rho\Omega^2 r_s^5} = \frac{|\dot{m}|}{\rho\Omega r_s^3}0.333\left[1 - \left(\frac{r}{r_s}\right)^2\right]. \tag{5.2.16}$$

These solutions are relevant to flow regions in rotor-stator and corotating disc cavities where the supplied fluid is being entrained into disc boundary layers. Such flows are further discussed further.

5.2.5 Rotor-Stator Cavity Flow

The integral methods described above have been applied to rotor-stator disc and cone cavity flows by Chew (1991) and May *et al.* (1994). Rotationally dominated flow was assumed and the flow structure is illustrated in Figure 4.3 for an enclosed disc and in Figure 5.7 for a conical rotor-stator configuration with net-positive radial outflow. The integral equations (5.1.8)–(5.1.10), with equation (5.1.12) and the coefficients given in Table 5.1 for $n = 7$, were solved numerically for the rotor boundary layer. The boundary-layer solution on the stationary cone was also obtained numerically, but using the modified integral method described in Section 5.2.3. This replaces equation (5.1.9) with a specified limiting flow angle and modifies the coefficient I_3. The conditions applied to the momentum-integral solutions and treatment of other flow regions are summarised as follows.

Source Region

The source region extends from the inlet to the radius r_e at which all the supplied flow has been entrained into the rotor boundary layer. The circumferential velocity component was calculated for a specified inlet value assuming a free vortex, so rv_ϕ = *constant*. The stator boundary layer in the source region was neglected for the conditions considered. For higher inlet values of rv_ϕ, this could be reconsidered. For inlet swirl fractions of 0 or 1, the rotor boundary-layer solution in the source region corresponds to solutions given above in Sections 5.2.1 and 5.2.4.

Interior Core

In the core flow, the pressure gradient was assumed to balance the centrifugal force, radial velocity was taken to be zero and axial and circumferential velocity components were taken as functions of radius only, as in equation 3.3.7. In the numerical solution, the values of v_ϕ were initially specified on a radial grid and then iteratively improved. In the final solution, the mass flows in the rotor and stator boundary layers are matched at the radial grid points to give the correct net mass flow rate.

Rotating Cone Boundary Layer

The numerical solution integrated the first order differential integral equations starting from conditions of zero boundary-layer flow and thickness at $r = a$. In the source region, $v_{\phi,\delta}$ was obtained from the assumed free vortex. At higher radii, $v_{\phi,\delta}$ was obtained from the interior core solution.

Shroud Boundary Layer

In the configuration shown in Figure 5.7, the flow entrained by the rotor feeds a boundary layer on the shroud. The mass flow in this layer is assumed constant and the rate of change in bulk (or mean) angular momentum with axial distance was estimated using an assumed friction factor and the bulk circumferential velocity. A simple differential equation for the bulk swirl velocity was obtained, and this can be solved analytically for a constant friction factor. In the references given, the shroud friction factor was set to equal that for the rotor boundary layer at the outer radius. Mass and angular momentum were conserved in calculating the starting conditions for the shroud boundary layer from the solution for the rotor boundary layer.

Outlet

For an outlet situated between the shroud and the rotor, as in Figure 5.6, a portion of the 'mixed out' boundary-layer flow (equal to the inlet flow rate) was assumed to leave the cavity. In configurations with the outlet between the rotor and shroud boundary layers, the outlet flow could be assumed to be fed from the portion of the rotor boundary layer closest to the rotor. In considering the effect of ingestion of external flow into the cavity allowance was also made for the angular momentum of the external flow. In all cases, mass and angular momentum were conserved in the flow interchanges.

Stationary Cone Boundary Layer

Integration of the integral momentum equation for the stationary cone boundary layer started at the outer radius with boundary-layer thickness, and $v_{\phi,\delta}$ chosen to give the required flux of mass and angular momentum. Apart from the starting point $r = b$, the solution used $v_{\phi,\delta}$ from the interior core solution. Thus, the mass flow balance for the outer point on the radial grid for the core solution was automatically satisfied and was replaced in the iterative solution by a requirement that $v_{\phi,\delta}$ at $r = b$ was matched in the core and boundary-layer solutions.

Further details of the model and numerical solutions are given in the referenced papers. The method was applied in studying rim seal flow ingestion and gave reasonable agreement with CFD studies and torque measurement for Daily and Nece's (1960) enclosed disc case and Daily *et al.*'s (1964) correlation for a disc cavity with throughflow. May *et al.* compared the model with CFD studies and torque measurement for Yamada and Ito's (1975, 1979) enclosed rotating cone configurations. For cone half angles greater than 60°, they showed excellent agreement of the integral method experimental moment coefficients. An increase in C_M as ε was reduced from 60° to 45° in the closed cavity experiments was not reproduced by the integral method. This discrepancy might be associated with 'Taylor-type vortices' observed in the experiment at lower cone angles. For a 45° half-angle cone, the agreement between experiment and the integral method improved as the throughflow rate increased, consistent with Yamada and Ito's observation that throughflow tended to suppress the Taylor-type vortices.

May *et al.*'s results for an enclosed cone without throughflow are illustrated in Figure 5.8. This includes their comparison with Yamada and Ito's data and steady CFD solutions using a mixing-length model of turbulence. In addition, results from equation (5.2.17) are included. This

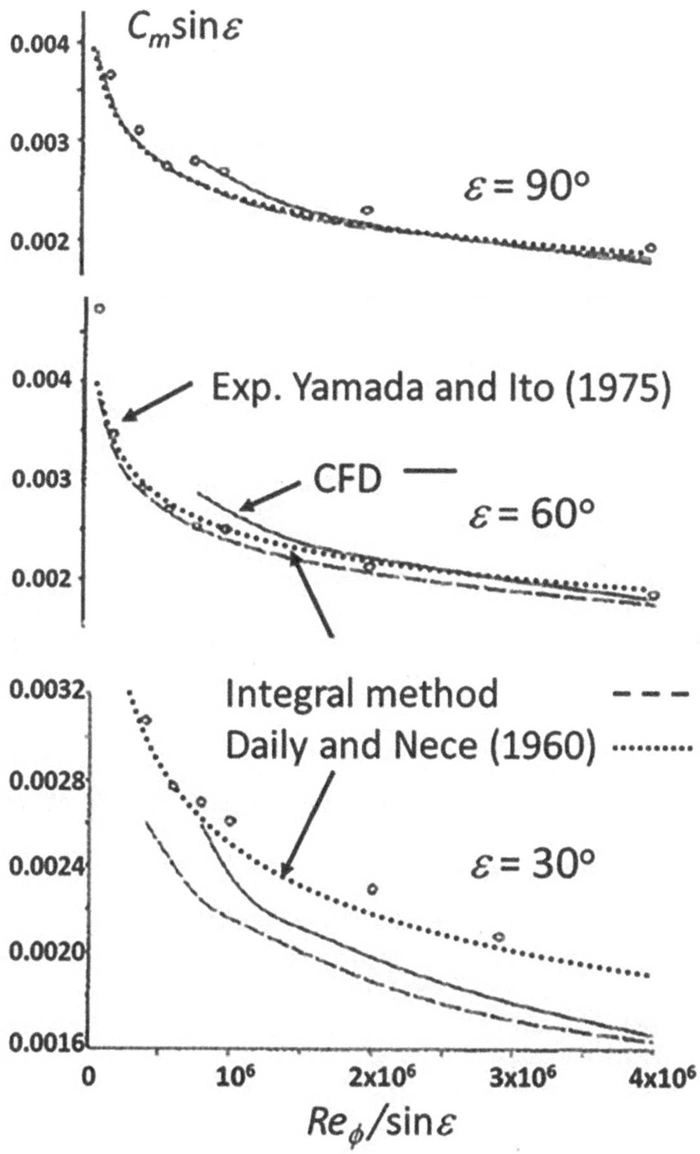

Figure 5.8 Enclosed rotating cone moment coefficients for $s/b = 0.08$. *Source:* Adapted from May *et al.* (1994).

extends Daily and Nece's experimental correlation for enclosed disc moment with turbulent, separated boundary layers assuming scaling with cone angle as deduced above:

$$C_M = 0.051(s/b)^{0.1}Re_\phi^{-0.2}(\sin\varepsilon)^{-0.8}. \tag{5.2.17}$$

Here s is taken as the minimum clearance between the rotating and stationary cones.

The results in Figure 5.8 for $\varepsilon = 90°$ and 60° show generally good agreement of the experimental data, integral solution, CFD and equation 5.2.17. For $\varepsilon = 30°$, both the integral method and CFD significantly underpredict the measurement. However, equation (5.2.17) remains in reasonable agreement with the cone measurements. This suggests that the unsteady and/or 3-dimensional flow features, not captured by the CFD or integral methods, significantly affect the mean flow for narrow cones.

Incompressible, uniform property flow was assumed in the above calculations. The model indicates that the nondimensional solution for a given geometry with no ingestion at the outlet can be obtained as a function of nondimensional throughflow rate $\lambda = C_W/[Re_\phi/\sin\varepsilon]^{0.8}$ and inlet swirl fraction $v_{\phi,\delta}/\Omega a$ only. Thus, with the assumptions made for the integral method, the number of independent variables has been reduced by 2 from that given by the application of the Buckingham Pi theorem, as discussed in Chapter 2. It can be seen that the effects of C_W, Re_ϕ and ε are combined in a single parameter. In their experimental correlation, Daily *et al.* (1964) used the same scaling as given by the integral solution, normalising the cavity throughflow rate with von Kármán's solution for free disc flow entrainment. Yamada and Ito's torque measurements for higher cone angles also show scaling close to that given by the integral equations.

Integral methods for rotor-stator disc cavities have also been considered by other workers. The method developed by Kurokawa and Sakuma (1988) may be of particular interest. In this approach, the integral angular momentum equations for both the rotor and stator boundary layers are combined with assumptions regarding the limiting flow angle at the surfaces. Correlations for the flow angle were deduced from experimental data. For the stator, these indicate (with some scatter) angles reducing from ~27° at $Re_\phi(r/b)^2 = 5\times10^5$ to ~12° at $Re_\phi(r/b)^2 = 5\times10^6$. This is reasonably consistent with the angle of 20° (independently) assumed in the treatment above. Will (2011) reported that Kurokawa and Sakuma's method has had considerable use in industry and developed an extension to the model using logarithmic scaling rather than the 1/7*th* power law. Comparisons of the extended model with pressure data for a radial compressor front cavity at $Re_\phi = 5\times10^7$ showed good agreement.

5.3 Turbulent Ekman Layers and Corotating Disc Cavity Flows

5.3.1 The Turbulent Ekman Layer Solution

As for laminar flow, the boundary-layer equations may be simplified when, in a rotating coordinate system, the Coriolis terms dominate over the nonlinear, mean inertial terms. Writing the boundary-layer equations in the rotating coordinate system (s, θ, n) and neglecting nonlinear terms in the relative velocity gives

$$\frac{\partial(\rho r u_s)}{\partial s} + \frac{\partial(\rho r u_n)}{\partial n} + \frac{\partial(\rho u_\theta)}{\partial\theta} = 0, \tag{5.3.1}$$

$$-2\sin(\varepsilon)\rho\Omega u_\theta = -\frac{\partial p_{red}}{\partial s} + \frac{\partial\tau_{sn}}{\partial n}, \tag{5.3.2}$$

$$2\sin(\varepsilon)\rho\Omega u_s = -\frac{\partial p_{red}}{r\partial\theta} + \frac{\partial\tau_{n\theta}}{\partial n}. \tag{5.3.3}$$

These are similar to equations 3.3.11–3.3.13 for laminar flow and, again, pressure gradients in the normal direction are negligible.

Introducing local coordinates (x, y) to replace (s, θ) such that x aligns with the flow at the boundary-layer edge, these equations may be written as follows:

$$\frac{\partial(\rho u_x)}{\partial x} + \frac{\partial(\rho u_n)}{\partial n} + \frac{\partial(\rho u_y)}{\partial y} = 0 \tag{5.3.4}$$

$$-2\sin(\varepsilon)\rho\Omega u_x = -\frac{\partial p_{red}}{\partial y} + \frac{\partial \tau_{yn}}{\partial n} \tag{5.3.5}$$

$$2\sin(\varepsilon)\rho\Omega u_y = -\frac{\partial p_{red}}{\partial x} + \frac{\partial \tau_{nx}}{\partial n}. \tag{5.3.6}$$

At the boundary-layer edge, the shear stresses and u_y are zero, giving the geostrophic flow solution (as in Section 3.3.1) with $2\sin(\varepsilon)\rho\Omega u_{x,\delta} = \partial p_{red}/\partial y$, $0 = \partial p_{red}/\partial x$. Using this result with assumed velocity profiles, as in equation (5.3.7), integration across the boundary layer from $n = 0$–δ gives equations (5.3.8)–(5.3.10):

$$u_y = u_{y0} f(\eta) \quad u_x = u_{x,\delta}\, g(\eta) \quad \text{where} \quad \eta = n/\delta \tag{5.3.7}$$

$$u_{n,\delta} = -\frac{I_1}{\rho r}\frac{d(\rho r v_{s0}\delta)}{d s} \tag{5.3.8}$$

$$2\sin(\varepsilon)\rho\Omega\delta\, u_{x,\delta}\, I_4 = -\tau_{y,0} \tag{5.3.9}$$

$$2\sin(\varepsilon)\rho\Omega\delta\, u_{y0} I_1 = -\tau_{x,0}. \tag{5.3.10}$$

Power law, von Kármán-type velocity profiles as in equations 5.1.7a and 5.1.7b are assumed with corresponding assumptions for surface shear stress giving

$$\tau_{x,0} = \rho K_n^2 \left(\frac{\mu}{\rho\delta}\right)^{2/(n+1)} u_{x,\delta}\left[u_{y0}^2 + u_{x,\delta}^2\right]^{(n-1)/2(n+1)}$$

$$\tau_{y,0} = \frac{u_{y0}}{u_{x,\delta}}\tau_{x,0}. \tag{5.3.11}$$

I_1, I_4 and K_n^2 are as defined in Section 5.1, with values tabulated in Table 5.1.

The second part of equation (5.3.11) assumes that the surface shear stress aligns with the velocity vector as $y \to 0$. With this relation, the integral equations (5.3.9) and (5.3.10) lead directly to $u_{y0}/u_{x,\delta} = -\sqrt{I_4/I_1}$, with the square root chosen to give a negative velocity ratio on physical grounds. The solution for other variables given below can then be found by substitution and algebraic manipulation. The form of the solution corresponds to that of equation (5.2.9), with $|u_{x,\delta}/\Omega r|$ replacing $|1 - k_\delta|$, and x and y coordinates replacing ϕ and s. The coefficients take their limiting values at $k_\delta = 1$. With $n = 7$, the solution is summarised as follows:

$$u_{y0}/u_{x,\delta} = -0.553, \quad \frac{\delta}{r} = 0.0983\left|\frac{u_{x,\delta}}{\Omega r}\right|^{0.6}\left(\frac{\rho\Omega r^2}{\mu}\right)^{-0.2}(\sin\varepsilon)^{-0.8},$$

$$\frac{|\tau_{x,0}|}{\rho\Omega^2 r^2} = 0.0444\left|\frac{u_{x,\delta}}{\Omega r}\right|^{1.6}\left(\frac{\rho\Omega r^2}{\mu}\right)^{-0.2}(\sin\varepsilon)^{0.2},$$

$$\frac{|\dot{m}|}{\rho\Omega r^3} = 0.140\left|\frac{u_{x,\delta}}{\Omega r}\right|^{1.6}\left(\frac{\rho\Omega r^2}{\mu}\right)^{-0.2}(\sin\varepsilon)^{-0.8}. \tag{5.3.12}$$

Note that here $\dot{m}$ is the mass flow in the y direction, that is, the flow normal to the mainstream induced by the Coriolis force. It has the opposite sign to $u_{x,\delta}$ so, for example, if the mainstream angular velocity component is greater than that of the disc (so that $u_{x,\delta}$ is positive), the induced flow will have a negative radial component. It is sometimes convenient to rearrange equation (5.3.12) to obtain the solution as a function of the induced mass flow:

$$\left|\frac{u_{x,\delta}}{\Omega r}\right| = 3.42\left(\frac{\rho\Omega r^2}{\mu}\right)^{1/8}\left(\frac{|\dot{m}|}{\rho\Omega r^3}\right)^{5/8}(\sin\varepsilon)^{0.5},$$

$$\frac{\delta}{r} = 0.206\left(\frac{|\dot{m}|}{\rho\Omega r^3}\right)^{3/8}\left(\frac{\rho\Omega r^2}{\mu}\right)^{-1/8}(\sin\varepsilon)^{-0.5},$$

$$\frac{|\tau_{x,0}|}{\rho\Omega^2 r^2} = \frac{|\dot{m}|}{\pi\rho\Omega r^3}\sin(\varepsilon). \tag{5.3.13}$$

Owen *et al.* (1985) derived the Ekman layer solution and compared equation (5.3.13) to velocity measurements for axisymmetric flow between corotating discs with known inflow and outflow rates. Good agreement was obtained for surprisingly high velocities $|u_{\theta,\delta}/\Omega r|$ of up to 0.67. It is also of interest to compare the Ekman solution with Newman's solution in Figure 5.4. In these graphs, the Ekman solution agrees with Newman's solution near $k_\delta = 1$, with differences becoming increasingly apparent for $k_\delta < 0.8$.

As for the laminar Ekman solution discussed in Section 3.3.3, the conditions under which the exact analogy between the angular momentum and energy equations holds are rather restrictive. However, it is of interest to note that the ratio of shear stress from the turbulent Ekman layer solution to that from the laminar case is $0.0444\{|u_{\theta,\delta}|/\sqrt{\nu\Omega\sin\varepsilon}\}^{0.6}$. This ratio might also be applied to heat transfer across an Ekman layer, as indicated for example, by equation 2.5.29. The ratio equals unity when the boundary-layer Reynolds number $|u_{\theta,\delta}|/\sqrt{\nu\Omega\sin\varepsilon} = 180$. As noted in Chapter 4, this is close to the transitional Reynolds number, suggesting that change in heat transfer at transition may be gradual rather than abrupt.

5.3.2 Corotating Disc Cavities with Radial Outflow

Solutions of the integral equations for radial outflow between corotating discs have been presented and discussed in detail by Owen *et al.* (1985) and Chew and Rogers (1988). Qualitatively, the flow structure, as illustrated in Figure 5.9, and heat transfer are very similar to those in laminar flow, as discussed in Section 3.3. Numerical solutions of the integral equations (with $n = 7$) show good agreement with measurements of velocity in the rotating flow core and disc heat transfer. The Ekman layer approximation for the core velocity applies at lower values of $|u_{\theta,\delta}/\Omega r|$, as mentioned above. The numerical solution of the integral equations given by Chew and Rogers started with zero boundary-layer thickness and entrained mass flow at $r = a$, and advanced radially outwards. In the source region, $u_{\theta,\delta}$ was specified assuming free vortex flow outside the boundary layer. The source region extended to the radius at which all the supplied flow was entrained in the boundary layers. At radii beyond this point, boundary-layer mass flow was assumed constant, and the core velocity was obtained from the integral equations. Good agreement with Northrop and Owen's (1988b) heat transfer measurements was demonstrated using a slightly more complex treatment of the energy equation than described above. Some sensitivity to the assumptions made to account for the effect of a nonquadratic disc temperature distribution on the Reynolds analogy, was found, and comparisons with heat transfer rates deduced from the Ekman layer solution were also given.

Figure 5.10 shows a numerical solution of the integral equations for a cavity radius ratio $a/b = 0.1$, an inlet swirl fraction $v_{\phi,in}/\Omega a = 1$ and a range of nondimensional flow rates. Fluid

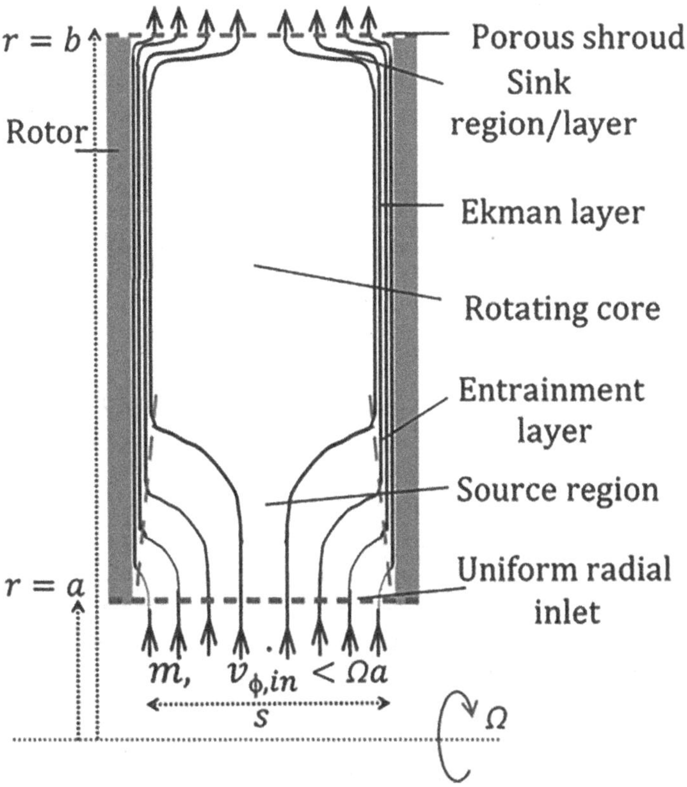

Figure 5.9 Schematic of flow for a rotating cavity with radial outflow.

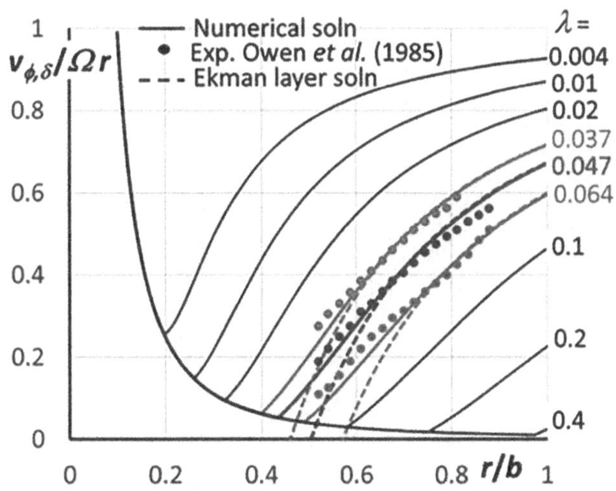

Figure 5.10 Solution for core velocity in a rotating cavity with radial outflow.

property variations are neglected. The mass flow coefficient λ is defined using the full mass flow rate into the cavity, so for $\lambda \approx 0.44$, the free disc entrainment for the two discs is equal to the supplied flow. At this point, the source flow region fills the cavity, and the core swirl velocity corresponds to a free vortex. As the flow rate reduces, so does the extent of the source region, and the rotating core separating the two constant mass flow disc boundary layers grows. Owen *et al.*'s measurements are shown on the figure for three conditions, corresponding to $C_W = 2500$ with $Re_\phi = 5.5 \times 10^5$, 8.2×10^5 and 1.1×10^6. The turbulent Ekman layer solution given by equation (5.3.13) is also included for these three conditions. Good agreement between the numerical solution of the full integral equations and measurements is shown. The Ekman layer solution agrees well with both the measurements and the full solutions for $v_{\phi,\delta}/\Omega r$ greater than about 0.3. This is better agreement than might be expected from comparisons of the Ekman and Newman solutions discussed above. This may be explained by the balancing out of some nonlinear terms for these radial outflow conditions.

5.3.3 Corotating Disc Cavities with Radial Inflow

As shown above for radial outflow, at low values of mass flow coefficient λ, a quiescent, rotating core with radial flow confined to Ekman layers develops between corotating discs with a forced radial inflow. In this case, the flow core rotates faster than the discs. A schematic of the flow structure in a rotating cavity with inflow is shown in Figure 5.11. Here the flow is assumed to enter the cavity through angled holes or nozzles in the rotating shroud so that the inlet swirl fraction $v_{\phi,in}/\Omega b < 1$, and leave through a porous inner cylinder. The quiescent core is bounded radially by source and sink regions in which the flow will depend on the inlet and exit geometry and conditions.

In the absence of surface friction, the flow into the cavity would form a free vortex with v_ϕ increasing as the radius reduces. This is modified by the disc torque and associated boundary-layer flows. At higher radii, where the discs and shroud rotate faster than the fluid, the rotor torque increases fluid swirl and radial outflow is induced in the disc boundary layers. Mixing of the disc pumping flows and the inflow may be assumed, with the mean flow swirl outside the boundary layer in the mixing/source regions assumed to be a free vortex. This vortex may have higher

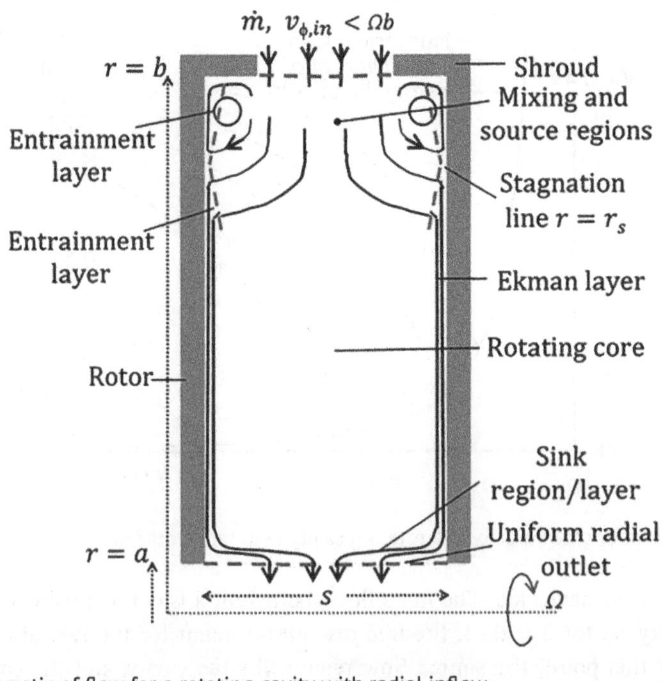

Figure 5.11 Schematic of flow for a rotating cavity with radial inflow.

angular momentum than the inflow due to the positive disc moment at outer radii. At the radius in the source region where the free vortex and disc circumferential velocities are equal, the flow in the disc boundary layers will reverse. This is indicated as a stagnation line on the disc at radius r_s in Figure 5.11 and marks the start of the radial outflow and inflow entrainment layers. The radius at which mass flow in the inflow entrainment layers matches the supplied flow marks the start of the quiescent core and constant mass flow disc boundary layers. The swirl velocity in the quiescent core continues to increase as radius reduces but at a slower rate than for a free vortex.

Using numerical methods similar to those described above for the radial outflow case, Chew and Snell (1988) presented solutions of the integral equations for inflow with an inlet swirl ratio $v_{\phi,in}/\Omega b = 1$. In this case, $r_s = b$ and the inflow entrainment layers start at the cavity outer radius. Good agreement with most of the available pressure and velocity measurements was found. For uniform property flows, the solution for nondimensional pressure drop across the cavity is a function of just the flow coefficient λ and the radius ratio a/b, and is encapsulated in Figure 5.12. All solutions follow the free vortex solution in the source region and for high enough λ or ratio a/b, the source region will fill the cavity. For low enough values of λ, the viscous drag on the discs dominates and the solution approaches that for a forced vortex. Figure 5.12 shows the transition between the forced and free vortex limits as λ increases. Very high-pressure drops are predicted at low values of a/b, and it should be noted that the solution neglects compressibility and viscous effects outside the boundary layers. Chew and Snell give further solutions showing the effect of compressibility for gas flows and found this to be relatively small (<5%) if the pressure drop across the cavity was less than 10% of the inlet static pressure.

Farthing *et al.* (1991) presented integral solutions and pressure measurements for a rotating cavity with deswirled radial inflow, produced by nozzles angled at 30° to the circumferential direction fitted in a rotating shroud. The integral solution assumed the flow structure in

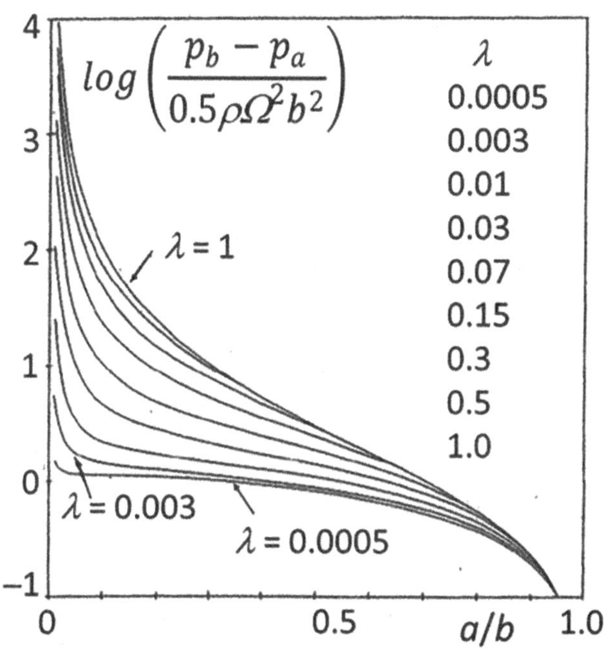

Figure 5.12 Integral solution for a pressure drop in a rotating cavity with radial inflow for an incompressible fluid with $v_{\phi,in}/\Omega b = 1$.

Figure 5.11. The first stage of the integral solution involved calculation of the source flow vortex strength using an iterative procedure. An estimate of the free vortex strength was used to find r_s and $v_{\phi,\delta}$ in the source region. The solution of the integral equations for the outward entrainment layers and a simple treatment of the shroud boundary layers, as described above for the rotor-stator disc cavity, then provided an estimate of the recirculating flow strength and its angular momentum. The estimate of the vortex strength could then be adjusted, if necessary, until it matched the value obtained from the mixing of the recirculating and nozzle flows with conservation of angular momentum. Once the source/mixing region free vortex strength was known, the integral for the radial inflow disc boundary layers for $r < r_s$ could be obtained. Farthing *et al.* used numerical solutions of the integral equations, but deswirled inflow solutions could also be obtained using equation (5.2.14) for the outflow entrainment layers and tabulated or graphical solutions for the case $v_{\phi,in}/\Omega b = 1$ (with adjustment to account for the starting radius r_s). This is discussed in Chapter 8.

As shown by the example in Figure 5.13, good agreement has been found between measurements and integral solutions. The figure shows variation with flow rate of the pressure difference between $r/b = 0.52$ and $r/b = 0.985$ divided by the disc dynamic head $0.5\rho\Omega^2 b^2$ at $Re_\phi = 6.1 \times 10^5$. At zero flow rate the integral solution reduces to a forced vortex. As flow rate increases, deswirl is initially weak and the inflow causes an initial increase in pressure drop across the cavity. As the flow through the nozzles continues to increase, the inlet swirl fraction reduces, causing a reduction in pressure drop. For the conditions of Figure 5.13, the absolute swirl velocity at the nozzle exits will be ~0 when $C_W = 10^4$. At this point ,there is still a small pressure drop due to the swirl imparted by the

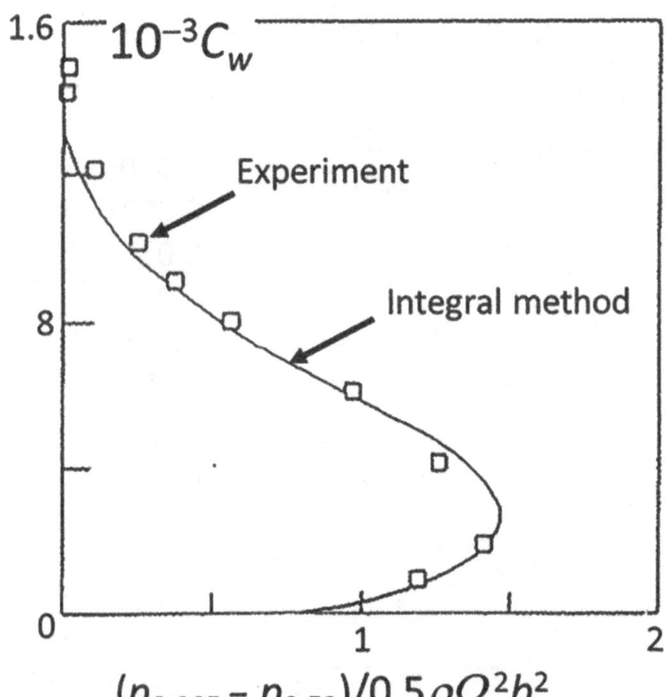

Figure 5.13 Pressure drop in a rotating cavity with deswirled radial inflow for $Re_\phi = 6.1 \times 10^5$. *Source:* Farthing *et al.* (1991) / with permission of American Society of Mechanical Engineers (ASME).

discs and shroud. At slightly higher flow rates, the pressure drop becomes negligible as negative swirl imparted by the nozzles is balanced by the rotor moment to give an effective swirl of zero. Experiments show that at even higher flow rates, the pressure drop can increase again as the deswirl drives a vortex rotating in the opposite direction to the discs. This condition was not modelled in the integral solution.

Figure 5.13 shows that imposing a specific pressure drop across a deswirled rotating cavity with inflow does not uniquely define the flow rate. This characteristic can lead to non-unique steady state solutions with operating conditions achieved depending on other elements of the flow system and how the steady point is approached. Further illustration and discussion of this effect is given in Chapter 8 and by May *et al.* (2013).

5.3.4 Buoyancy Effects

In the examples given above, uniform fluid properties have been assumed. The effects of density and viscosity variations in the streamwise s direction, as included in the integral equations discussed in Section 3.1, can readily be incorporated in the numerical integration of the boundary-layer flows. However, in some circumstances, centrifugal buoyancy effects may change the flow outside the boundary layers. This is yet to be fully investigated.

As discussed for laminar flow in Section 3.3, if two discs in a closed rotating cylindrical cavity are at different temperatures, buoyancy may generate Ekman layers with radial outflow on the cooler disc and radial inflow on the hotter disc. Consideration of the laminar flow analysis for infinite, corotating discs together with the criterion for Ekman layer transition to turbulence given in Chapter 4, suggests that turbulent flow is likely for $(\beta\Delta T)^2 Re_\phi$ of order 10^6 and higher where ΔT is the temperature difference between the discs. For such extreme unexplored conditions, the analysis can be extended to use turbulent Ekman layer solutions, but the extent to which turbulence would modify the heat transfer in the core is not clear. With an imposed radial outflow or inflow through the cavity, the mass flows in the two-disc boundary layers will differ (depending on the strength of the buoyancy and net-flow rate). In these cases, the loss of symmetry between the two discs leads to a weak axial flow in the rotating core. The strength of this effect may be gauged by considering the variation in swirl velocity needed to balance the change in centrifugal force due to buoyancy. This indicates that axial buoyancy effects will be small when $\beta\Delta T/4 \ll |u_{\theta,\delta}/\Omega r|$ and can often be neglected.

In turbomachinery applications, buoyancy effects due to a positive radial temperature gradient are known to be important in rotating cavities with no net-radial flow. As discussed in Chapter 4, centrifugal Rayleigh-Bénard instability leads to turbulence and large-scale flow unsteadiness in the flow. Such effects are not included in the inviscid core treatments used above with the integral boundary-layer equations.

5.4 Concluding Remark

Integral boundary-layer methods have played an important part in developing the understanding of rotating flows over the last century and are especially useful for quick engineering estimates and scaling between different conditions. The examples given above show considerable success of methods based on von Kármán's $1/7th$ ($n = 7$) power-law-velocity profiles for boundary-layer-dominated flow in rotating disc and cone cavities. It should be recognised that engine operating Reynolds numbers are often higher than for laboratory measurements used in validating the integral solutions.

There is evidence for the free rotating disc that logarithmic or lower power-law-velocity profiles are more appropriate at very high Reynolds numbers. However, straightforward extension of the $1/7th$ power-law-velocity profiles to $n = 8$ and higher may introduce other errors in representing the radial velocity for this case.

An important limitation of the integral solutions described above is the assumption of axisymmetric rotor and stator surfaces. Features such as boltheads or holes may dominate the rotor and stator moments and centrifugal pumping. This changes the scaling with Reynolds numbers. Some studies have included drag on non-axisymmetric features in integral solutions through additional terms in the equations representing circumferentially averaged effects. An example is given by Chew *et al.* (1989), and further discussion of the effect of 3D features is given in Chapter 7.

Difficulties in applying integral methods occur with recirculating flows and complex geometries. The examples above show that such issues can sometimes be overcome using careful, ad hoc assumptions and numerical, iterative solution methods. This is challenging in more complex cases and considerable engineering judgement or adjustment to match experimental data may be required for successful modelling.

References

Cham, T. and Head, M., 1969. Turbulent boundary-layer flow on a rotating disk. J. Fluid Mech., 37, 129–147. https://doi.org/10.1017/S0022112069000450.

Chew, J. W., 1988. The effect of hub radius on the flow due to a rotating disk. ASME J. Turbomach., 110, 417–418. https://doi.org/10.1115/1.3262213.

Chew, J. W., 1991. A theoretical study of ingress for shrouded rotating disk systems with radial outflow. ASME J. Turbomach., 113, 91–97. https://doi.org/10.1115/1.2927742.

Chew, J. W., Farthing, P. R., Owen, J. M., and Stratford, B., 1989. The use of fins to reduce the pressure drop in a rotating cavity with a radial inflow. ASME J. Turbomach., 111, 349–356. https://doi.org/10.1115/1.3262279.

Chew, J. W. and Rogers, R. H., 1988. An integral method for the calculation of turbulent forced convection in a rotating cavity with radial outflow. Int. J. Heat Fluid Flow, 9, 37–48. https://doi.org/10.1016/0142-727X(88)90028-8.

Chew, J. W. and Snell, R. J., 1988. Prediction of pressure distribution for radial inflow between co-rotating discs. ASME paper 88-GT-127, https://doi.org/10.1115/88-GT-61.

Childs, P. R. N., 2011. Rotating flow, Elsevier.

Cobb, E. C. and Saunders, O. A., 1956. Heat transfer from a rotating disk. Proc. R. Soc. London, Ser. A, 236, 343–351. https://doi.org/10.1098/rspa.1956.0141.

Daily, J. W., Ernst, W. D. and Asbedian, V. V., 1964. Enclosed rotating disks with superposed throughflow. Report 64, M.I.T, Department of Civil Engineering. https://hdl.handle.net/1721.1/142983.

Daily, J. W. and Nece, R. E., 1960. Chamber dimension effects on induced flow and frictional resistance of enclosed rotating disks. ASME J. Basic Eng., 217–230. https://doi.org/10.1115/1.3662532.

Dorfman, L. A., 1963. Hydrodynamic resistance and the heat loss of rotating solids, First English edition, London: Oliver and Boyd.

Farthing, P. R., Chew, J. W., and Owen, J. M., 1991. The use of deswirl nozzles to reduce the pressure drop in a rotating cavity with a radial Inflow. ASME J. Turbomach., 113, 106–114. https://doi.org/10.1115/1.2927727.

Itoh, M. and Hasegawa, I., 1994. Turbulent boundary layer on a rotating disk in an infinite quiescent fluid. JSME Int. J., Ser. B, 37, 449–456. https://doi.org/10.1299/jsmeb.37.449.

Kurokawa, J. and Sakuma, M., 1988. Flow in a narrow gap along an enclosed rotating disk with through-flow. JSME Int. J., Ser. 2, 31, 243–251. https://doi.org/10.1299/jsmeb1988.31.2_243.

May, N. E., Chew, J. W., and James, P. W., 1994. Calculation of turbulent flow for an enclosed rotating cone. ASME J. Turbomach., 116, 548–554. https://doi.org/10.1115/1.2929444.

May, D., Chew, J. W., and Scanlon, T. J., 2013. Prediction of deswirled radial Inflow in rotating cavities with hysteresis. ASME J. Turbomach., 135, 041025. https://doi.org/10.1115/GT2012-68647.

Newman, B. G., 1983. Flow and heat transfer on a disk rotating beneath a forced vortex. AIAA J., 21, 1066–1070. https://doi.org/10.2514/3.8205.

Northrop, A. and Owen, J. M., 1988a. Heat transfer measurements in rotating-disc systems part 1: The free disc. Int. J. Heat Fluid Flow, 9, 19–26. https://doi.org/10.1016/0142-727X(88)90026-4.

Northrop, A. and Owen, J. M., 1988b. Heat transfer measurements in rotating-disc systems part 2: The rotating cavity with a radial outflow of cooling air. Int. J. Heat Fluid Flow, 9, 27–36. https://doi.org/10.1016/0142-727X(88)90027-6.

Owen, J. M., Pincombe, J. R., and Rogers, R. H., 1985. Source–sink flow inside a rotating cylindrical cavity. J. Fluid Mech., 155, 233–265. https://doi.org/10.1017/S0022112085001793.

Owen, J. M. and Rogers, R. H., 1989. Flow and heat transfer in rotating disc systems, vol. 1: Rotor-Stator Systems, Taunton: U. K. Research Studies Press.

Owen, J. M. and Rogers, R. H., 1995. Flow and heat transfer in rotating disc systems, vol. 2: Rotating Cavities, Taunton: U. K. Research Studies Press.

Schlichting, H., 1979. Boundary-layer theory, Seventh edition, McGraw-Hill.

Shevchuk, I. V., 2009. Convective heat and mass transfer in rotating disc systems, Springer.

Theodorsen, T. and Regier, A., 1944. Experiments on drag of revolving disks, cylinders, and streamline rods at high speeds. NACA Report 793. https://ntrs.nasa.gov/api/citations/20050241738/downloads/20050241738.pdf.

Tien, C. and Campbell, D., 1963. Heat and mass transfer from rotating cones. J. Fluid Mech., 17(1), 105–112. https://doi.org/10.1017/S0022112063001142.

von Kármán, T., 1921. Über laminare und turbulente Reibung. ZAMM, 1, 233–252. https://doi.org/10.1002/zamm.19210010401.

Will, B. C., 2011. Theoretical, numerical and experimental investigation of the flow in rotor-stator cavities with application to a centrifugal pump. Duisburg, Essen, Universität Duisburg-Essen, Diss. HYPERLINK "https://bibliographie.ub.rub.de/"bibliographie.ub.rub.de.

Wormley, D. N., 1969. An analytical model for the incompressible flow in a short vortex chambers. ASME J. Basic Eng., 264–276. https://doi.org/10.1115/1.3571091.

Yamada, Y. and Ito, M., 1975. On the frictional resistance of enclosed rotating cones. Bull. JSME, 1026–1034. https://doi.org/10.1299/jsme1958.18.1026.

Yamada, Y. and Ito, M., 1979. Frictional resistance of enclosed rotating cones with superposed throughflow. ASME J. Fluids Eng., 101, 259–264. https://doi.org/10.1115/1.3448948.

6

Computational Fluid Dynamics

Many of the limitations of the integral methods described in Chapter 5 can be overcome using computational fluid dynamics (CFD). The general aim in CFD is to obtain numerical solutions to exact or approximate governing equations for the flow with fully specified boundary conditions. The full Navier–Stokes equations can be considered exact for the conditions considered here. Numerical solution of these equations, known as direct numerical simulation (DNS), is ostensibly very attractive until the computational resources required for high Reynolds number flows are considered. Turbulence length and time scales reduce as flow Reynolds number increases, roughly in proportion to $Re^{-3/4}$ and $Re^{-1/2}$, respectively. This requires finer spatial and temporal resolution in the numerical solution so that computational requirements increase roughly with $Re^{11/4}$. Assuming increases in available computing power by a factor of 5 every five years, Spalart (2000) estimated that DNS of flow over a full aircraft or car would be possible around 2080. On this basis, Chew and Hills (2009) suggested that DNS for a full aeroengine simulation may not be possible before 2100. Such predictions are speculative but illustrate an important point. Compromises are required in the choice of modelling strategy. These should be guided by knowledge of underlying flow mechanisms, capability of the models and the application in question.

Owing to computational limitations, most current applications of CFD to rotating disc cavity flows consider the Reynolds-averaged Navier–Stokes (RANS) equations as introduced in Section 2.4 of Chapter 2. However, large eddy simulation (LES), in which larger scale turbulence is resolved numerically but smaller scales are modelled, is established in research studies and is attracting increasing interest for industrial application. DNS has been used in research studies for rotational Reynolds numbers up to 10^6 and has given new insight into the flow physics in some cases. General introductions to CFD methods and practices are available in other books, such as that by Hirsch (2007). In this chapter, some prior knowledge is assumed, and attention focusses on aspects that are particularly relevant or specific to rotating flows.

Section 6.1 of this chapter gives a brief discussion of CFD solution methods. The following sections then focus on modelling approaches, illustrating performance for rotating flow test cases, and indicating model strengths and weaknesses.

6.1 Computational Fluid Dynamics (CFD) Solution Methods

Typically, but not exclusively, general purpose finite volume CFD codes are applied in industry and for research in analysing disc cavity flow and heat transfer. Discussion here focusses largely on such methods. As in other CFD applications, mesh resolution and quality are crucial factors. While there is some dependency of solution accuracy on the discretisation scheme used, all methods rely on appropriate choice of mesh. Requirements depend on the modelling approach selected and include careful consideration of the boundary layer resolution, as in many cases it is the disc boundary layers that drive the flow. Ideally, mesh sensitivity tests might be undertaken for all solutions, or mesh generation might be automated to achieve a required accuracy. A currently more practical approach is to develop meshing and modelling guidelines for specific classes of flow and apply these to more routine applications. Guidance is available from results and examples in this book and reported in the open literature, with some key points related to turbulence modelling being discussed further in the following sections.

Both pressure-based and density-based solution methods are successfully used for rotating disc flows. The former enforce conservation of mass through adjustment of the pressure field, while the latter adjust the density to achieve the same aim. For incompressible fluids or low Mach number conditions pressure-based methods may be preferred whilst density-based methods may be preferred at higher Mach numbers. In turbomachinery, applications where the flow in the disc cavities at relatively low Mach number interacts with main gas path or seal flows at higher Mach number are common. The computing time required for convergence to a steady CFD solution and the simulation time required for an unsteady flow solution to reach a statistically steady state are affected by the range of time scales for flow and temperature fields in different regions of the flow. Typically, the timestep required to resolve convective terms is of order of the grid size divided by the flow velocity, and this sets an upper limit on the timestep. The simulation time required may be related to the time for the flow to pass through the solution domain and the diffusion, or viscous, timescales. The examples considered in earlier chapters include regions where the flow is rotating with little or no axial and radial convection. In these regions, diffusion may dominate, requiring relatively long simulation periods to obtain convergence.

Further issues affecting convergence and computing times for rotating flows are the strong spatial coupling and interaction between velocity components in rotating flows. The former is illustrated by the Taylor–Proudman theorem showing that axial gradients of velocity and pressure are suppressed by rotation. The latter is apparent from the Coriolis forces in the momentum equations and the dominance of these when considering flow as a perturbation from forced vortex rotation. These two effects are interrelated. They result in mathematically stiff systems of equations and can destabilise iterative solutions that neglect or approximate the coupling terms in advancing the solution. Practical measures taken to address this include the use of underrelaxation, more accurate solution of the equations at each iteration and the use of iterative schemes that better account for the coupling.

As with other CFD applications the convergence of solutions needs to be examined carefully. A solution that is slowly varying may still be some way from convergence, especially if, for example, strong underrelaxation is used. Checking of overall mass, angular momentum and energy balances (in addition to other convergence criteria) has been found to be a particularly useful check for rotating flows. These are given by equations 2.1.1, 2.1.7 and 2.1.16 in Chapter 2. Typically, in engineering calculations errors of less than 1%, 3% and 5% (of representative flow rate, moments and energy transfer) in mass, angular momentum and energy are considered satisfactory. The levels achieved may depend on modelling approach, mesh size and quality, flow complexity and numerical convergence levels. The accuracy levels above may well be difficult to achieve for large models that approach the limits of feasibility with the available computing resources. In such cases, the solutions may still be useful provided the results are interpreted carefully. An analogy might be made with experimental studies where the accuracy achievable in more engine-representative conditions may be limited but effects that are missed in simpler laboratory testing may be captured.

6.2 Reynolds-Averaged Navier–Stokes (RANS) Models

6.2.1 The Eddy Viscosity Hypothesis

As described in Section 2.4, the RANS equations provide a basis for estimation of mean flow fields but require additional assumptions to be made regarding the Reynolds stresses. The Boussinesq eddy viscosity hypothesis, given by equations 2.4.12, is widely used and is adequate for many, but not all, turbulent disc cavity flows. This approach is popular because it is easy to implement in flow solvers, the additional viscosity helps to stabilise solution algorithms, it has some theoretical foundation in simple shear flows, and because it is found useful in many engineering calculations. With this model specifying turbulence is reduced to finding values for a single scalar. However, its physical realism in complex flows is questionable. It is most appropriate for flow regions where there is a single dynamically significant Reynolds stress (e.g. $\overline{\acute{v}_x \acute{v}_y}$), such as attached boundary layers or other simple shear flows.

Considering a simple shear flow in which velocity gradients in the y direction dominate and the streamwise direction is x, the effects of turbulence are dominated by the Reynolds stress component $\overline{\acute{v}_x \acute{v}_y}$. The eddy viscosity hypothesis then reduces to

$$-\overline{\acute{v}_x \acute{v}_y} = \nu_t \frac{\partial \overline{v_x}}{\partial y}. \tag{6.2.1}$$

This is given in the stationary frame of reference but a similar equation applies in rotating coordinate systems. As in Chapter 2, $\nu_t = \mu_t/\rho$ is the kinematic turbulent or eddy viscosity. The kinematic viscosity has dimensions of length x velocity, and eddy viscosity models can be viewed as treating ν_t as the product of a length characteristic of the size of the turbulent eddies and a velocity reflecting the magnitude of the turbulent fluctuations.

6.2.2 Mixing Length Models

As discussed, for example, in the book by Davidson *et al.* (2011) Prandtl's, c1925, mixing length hypothesis draws an analogy between turbulent and molecular viscosity, and relates the turbulent velocity to the product of a turbulent (or mixing) length scale times the mean velocity gradient. For the simple shear layer this results in the expression

$$-\overline{\acute{v}_x \acute{v}_y} = l_m^2 \left| \frac{\partial \overline{v_x}}{\partial y} \right| \frac{\partial \overline{v_x}}{\partial y}. \tag{6.2.2}$$

where l_m is the mixing length. This model has had significant success, particularly in boundary layers where l_m is usually assumed to scale with boundary layer thickness δ and have a basic ramp distribution rising linearly from zero at the wall to a constant of ~0.085δ in outer parts of the layer. Refinements include further exponential damping very near the wall and Richardson number correction to account for some rotational effects. For comparison, in free mixing layers $l_m \sim 0.07$ of the layer thickness, in plane jets $l_m \sim 0.09$ of the jet width, in round jets $l_m \sim 0.075$ of the jet diameter and in plane wakes $l_m \sim 0.16$ of the wake thickness.

Koosinlin *et al.* (1974) and Cebeci and Abbott (1975) show success of mixing length models in calculating flow on free rotating discs, cones and cylinders. Chew (1985) found good agreement between mixing length predictions and measurements for the problem described in Section 5.3.2 of radial outflow between corotating discs. Presenting results for a rotor–stator disc cavity with and without a net radial outflow Chew and Vaughan (1988) concluded that the mixing length model was suitable for many engineering calculations of boundary layer dominated flows in rotating disc

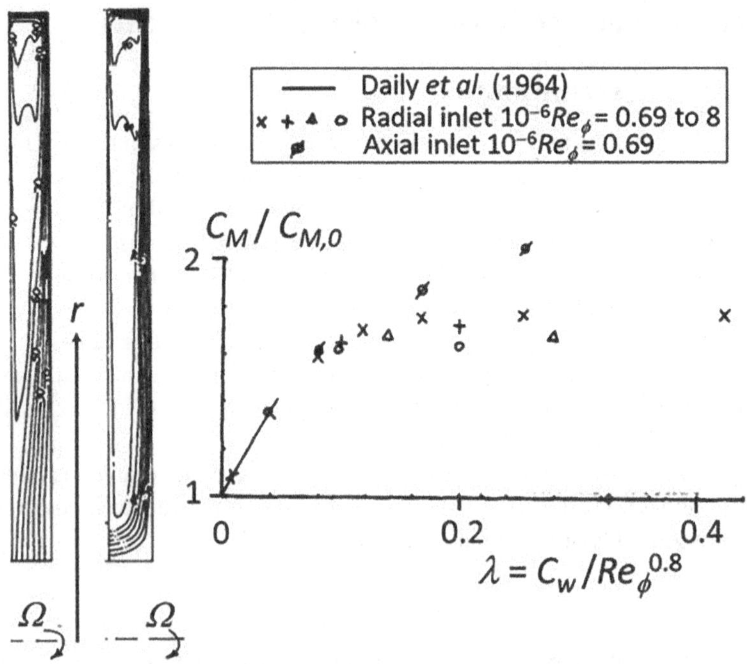

Figure 6.1 Computational fluid dynamics (CFD) results for a rotor–stator disc cavity with outflow, showing streamlines for radial and axial inlets with $Re_\phi = 0.69 \times 10^5$, $C_W = 3795$ and normalised disc moment coefficient. *Source:* Chew and Vaughan (1988) / with permission of American Society of Mechanical Engineers (ASME).

systems. Hart and Turner (1994) gave further comparisons of the mixing length model with flow measurements for a rotor–stator cavity with radial inflow or outflow. Generally good agreement was found with some discrepancies and uncertainties associated with inflow regions.

Figure 6.1 shows calculated streamlines and disc moment coefficients for the configuration studied by Chew and Vaughan, comparing with Daily *et al.*'s (1964) experimental correlation. The streamlines (for mass flow parameter $\lambda = 0.081$) show the expected separate boundary layers on the rotor and stator with a rotating core in the outer part of the cavity. The flow in the inner part of the cavity differs for the two inlet configurations shown, reflecting weaker rotational effects in this region. The moment coefficients are normalised by the zero flow value given by Daily and Nece's (1960) correlation. Von Kármán's free disc solution corresponds to a value of 1.87 on this scale. At lower values of λ the flow can be said to be rotationally-dominated and CFD agrees well with measurements irrespective of inlet configuration. At higher values of λ the CFD results for the radial inlet level off at about 10% below the free disc value. This may be due to the influence of the outer shroud. The radial inlet results have also been shown by Chew (1990) to agree reasonably well with the integral method described in Section 5.2.5. With an axial inlet, impingement of the inflow onto the rotating disc leads to higher moments. This illustrates the value of the CFD in capturing effects that are beyond the capability of integral methods.

In more general flows the eddy viscosity for the mixing length model can be taken as $\nu_t = l_m^2\sqrt{\Phi/\mu}$ where Φ is has the form of equation 2.1.41 (for viscous energy dissipation) and is evaluated using the mean velocity field. However, specification of the mixing length becomes difficult and so alternative turbulence models have been adopted as CFD capability has developed for more complex flows.

6.2.3 The $k - \varepsilon$ Model

6.2.3.1 Model Description

The $k - \varepsilon$ model was developed in the 1970s and has played a prominent role in industrial CFD over several decades. In this model, the turbulence length and velocity scales are estimated from transport equations for turbulent kinetic energy k and turbulent energy dissipation rate ε, with $\nu_t \propto k^2/\varepsilon$. The equations describing the model in regions of fully turbulent flow may be written as

$$\frac{\partial(\rho k)}{\partial t} + \frac{\partial(\rho r \bar{\upsilon}_r k)}{r\partial r} + \frac{\partial(\rho \bar{\upsilon}_\phi k)}{r\partial \phi} + \frac{\partial(\rho \bar{\upsilon}_z k)}{\partial z} = \frac{\partial}{r\partial r}\left(\frac{r\mu_t \partial k}{\sigma_k \partial r}\right) + \frac{\partial}{r\partial \phi}\left(\frac{\mu_t \partial k}{\sigma_k r\partial \phi}\right) + \frac{\partial}{\partial z}\left(\frac{\mu_t \partial k}{\sigma_k \partial z}\right) + P_k - \rho\varepsilon \tag{6.2.3}$$

$$\frac{\partial(\rho \varepsilon)}{\partial t} + \frac{\partial(\rho r \bar{\upsilon}_r \varepsilon)}{r\partial r} + \frac{\partial(\rho \bar{\upsilon}_\phi \varepsilon)}{r\partial \phi} + \frac{\partial(\rho \bar{\upsilon}_z \varepsilon)}{\partial z} = \frac{\partial}{r\partial r}\left(\frac{r\mu_t \partial \varepsilon}{\sigma_\varepsilon \partial r}\right) + \frac{\partial}{r\partial \phi}\left(\frac{\mu_t \partial \varepsilon}{\sigma_\varepsilon r\partial \phi}\right) + \frac{\partial}{\partial z}\left(\frac{\mu_t \partial \varepsilon}{\sigma_\varepsilon \partial z}\right) + (C_{\varepsilon 1} P_k - C_{\varepsilon 1}\rho\varepsilon)\frac{\varepsilon}{k} \tag{6.2.4}$$

$$\mu_t = \rho C_\mu \frac{k^2}{\varepsilon}. \tag{6.2.5}$$

In the above P_k is the production rate of turbulent kinetic energy, arising from turbulent dissipation of mean flow kinetic energy, and is given by

$$P_k = -\left(\overline{\rho \acute{\upsilon}_r \acute{\upsilon}_r}\frac{\partial \bar{\upsilon}_r}{\partial r} + \overline{\rho \acute{\upsilon}_r \acute{\upsilon}_\phi}\frac{\partial \bar{\upsilon}_\phi}{\partial r} + \overline{\rho \acute{\upsilon}_r \acute{\upsilon}_z}\frac{\partial \bar{\upsilon}_z}{\partial r} + \overline{\rho \acute{\upsilon}_r \acute{\upsilon}_\phi}\frac{\partial \bar{\upsilon}_r}{r\partial \phi} + \overline{\rho \acute{\upsilon}_\phi \acute{\upsilon}_\theta}\frac{\partial \bar{\upsilon}_\phi}{r\partial \phi} + \overline{\rho \acute{\upsilon}_\phi \acute{\upsilon}_z}\frac{\partial \bar{\upsilon}_z}{r\partial \phi} + \overline{\rho \acute{\upsilon}_r \acute{\upsilon}_z}\frac{\partial \bar{\upsilon}_r}{\partial z} + \overline{\rho \acute{\upsilon}_\phi \acute{\upsilon}_z}\frac{\partial \bar{\upsilon}_\phi}{\partial z} + \overline{\rho \acute{\upsilon}_z \acute{\upsilon}_z}\frac{\partial \bar{\upsilon}_z}{\partial z} + \frac{\bar{\upsilon}_r \overline{\rho \acute{\upsilon}_\phi \acute{\upsilon}_\theta}}{r} - \frac{\bar{\upsilon}_\phi \overline{\rho \acute{\upsilon}_r \acute{\upsilon}_\phi}}{r}\right) \tag{6.2.6}$$

where the Reynolds stresses are obtained using the eddy viscosity hypothesis, equation 2.4.12. C_μ, $C_{\varepsilon 1}$ and $C_{\varepsilon 2}$ $C_{\varepsilon 2}$ are model constants and σ_k and σ_ε are turbulent Prandtl or Schmidt numbers for k and ε, for both of which diffusion is assumed proportional to spatial gradients. Typical values for model constants are given below following Launder and Spalding (1974).

$$C_\mu = 0.09 \quad C_{\varepsilon 1} = 1.44 \quad C_{\varepsilon 2} = 1.92 \quad \sigma_k = 1 \quad \sigma_\varepsilon = 1.3 \tag{6.2.7}$$

Apart from the modelled diffusion term, the k transport equation 6.2.3 is essentially as derived from the Navier–Stokes equations. Equation 6.2.4 for ε relies on further assumptions with limited validity. The choice of model constants is based on matching to measurements for relatively simple shear flows and decaying isotropic turbulence. The basic model given above and several variants of this have achieved wide popularity in general engineering over several decades.

Boundary conditions on k and ε are required to close the problem. At inlets, it is common practice to specify values of k and ε calculated from representative values of the turbulence intensity = $\sqrt{2k/3}/V$ (typically 1% to 10%), and turbulent length scale = $C_\mu^{3/4}k^{3/2}/\varepsilon$ (e.g. 0.07 times duct diameter). Zero streamwise derivatives are usually used at exit boundary conditions, as for other variables.

No-slip conditions at walls require special consideration. The basic $k - \varepsilon$ model above does not attempt to model regions of low turbulence (where molecular viscosity and turbulent viscosity have comparable magnitudes) that inevitably occur in the boundary layers. A common modelling approximation is to assume the flow near the wall follows a classical logarithmic behaviour as established for fully developed turbulent boundary layers on flat plates and also found to be representative in many other attached boundary layer flows. Assuming velocity u to be zero at the wall and defining nondimensional near-wall velocity $u^+ = u/\sqrt{\tau_0/\rho}$ and normal distance from the wall $n^+ = n\sqrt{\rho\tau_0}/\mu$ the 'log law' gives $u^+ = \ln(En^+)/\kappa$ where E and κ are constants. Typically for smooth walls $E \sim 9$ and the von Kármán constant $\kappa \sim 0.4$. The log law relation is often assumed to hold in the region $30 \leq n^+ \leq 100$. The upper limit of n^+, in particular, will depend on the flow case considered. A laminar sublayer in which $u^+ = n^+$ is expected for $n^+ \leq 5$ and a buffer layer in which both laminar and turbulent diffusion are significant is expected for $5 \leq n^+ \leq 30$. The log law relates wall shear stress to the near-wall distance and velocity. It can be implemented in CFD to replace equation 6.2.5 for eddy viscosity in the near-wall region. Noting that the velocity profiles for the laminar sublayer and logarithmic region intersect at $n^+ \sim 11.5$, a possible implementation is

$$\mu_{eff} = \mu_t + \mu = \frac{\mu n^+ \kappa}{\ln(En^+)} \quad \text{if } n^+ \geq 11.5$$

$$\mu_{eff} = \mu \quad \text{if } n^+ \leq 11.5. \tag{6.2.8}$$

A common refinement is to replace n^+ and u^+ with $n^* = \rho n C_\mu^{1/4} k^{1/2}/\mu$ and $u^* = u C_\mu^{1/4} k^{1/2}/\sqrt{\tau_0/\rho}$. For equilibrium boundary layers in which turbulence production balances turbulence dissipation $P_k = \rho\varepsilon$, and this modification makes no difference. In practice, the wall functions are applied at near-wall mesh points and n^+ or n^* values provide a measure of near-wall mesh resolution. The wall functions are supplemented with near-wall relations for k and ε, with a relatively simple treatment assuming $\partial k/\partial n = 0$ and $\varepsilon = C_\mu^{3/4} k^{3/2}/\kappa n$. Extensions of wall functions to the energy equation are also widely used, drawing on the analogy between momentum and energy transfer with correction for Prandtl number and other effects, as was illustrated for integral methods in Section 5.2.1.

Attempting to overcome limitations of wall functions a number of alternative treatments allowing full resolution of the mean flow in boundary layers have been devised. Some of the most successful blend the $k - \varepsilon$ model in the mainstream to a near-wall treatment assuming a turbulence length scale related to distance from the wall (as in the mixing length model). Some implementations automatically select wall functions or the alternative treatment depending on the (appropriately scaled) near-wall mesh spacing. In many variations on the basic model, developers have adjusted the equations to improve specific aspects of performance. As these are often based on limited comparison with experimental data the improvements achieved for specific flow cases may not be maintained across a wider range of flows.

6.2.3.2 Model Evaluations for Rotating Disc Flows

As in many other engineering applications the $k - \varepsilon$ model has had remarkable success for rotating disc flows despite its undoubted limitations. Early studies applying $k - \varepsilon$ models to rotating disc flows showed mixed results. This is discussed by Virr *et al.* (1994) who evaluated and compared $k - \varepsilon$ model predictions using both wall functions and a near-wall resolved model which merged the $k - \varepsilon$ treatment to a $k - l_m$ treatment for $n^+ < 100$. For radial outflow between corotating discs, it was found that wall functions performed poorly at lower Reynolds numbers where it was not possible to resolve the radial velocity profile with n^* in the appropriate range for the log law to apply. At higher Reynolds numbers, this problem did not arise as the peak radial velocity occurred at higher values of n^*. Apart from this issue the two near-wall treatments gave similar results, and reasonable agreement with measurements, for a range of test cases. Some care was needed in mesh generation. With wall functions some dependency on near-wall mesh spacing is inevitable. Typically, for a range of meshes and for the two models, moment coefficients for a closed rotor–stator disc cavity agreed with Daily and Nece's correlation to within about 10% (which is also representative of experimental scatter).

Virr *et al.*'s comparison of CFD heat transfer prediction with measurements from Millward and Robinson's (1989) industrial research rig are shown in Figure 6.2. This is a rotor–stator configuration with radial outflow with $Re_\phi = 7 \times 10^6$ and $C_W = 36800$ (giving $\lambda = 0.12$). A high speed condition with $\Omega^2 b^2/2C_p \sim 50$ K and disc windage heating of similar magnitude to the disc heat flux is considered. Further results in the paper show that the CFD prediction for the disc moment from the wall function model is 4.5% higher than that for the wall resolved model. The latter gives a moment coefficient 1.9 times the value given by Daily and Nece's closed cavity correlation. This is higher than might be expected from the results in Figure 6.1 but might be explained by the difference between 1/7th power law and logarithmic scaling. Daily and Nece's correlation follows the 1/7th power law scaling. Equations 5.2.2 and 5.2.3 for the free disc indicate a 10% higher moment using logarithmic scaling at this Reynold number. Applying a 10% adjustment to the present condition would bring the moment ratio result close to the lower Reynolds number results in Figure 6.1.

The predicted streamlines in Figure 6.2 follow expectations. Von Kármán's free disc solution suggests that the supplied flow will match disc pumping at $r/b \sim 0.7$ and this is reflected in the flow pattern. For heat transfer, wall function results are given for two values of the recovery factor R_{fac} used in the energy equation wall function. This factor plays a similar role to that of R_f in equation 5.1.18. With $R_{fac} = 0.9$, corresponding to the turbulent Prandtl number, good agreement with the wall resolved model is shown, demonstrating that with careful formulation the energy equation wall function can work well. Using $R_{fac} = 1$ gives significantly higher heat transfer, reflecting the importance of frictional heating for this test case. The comparison with measurements was considered very encouraging. The wall resolved results are within 11% of the experiments at the two inner measurement points. The discrepancy of 22% at the outer point may be associated with uncertainties in extrapolating temperature measurements to define boundary conditions for the calculations.

Many other studies have confirmed the utility of $k - \varepsilon$ models and have also illustrated challenges in applying CFD to more complex situations. An interesting example, related to the heat transfer results shown above, is given by Roy *et al.* (2001). In their relatively low speed, rotor–stator experiment Roy *et al.* had more extensive instrumentation than Millward and Robinson. They compared measurements and CFD calculations of disc heat transfer and fluid temperatures for $Re_\phi = 4.6 \times 10^5$ to 8.6×10^5, and $C_W = 1500$ to 7520 (corresponding to $\lambda = 0.027$ to 0.222). In the test configuration cavity purge flow was supplied axially at inner radii, left the cavity through a rim seal at the disc

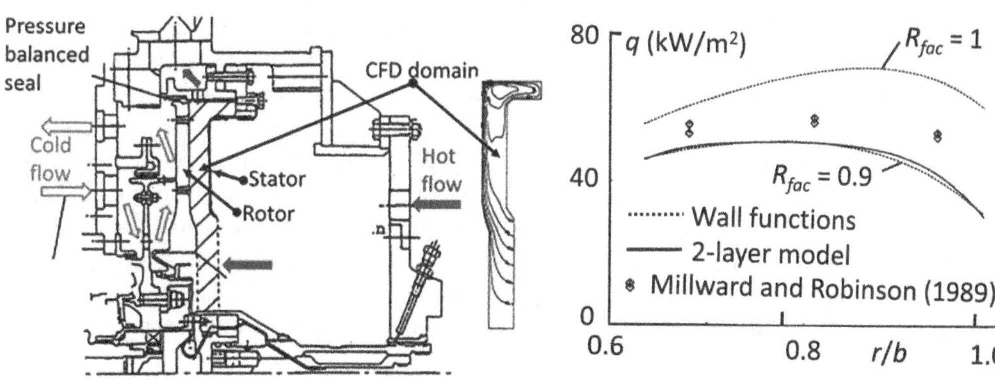

Figure 6.2 Computational fluid dynamics (CFD) result for a rotor–stator disc cavity with outflow, showing Millward and Robinson's (1989) test rig, predicted streamlines, and comparison of disc heat transfer measurements with prediction for $Re_\phi = 7 \times 10^6$, $C_W = 36800$. *Source:* Virr et al. (1994) / with permission of American Society of Mechanical Engineers (ASME).

outer radius and mixed with a turbine-representative annulus flow. At all except the higher purge flow rate tested some ingestion of the annulus flow into the cavity was reported and the CFD models used were not expected to capture this. Based on evaluations against Daily and Nece's enclosed disc measurements, Roy *et al.* selected the renormalisation group (RNG) $k - \varepsilon$ model with a $k - l_m$ near-wall treatment for their axisymmetric CFD modelling. The main difference between the RNG $k - \varepsilon$ model and the standard model lies in modifications to the ε equation.

Comparisons of local disc heat transfer coefficient, based on the disc-to-boundary layer edge temperature difference, were reported by Roy *et al.* to be good in some cases and to show differences of up to 25% for other cases. Annulus flow ingestion and thermal boundary conditions on the stator disc were noted as sources of uncertainty. It may be significant that some of the best agreement between CFD and measurement was for the highest Re_ϕ and C_W considered, for which $\lambda = 0.13$ is close to that for the results in Figure 6.2. For these conditions ingestion was not detected and any low Reynolds number effects would be reduced. Fluid temperatures within the cavity also show variable levels of agreement between CFD and measurement. For a condition with maximum C_W and $\lambda = 0.16$, the measured fluid core temperature at $r/b \sim 0.88$ is about 45% of the disc-to-inlet temperature difference higher than inlet temperature. The CFD prediction for this value is about 35%. This could reflect sensitivities to thermal boundary conditions or deficiencies in modelling. The CFD appears to predict considerably thinner boundary layers on the stator disc than were measured. Overestimation of heat loss to the stator might account for significant underestimation of core temperature by the CFD in a number of cases. It may be noted that local Reynolds numbers for the stator boundary layer will be considerably lower than those for the rotor and that stator heat transfer prediction has not been tested to the same degree as that for the rotor. A further interesting point is that the measurements and CFD show comparable levels of positive radial temperature gradients in the core flow. This suggests that destabilising effects of centrifugal buoyancy (which are unlikely to be captured in the axisymmetric CFD) are overcome by stabilising effects of the radially increasing circumferential velocity and axial flow across the cavity in this case.

A further evaluation of the $k - \varepsilon$ model for prediction of disc heat transfer is illustrated in Figure 6.3. This is taken from a study by Javiya *et al.* (2012) evaluating sensitivity to turbulence models and other assumptions for a rotor–stator flow with highly swirling inflow. The CFD model of the experimental configuration is shown in the figure. This represents a simplified engine preswirl chamber in which the air is given a strong circumferential velocity component, as it accelerates through angled nozzles before transferring to cooling passages in the rotor. In the experiment, described by Kakade *et al.* (2009), disc heat transfer rates were deduced from disc temperature changes detected by liquid crystals as the hotter air heated the initially cold disc. The 3D CFD models approximate 60 discrete receiver holes in the rotor by an axisymmetric slot to allow a steady solution in the stationary frame and use the $k - \varepsilon$ model with wall functions. For one model (denoted 'slot inlet' in the figure) a steady solution in the rotating frame was obtained with the 24 inlet nozzles modelled as a slot. The rotor disc was assumed isothermal at its initial temperature while the stator disc was either assumed adiabatic or isothermal at the initial temperature of the rotor. The free disc integral solution given by equation 5.3.4 with $Pr = 0.7$ and $p = 2$ (for a quadratic disc temperature distribution) is included in the figure as a reference. The Nusselt number in the figure is based on the inlet-to-disc temperature difference and the local radius.

The agreement between CFD and measurements at the higher flow rate ($\lambda = 0.369$) in Figure 6.3 is good, especially when the stator is assumed to be isothermal. The levels of sensitivity to thermal conditions on the stator and the treatment of the inlet and outlet as slots are similar. At lower radii the heat transfer is dominated by impingement of the inlet jets and the heat transfer exceeds the

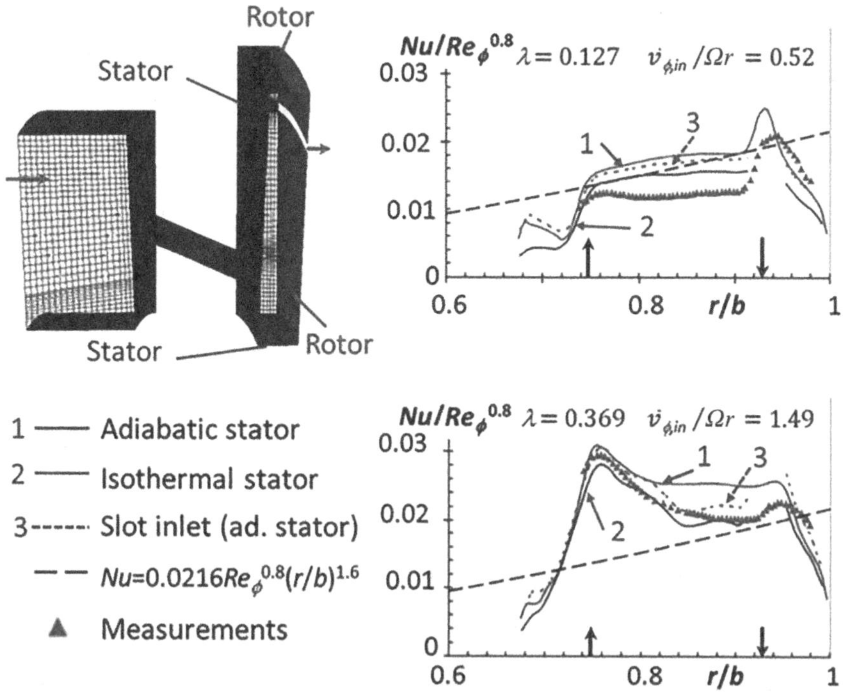

Figure 6.3 Results for a heat transfer on a rotating disc in a preswirl chamber, showing computational fluid dynamics (CFD) domain, measurements, computational fluid dynamics (CFD) results and the integral free disc solution. $Re_\phi = 7.8 \times 10^5$. *Source:* Javiya *et al.* (2012) / with permission of American Society of Mechanical Engineers (ASME).

free disc level throughout. At the lower flow rate ($\lambda = 0.127$), the measured heat transfer is generally below the free disc level. The CFD predicts significantly higher values than measured, although it shows a similar, flat distribution over the main part of the disc. The reasons for this discrepancy are not clear. Comparisons of CFD predictions with core velocity measurements show reasonable agreement for this condition, better than for the higher flow rate. Possible causes of the discrepancy in heat transfer prediction include the assumption of uniform disc temperature, assumptions of steady flow and circumferential periodicity over a small sector, neglect of the gap between the rotor and stator shrouds, turbulence modelling in the CFD and experimental uncertainty. Further investigations of this test case, as discussed later in this chapter, have not fully resolved this issue.

Further results from Javiya *et al.*, assuming steady flow in the rotating frame, examined the surface heat transfer around the receiver holes in the rotor. Differences between results from the $k - \varepsilon$ model with wall functions and a $k - \varepsilon/k - l_m$ were observed. Neither model matched the experimental results well, but the wall function results appeared closer to the measurements. Sensitivity to turbulence modelling is perhaps to be expected in this region as boundary layer separation is likely.

6.2.4 The $k - \omega$ SST Model

The $k - \omega$ SST model, with ω having units 1/time and SST standing for shear stress transport, was introduced by Menter (1994) and has since gained considerable popularity for general engineering use. ω is described as the rate of energy dissipation per unit turbulent kinetic energy or a turbulence frequency. The model combines (Wilcox's, 1988) $k-\omega$ model near walls with the $k-\varepsilon$ model elsewhere and introduces the SST modification to improve the prediction of flow in adverse pressure gradients. The development was on an empirical basis with emphasis on aerodynamic applications. The $k-\omega$ model was selected for its numerical stability near the wall and the $k-\varepsilon$ model for its performance in outer boundary layer regions and free shear flows. Menter evaluated performance for a flat plate boundary layer, free shear layers, a backward facing step, 2D adverse pressure gradient flows with separation and a transonic bump. Further testing for complex 3D flows was recommended.

The transport equations for the model are as follows:

$$\frac{\partial(\rho k)}{\partial t}+\frac{\partial(\rho r\overline{v}_r k)}{r\partial r}+\frac{\partial(\rho\overline{v}_\phi k)}{r\partial\phi}+\frac{\partial(\rho\overline{v}_z k)}{\partial z}=\frac{\partial}{r\partial r}\left(\left(\mu+\mu_t\sigma_k^{-1}\right)\frac{r\partial k}{\partial r}\right)$$
$$+\frac{\partial}{r\partial\phi}\left(\left(\mu+\mu_t\sigma_k^{-1}\right)\frac{\partial k}{r\partial\phi}\right)+\frac{\partial}{\partial z}\left(\left(\mu+\mu_t\sigma_k^{-1}\right)\frac{\partial k}{\partial z}\right)+P_k-\beta^*\rho\omega k \tag{6.2.9}$$

$$\frac{\partial(\rho\omega)}{\partial t}+\frac{\partial(\rho r\overline{v}_r\omega)}{r\partial r}+\frac{\partial(\rho\overline{v}_\phi\omega)}{r\partial\phi}+\frac{\partial(\rho\overline{v}_z\omega)}{\partial z}=\frac{\partial}{r\partial r}\left(\left(\mu+\mu_t\sigma_\omega^{-1}\right)\frac{r\partial\omega}{\partial r}\right)$$
$$+\frac{\partial}{r\partial\phi}\left(\left(\mu+\mu_t\sigma_\omega^{-1}\right)\frac{\partial\omega}{r\partial\phi}\right)+\frac{\partial}{\partial z}\left(\left(\mu+\mu_t\sigma_\omega^{-1}\right)\frac{\partial\omega}{\partial z}\right)+\frac{\rho\gamma}{\mu_t}P_k-\beta\rho\omega^2$$
$$+2(1-F_1)\frac{\rho\sigma_{\omega2}^{-1}}{\omega}G_{k\omega} \tag{6.2.10}$$

In the above, the subscript 2 in the inverse Prandtl or Schmidt number $\sigma_{\omega2}^{-1}$ denotes a value from the $k-\varepsilon$ model. P_k is the production rate of turbulence energy as given by equation (6.2.6) with addition of laminar stresses to the Reynolds stress to extend into the near-wall region. $G_{k\omega}$, occurring in the cross diffusion term arising from transformation of the ε equation, is given by

$$G_{k\omega}=\frac{\partial k}{\partial r}\frac{\partial\omega}{\partial r}+\frac{\partial k}{r\partial\phi}\frac{\partial\omega}{r\partial\phi}+\frac{\partial k}{\partial z}\frac{\partial\omega}{\partial z}. \tag{6.2.11}$$

F_1 is a scaling factor taking values of 0 and 1 in the far field and near-wall regions, so that the cross diffusion term goes to zero near walls.

The SST modification is implemented in the expression for eddy viscosity given below:

$$\mu_t=\rho\frac{a_1 k}{\max\left(a_1\omega,\overline{\Omega}F_2\right)}. \tag{6.2.12}$$

Menter (1994) takes $\overline{\Omega}$ to be the absolute value of vorticity in the above. In other versions of the model $\overline{\Omega}$ is taken to be the magnitude of the rate of strain tensor. F_2 is designed to be 0 in free shear flows and 1 in boundary layers, so that the SST modification only applies in boundary layers.

The model parameters $\sigma_k^{-1}, \sigma_\omega^{-1}, \beta^*, \gamma, \beta$ and a_1 are scaled linearly in F_1 between their values for the $k-\omega$ model near the wall and their values for the $k-\varepsilon$ model in the outer boundary layer and far field. Both F_1 and F_2 are functions of the distance to the nearest wall. Further details are given by Menter and others. Wall functions, valid for near-wall mesh points in the buffer and logarithmic layers, can also be implemented in the model, with automatic activation on coarser meshes.

Considering the close relationship of the $k-\varepsilon$ and $k-\omega$ SST models, they can be expected to give similar results in many cases and particularly when the boundary layers are fully attached and conform with the standard log law. Comparison of the $k-\omega$ SST model with Daily and Nece's rotor–stator cavity moment and velocity measurements as reported, for example, by Burden *et al.* (2022) for a case with $Re_\phi = 4.4\times10^6$, confirms this. From comparisons of various turbulence models with velocity measurements (Poncet *et al.*, 2010) recommend the $k-\omega$ SST model for industrial calculation of rotor–stator disc cavity flows. Their test cases included a closed cavity, radial outflow (with $\lambda \sim 0.08$) and radial inflow (with $\lambda \sim -0.15$) with $Re_\phi \sim 10^6$. Calculations and measurements for the closed cavity are noticeably closer than for the throughflow cases. A further study by Will (2011) compares CFD models with (Itoh *et al.*'s, 1992) closed rotor–stator disc cavity velocity measurements at $Re_\phi \sim 10^6$. The $k - \omega$ SST model was found to improve slightly on the realizable $k - \varepsilon$ model. This was attributed to the effects of transition in the rotor boundary layer. For application of CFD to a full radial pump including front and rear rotor–stator disc cavities. Will selected a $k - \omega$ SST model for overall performance and computational efficiency. A 'curvature correction' was included in the model used. This was reported to have no significant effect on results but to improve numerical convergence.

Da Soghe *et al.* (2018) and Javiya *et al.* (2012) give results from the $k - \omega$ SST models for the preswirl chamber shown in Figure 6.3. Da Soghe *et al.* confirmed some differences between RANS model predictions and measurements shown by Javiya *et al.* although their heat transfer predictions for the lower flow rate case shown in Figure 6.3 are closer to the measurements than the $k - \varepsilon$ results in the figure. Da Soghe *et al.*'s model included both discrete nozzles and discrete receiver holes with a 'frozen rotor' assumption allowing a steady solution. With this approach, the rotor and CFD mesh have a fixed position but the rotor velocity is applied on rotor surfaces. Comparison of frozen rotor and fully unsteady solutions assuming periodicity in a 30° sector, including two nozzles and five receiver holes, for the high flow rate condition of Figure 6.3 showed only weak effects. Javiya *et al.* also mention that their preliminary unsteady CFD solution showed little difference from a steady model.

6.2.5 Other Eddy Viscosity Models

While the $k - \varepsilon$ and $k - \omega$ SST models were selected for special mention above, there are numerous variations of these and other eddy viscosity models employing transport equations for turbulence properties. Many of the alternative formulations will show similar performance for the high Reynolds number, disc boundary layer dominated flows illustrated above. This might be expected when models are calibrated using standard boundary layer flows, as integral methods and mixing length models have shown success in applying methods from canonical 2D cases to the skewed disc boundary layers.

An example of a rotor–stator cavity calculation using the Spalart–Allmaras (SA) one-equation model of turbulence is given by Palermo *et al.* (2021) and illustrated in Figure 6.4. This turbulence model was adopted here for consistency with modelling of the main turbine annulus flow. The cavity geometry here departs from the simpler geometries used in many laboratory studies, but good agreement of the CFD with measurements for the radial pressure distribution is still found. The experimental data suggest the presence of an outer core vortex rotating at ~45% of the rotor disc speed in the outer part of the cavity and an inner cavity vortex rotating at ~34% of the disc rotor speed. Uncertainties in inlet swirl could account for some differences between CFD and measurements. A further point of interest is that (Bru Revert *et al.*, 2023) have measured flow unsteadiness in this cavity with a fast response pressure transducer close to the corner on the stator at $r/b \sim 0.93$ showing the greatest amplitude. It was suggested that this might be associated with shear of the two coexisting vortices or flow separation of the radial inflow on the stator.

Exceptions to the general agreement between various eddy viscosity models indicated above may be found for flows at lower Reynolds numbers and in models that attempt to predict laminar-turbulent transition. In studying rotor–stator cavities Will (2011) included calculations from the transitional SST model. This extends the $k - \omega$ SST model with the addition of two more

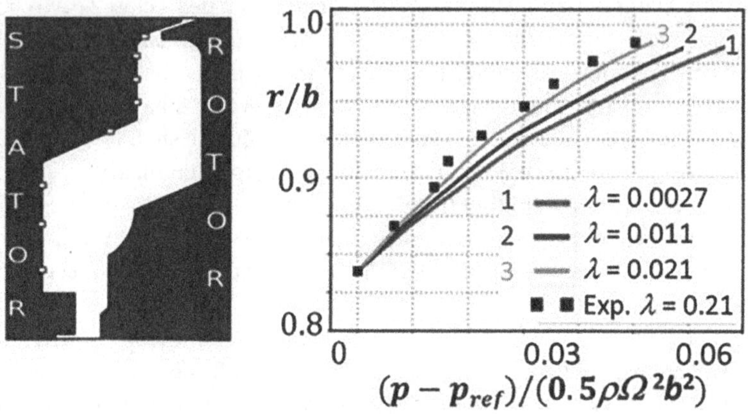

Figure 6.4 Computational fluid dynamics (CFD) results from the Spalart–Allmaras (SA) turbulence model for a rotor–stator disc cavity with weak outflow, showing cavity geometry and comparison of cavity pressure with measurements, for $Re_\phi = 3.3 \times 10^6$. *Source:* Palermo *et al.* (2021) / with permission of American Institute of Aeronautics and Astronautics.

transport equations for intermittency (i.e. a measure of the degree to which the flow is laminar and turbulent) and a transition onset criterion. Will reports some improvements using the transition SST model for a rotor–stator cavity with $Re_\phi \sim 10^6$ but results for a free disc were not so encouraging. Considering the inherent sensitivity of boundary layer transition to environmental conditions such as freestream turbulence and vibration, this might be expected. The results reinforce the need for caution in adopting unproven models and predicting lower Reynolds number flows.

The examples discussed above show that eddy viscosity models can give generally reliable predictions for flows dominated by fully turbulent disc boundary layers, such as occur in rotor–stator disc cavities with weak throughflow. Results for conditions with an imposed radial inflow or outflow are less conclusive than those for closed cavities. This may be associated with sensitivity to inlet geometry and flow conditions, the occurrence of low local Reynolds numbers on some regions of the stator and rotor, and turbulence modelling in the cavity interior. Javiya *et al.* (2012) considered some of these issues for the reasonably complex rotor–stator flow configuration shown in Figure 6.3. Although they concluded that eddy viscosity models are useful, some significant differences with measurements were identified. A further issue regarding turbulence modelling is that, with radial inflow, strong vortices develop in the core flow and eddy viscosity models tend to overestimate turbulent diffusion. This is discussed further in the next section.

6.2.6 Other Reynolds-Averaged Navier–Stokes (RANS) Models

Reynolds stress models (RSMs) that introduce transport models for the six independent Reynolds stresses are often proposed as improvements on eddy viscosity models for rotating flows. There is no doubt that they offer improved modelling in some cases. This is illustrated by Kumar *et al.*'s (2013) results for radial inflow in stationary and rotating cavities. Integral methods have had considerable success in this class of flow, as discussed in Section 5.3.3, but eddy viscosity models produce poor results (when compared to measurements) in some cases. A rather extreme example is given by the wide vortex chamber shown in Figure 6.5. A RSM gives good agreement with velocity measurements. The integral solution, which neglects diffusion in the core, also agrees reasonably well. The $k - \varepsilon$ and SA models severely underpredict core velocity. This is attributed to overprediction of turbulence kinetic energy in the core which is particularly severe for the standard $k - \varepsilon$ model.

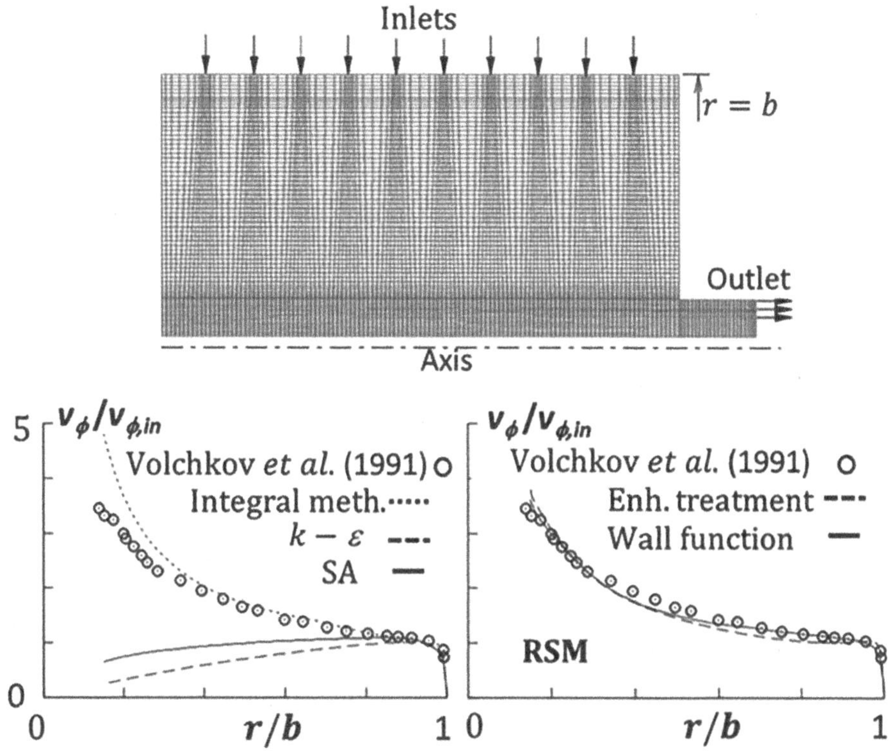

Figure 6.5 Computational fluid dynamics (CFD) results from Spalart–Allmaras (SA), $k-\varepsilon$, Reynolds stress models (RSM) and integral method for a stationary cavity with swirling inflow, showing computational fluid dynamics (CFD) mesh and comparison of core velocity with Volchkov *et al.*'s, (1991) measurements, for $\rho b v_{\phi,\text{in}}/\mu = 22000$, $C_W = 2360$, $s/b = 2$. *Source:* Kumar *et al.* (2013) / with permission of American Society of Mechanical Engineers (ASME).

The trends of underprediction of core velocity by eddy viscosity models in swirling radial inflow has been confirmed in cases more representative of turbomachinery internal air systems but is usually less severe than shown in Figure 6.5. Modifications to the standard $k-\varepsilon$ model can alleviate this but do not appear to eliminate the problem.

A further example from Kumar *et al.* is deswirled radial inflow between corotating discs. The expected flow structure for this case is shown in Figure 5.11, and Figure 6.6 shows the 3D geometry and associated results considered by Kumar *et al.* CFD results from the commercial general purpose code ANSYS/FLUENT and the Rolls–Royce turbomachinery code Hydra are shown. Most cases use an axisymmetric model, approximating the nozzle inlets with a slot of the same area as the nozzles and matching the inlet swirl for Farthing *et al.*'s (1991) experiment. 3D calculations were conducted in Hydra only. For the case shown the rotating deswirl nozzles reduce the inlet swirl to zero in the absolute frame of reference. Considering the results in Figure 6.6 and those for a lower flow rate with $v_{\phi,\text{in}} = 0.4\Omega b$ presented by Kumar *et al.* all models capture the experimental trends to some extent. The comparisons reveal sensitivities to modelling assumptions and details of implementation in different codes. Agreement of models with experiment is generally better at the higher inlet swirl condition than for the case shown here, reflecting the increased significance of the nozzle jets and source flow regions at higher flow rates.

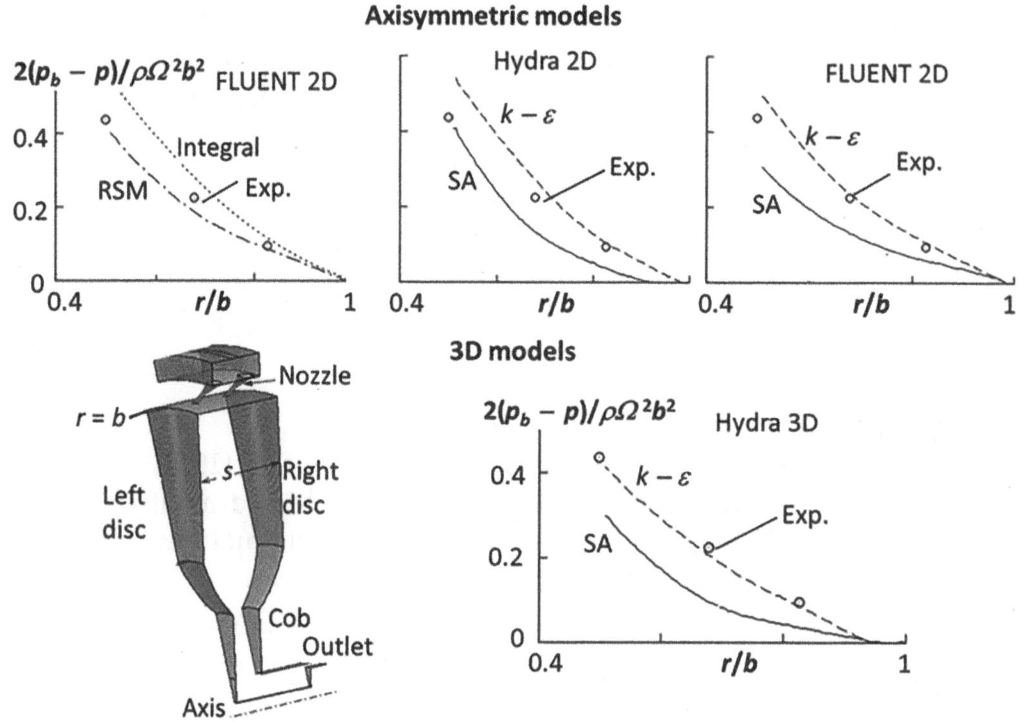

Figure 6.6 Computational fluid dynamics (CFD) results from Spalart–Allmaras (SA), $k - \varepsilon$, Reynolds stress models (RSM) and integral method for a rotating cavity with deswirled inflow, showing comparison of cavity pressure with (Farthing *et al.*'s, 1991) measurements, for $v_{\phi,\mathrm{in}} = 0$, $Re_\phi = 6.1 \times 10^5$, $\lambda = 0.14$, $s/b = 0.31$. *Source:* Kumar *et al.* (2013) / with permission of American Society of Mechanical Engineers (ASME).

The axisymmetric model results in Figure 6.6 and other test cases considered by Kumar *et al.* show a RSM giving the best agreement with experimental data. The integral method shows some divergence from the measurements at lower radii, but this could be due to approximations in the source region of flow where recirculation plays an important role for high deswirl. Differences between the Hydra and FLUENT code results have been associated with the way wall functions are implemented in the two codes. The SA model shows the greatest discrepancy with measurements at this condition but could be said to do slightly better than the $k - \varepsilon$ model at the lower flow rate. The 3D model results show a noticeable difference from the axisymmetric model counterparts but this is less than the difference between the two eddy viscosity models. Considering all results, the $k - \varepsilon$ model performs reasonably well.

Results for the overall radial pressure drop across the cavity are shown in Figure 6.7. This compares directly to Figure 5.13 which gave just the measurements and the integral method solution. The condition shown in Figure 6.6 corresponds to the point with $C_W \sim 6000$. Both the RSM and the $k - \varepsilon$ model capture the experimental trend well, showing more than one possible flow rate for a given pressure drop. From these overall results, there is little or no advantage of the RSM over the $k - \varepsilon$ model.

Despite their good performance in some rotating flows, RSMs are not widely adopted for industrial applications to turbomachinery internal air systems. This can be attributed to eddy viscosity models being more numerically stable and more computationally efficient. This effect has been observed by many workers and is reflected in recommendations by, for example, Will (2011) and

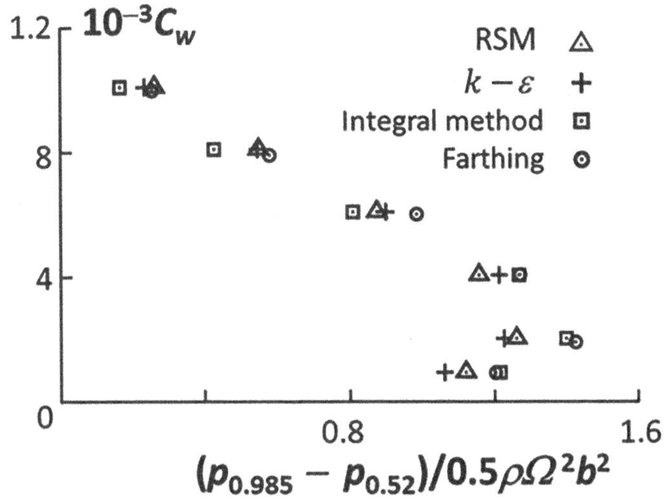

Figure 6.7 Pressure drop in a rotating cavity with deswirled radial inflow for $Re_\phi = 6.1 \times 10^5$, $s/b = 0.31$. *Source:* Kumar *et al.* (2013) / with permission of American Society of Mechanical Engineers (ASME).

Poncet *et al.* (2010). It may also be observed that for most applications the accuracy improvement from RSMs is likely to be small relative to other modelling uncertainties. The performance of RSMs is also dependent on modelling assumptions in their formulation.

In addition to RSM, other RANS models that go beyond the usual eddy viscosity approach have also been developed. These include algebraic stress models (ASMs) and nonlinear eddy viscosity models. With ASMs, the individual Reynolds stresses are estimated from algebraic modelling equations rather than transport equations. With nonlinear eddy viscosity models, stress is assumed to be a polynomial function of strain, rather than a linear function as in standard eddy viscosity models. As little evaluation of these models has been undertaken for rotating disc flows they are not considered further here.

6.3 Large Eddy Simulation (LES)

6.3.1 Modelling Approach

The idea of LES is to solve for the unsteady large, energy containing turbulent eddies that interact with the mean flow and can be captured on the calculation mesh. The effects of smaller eddies is modelled through the use of a sub-grid scale turbulence model. The sub-grid scale model thus depends on mesh size. LES is considerably less computationally demanding than DNS, and the escalation of computing requirements with Reynolds number is not so severe as for DNS. Nevertheless, LES is much more computationally intensive than RANS and this limits its suitability for engineering applications. However, for some applications, particularly where large-scale unsteadiness is significant, LES may be required to capture important flow mechanisms.

Theoretical justification for LES is given by considering the production and dissipation of turbulent kinetic energy in high Reynolds number turbulent flow. This is illustrated in Figure 6.8 and explained in more detail by, for example, Pope (2000). Most of the turbulent energy is contained in relatively large-scale eddies. These anisotropic eddies are continually produced from

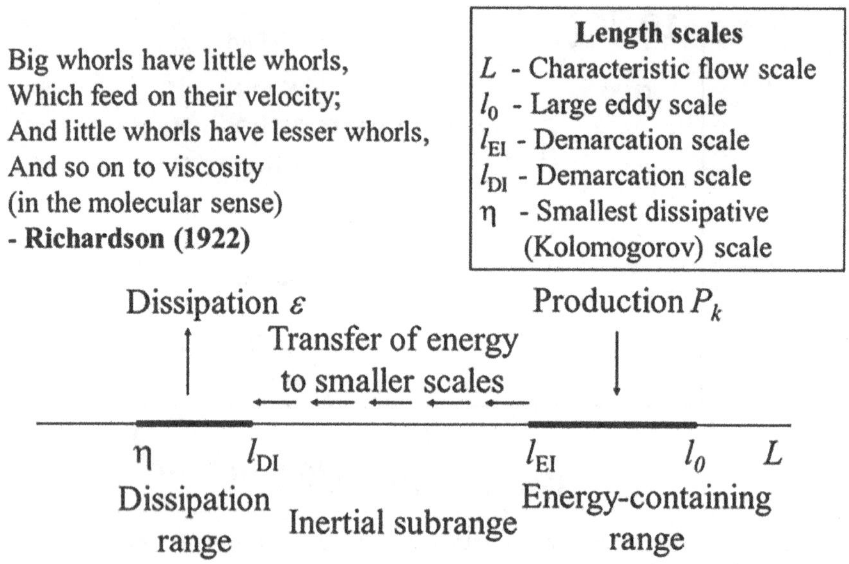

Figure 6.8 Illustration of turbulent energy cascade at very high Reynolds numbers.

the mean flow and break up into successively smaller eddies until fluid viscosity dissipates the energy of the smallest eddies. At high enough Reynolds numbers, an inertial subrange of isotropic eddies exists which transfer the energy from the energy containing range that interacts with the mean flow to the dissipation range. This energy cascade mechanism is expressed in Richardson's (1922) poem, quoted in the figure and summarising his (1920) paper on atmospheric turbulence. The ratio of large eddy length scale to the dissipation scale increases with Reynolds number and the mean flow becomes insensitive to the dissipation length scale. In LES, the turbulent kinetic energy is dissipated on a length scale related to the mesh resolution.

Sub-grid scale models usually use the eddy viscosity concept. A common formulation follows that proposed by Smagorinsky (1963) for atmospheric flow modelling. This essentially uses the mixing length approach (in generalised form) with sub-grid scale viscosity given by the following equation.

$$\mu_{t,SGS} = \rho C_{Smag} l_{mesh}^2 \sqrt{\Phi_{\mathrm{LES}}/\mu} \tag{6.3.1}$$

where C_{Smag} is a constant often taken to be ~ 0.1, l_{mesh} is the local mesh length scale and Φ_{LES} has the form of the viscous energy dissipation (given by equation 2.1.41) evaluated using the resolved velocity field. In practical applications, numerical diffusion (associated with the discretisation scheme) may also contribute to turbulent energy dissipation and with implicit LES (ILES) the numerical dissipation entirely replaces a sub-grid scale model. In LES schemes that use artificial viscosity (or dissipative smoothing) to stabilise the solution, the numerical smoothing may be reduced to levels that are just sufficient to avoid divergence of the solution. This illustrates a connection between numerical and physical stability. Without sufficient numerical dissipation to compensate for the failure to resolve the viscous dissipation the kinetic energy may build up and cause numerical divergence of the solution.

In weakly turbulent flow, the inertial subrange may not fully develop, the flow is Reynolds number dependent, and LES is less reliable. Thus further attention to modelling is needed to account for damping of turbulence in the near-wall region. Considerable research has been conducted and is ongoing in this area. Near-wall damping of $\mu_{t,SGS}$ is commonly used in wall resolved LES (WRLES), and wall-modelled LES (WMLES) in which turbulence in the inner boundary layer is not resolved is frequently used. WMLES can reduce computational cost considerably (compared to WRLES). Approaches available include the use of wall functions and hybrid LES/RANS methods which merge LES to established RANS models near the wall.

6.3.2 Performance for Boundary Layer Dominated Disc Cavity Flows

Considering the nature of flow in rotating disc systems, it is a requirement that LES should adequately capture boundary layer effects. Hence, before considering more complex flows where LES is likely to improve on RANS models, it is useful to examine LES performance for boundary layer dominated flows.

Gao and Chew (2021) report a comparison of DNS, WRLES, WMLES and RANS models for a closed rotor–stator cavity. Comparisons were made with Séverac *et al.*'s, (2007) velocity and Reynolds stress measurements at $Re_\phi = 10^6$ with cavity dimensions, $a/b = 0.286$ and $s/b = 0.143$. All LES used the Smagorinsky sub-grid scale model, with van Driest damping used near the wall in the WRLES. The WMLES used wall functions as these were relatively easy to implement and offered greater savings in mesh size than many other approaches. The RANS results were obtained with the SA model. The DNS solution is discussed further in the next section. Some of the results are shown in Figure 6.9.

Mesh density for the WRLES follows general guidelines and practice for LES. The mesh spacings (normalised using wall units) satisfy $\Delta n^+ = \Delta n\sqrt{\rho\tau_0}/\mu \sim 1$ near the wall, $\Delta r^+ < 40$, $r^+\Delta\phi < 80$. For

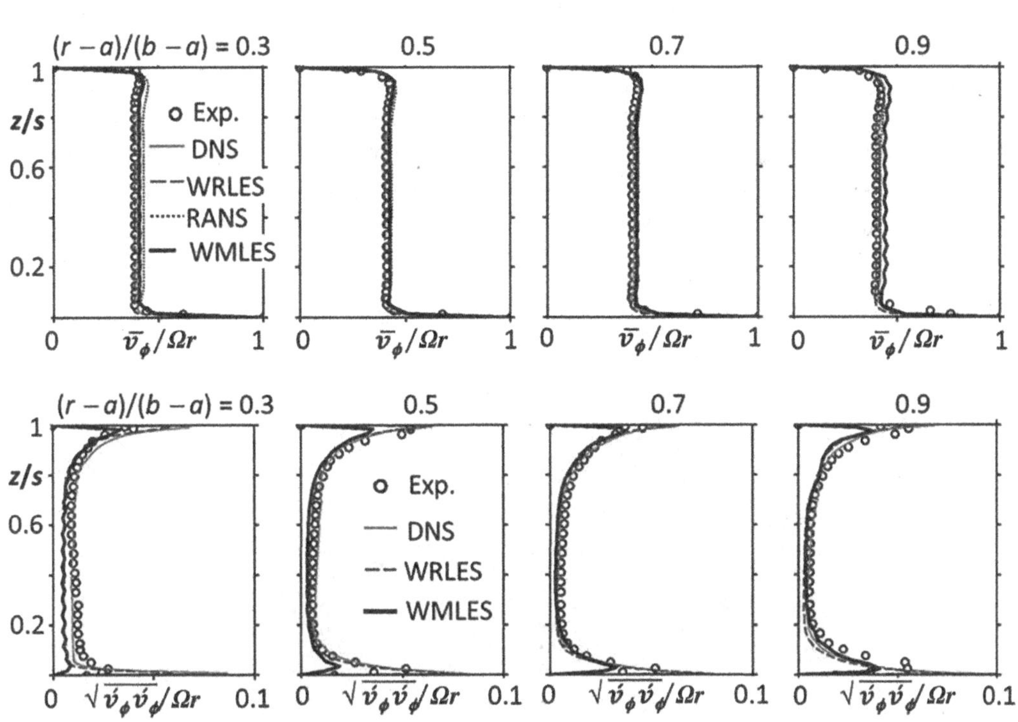

Figure 6.9 Comparison of wall resolved large eddy simulation (WRLES), wall-modelled large eddy simulation (WMLES), direct numerical simulation (DNS) and Reynolds-averaged Navier–Stokes (RANS) results with Séverac *et al.*'s, (2007) velocity and Reynolds stress measurements in a closed rotor–stator cavity at $Re_\phi = 10^6$. *Source:* From Gao and Chew (2021).

the WMLES $\Delta n^+ \sim 50$ near the wall, $\Delta r^+ < 100$, $r^+\Delta\phi < 100$. All the LES and DNS solutions assumed circumferential flow periodicity on a 30° sector. The resulting mesh size for WMLES (~0.42 million mesh points) was considerably lower than for WRLES (~21 million). Further WMLES calculations were conducted at $Re_\phi = 4.4 \times 10^6$ for one of Daily and Nece's (1960) test conditions but WRLES and DNS were not feasible at the higher Reynolds number.

The calculated mean circumferential velocity profiles in Figure 6.9 match the measurements well for all CFD models with only minor differences. Close inspection shows slight overprediction of the core rotation by RANS at the lower radius, and by WMLES at the outer radius, where a weak spatial oscillation of the solution is apparent. The oscillation is attributed to local numerical effects. The Reynolds stresses provide a sensitive test of model realism but are of less direct engineering relevance, and the RANS eddy viscosity model does not attempt to reproduce these. Reasonable agreement with the measured Reynolds stresses is shown with DNS giving the closest agreement, WRLES the next closest, and WMLES failing to resolve the near-wall peaks but otherwise being close to the WRLES. Some underprediction of Reynolds stresses in the core is seen in the WRLES and WMLES, particularly at the lower radius. More detailed results in the disc boundary layers are given by Gao and Chew.

An instantaneous isosurface of Q-criterion coloured by radial velocity from the WRLES solution is shown in Figure 6.10. This may be compared to the DNS results in Figure 4.4. As discussed in Section 4.3, transition has been observed to occur from $r/b \sim 0.4$ to 0.6 at this Reynolds number. The WRLES shows more distinct coherent structures than the DNS in Figure 4.4, but, as further discussed by Gao and Chew, indicates transition at a similar location to DNS. While LES should not be expected to predict transition exactly, the results indicate that LES can correctly predict turbulent flow at higher Reynolds numbers in a closed system with no turbulence introduced through boundary conditions.

As previously discussed, disc torque prediction is of direct engineering relevance and is generally well predicted for the closed rotor–stator cavity by RANS models. Table 6.1 gives Gao and Chew's comparison of moment coefficients with Daily and Nece's empirical correlation. The SA RANS model results are within the expected level of agreement, the WMLES results are also within the likely accuracy of the correlation, the WRLES result is significantly below the correlation, and the DNS is slightly lower. The DNS underprediction is within probable levels of combined numerical

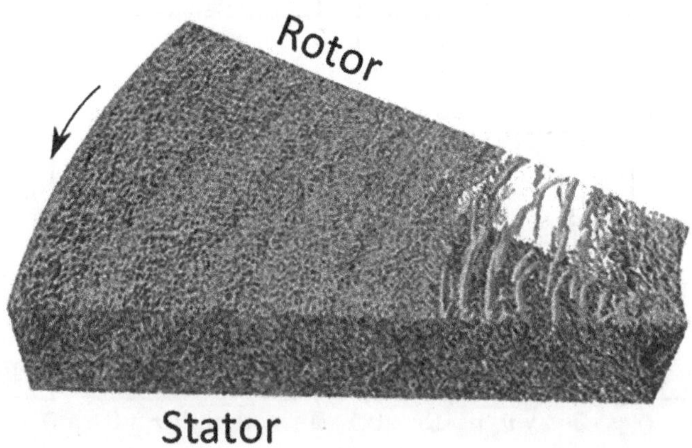

Figure 6.10 Isosurfaces of Q-criterion coloured by radial velocity from wall resolved large eddy simulation (WRLES) for flow in a closed rotor–stator cavity $Re_\phi = 10^6$. *Source:* Gao and Chew (2021) / with permission of SAGE Publications.

Table 6.1 Rotor moment coefficients calculated for a rotor–stator closed cavity.

	$Re_\phi = 10^6$		$Re_\phi = 4.4 \times 10^6$	
	10^{-3} C_M	Difference (%)	10^{-3} C_M	Difference (%)
D&N correlation	2.65	—	1.82	—
RANS-SA	2.51	5.2	1.76	3.2
WMLES	2.64	0.2	1.91	4.9
WRLES	2.12	20		
DNS	2.45	7.6		

and experimental uncertainty levels. The WRLES underprediction by 20% shows a weakness in fully resolved LES when using commonly adopted mesh refinement levels and an exponential near-wall damping. It corresponds to a 10% underprediction in friction velocity $\sqrt{\tau_0/\rho}$ that has been seen in LES for other flows by, for example, Amirante and Hills (2015), Piomelli *et al.* (1988), and Gao *et al.* (2015). Imposing the wall function in WMLES improves the shear stress and moment prediction.

Other studies have also shown that WRLES and WMLES can give good predictions of disc boundary layer flows. Séverac *et al.*'s study showing agreement of LES with velocity measurements used a spectral numerical approximation with a spectral vanishing viscosity (SVV) that only acts on high frequency fluctuations. Bridel-Bertomeu *et al.* (2016) noted the need for near-wall modelling for high Reynolds number applications. Paying particular attention to the formation of spiral flow structures, they concluded from simulations at $Re_\phi = 10^5$ that wall modelling produced some loss of accuracy but preserved many aspects of the rotor and stator boundary layers.

Da Soghe *et al.* (2018) compared RANS and LES/RANS hybrid models for the preswirl chamber illustrated in Figure 6.3. As described above, some differences between RANS eddy viscosity models and measurements have been identified for this configuration. Da Soghe *et al.* reported some improvements using hybrid methods at $Re_\phi = 7.8 \times 10^5$, although some difference with measurements remained. They suggested that low Reynolds number effects may be a factor and concluded that a scale-adaptive simulation (SAS), that automatically allows turbulent fluctuations to be resolved where the mesh is sufficiently fine, was a suitable hybrid model for this application. Kusbeci and Chew (2018) also applied LES to the high flow rate condition in Figure 6.3. Results from a two layer $k - \varepsilon/k - l_m$ model were compared to WMLES with a modified Smagorinsky subgrid scale model blended to a near-wall mixing length formulation. Differences between core velocity predictions from unsteady RANS and WMLES, shown in Figure 6.11, were modest, and no conclusive advantage of LES was found. It was noted that sensitivity to near-wall treatment, as consistent with (Javiya *et al.*'s, 2012) steady RANS studies, could account for differences in core velocity from the two models. None of the CFD models have reproduced the relatively sharp drop in core velocity between $r/b = 0.82$ and 0.85 shown by the experiments.

The results described above show that WRLES and WMLES can produce similar quality results to RANS models in high Reynolds number boundary layer dominated flows and suggest that LES may have advantages over RANS in modelling low Reynolds number effects. However, there are many variations in sub-grid scale models and wall modelling approaches and implementation. Too much emphasis should not be placed on evaluations against single test cases, without full investigation of all modelling assumptions. WMLES clearly has advantages over WRLES in terms of computing requirements and should perform similarly to WRLES in flow prediction away from boundaries.

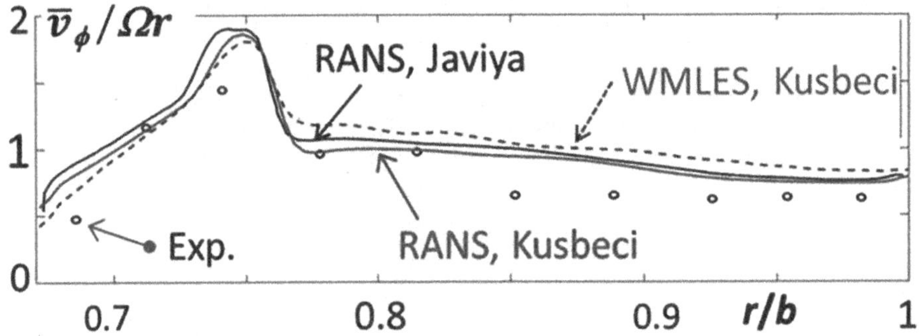

Figure 6.11 Comparison of Reynolds-averaged Navier–Stokes (RANS) and wall-modelled large eddy simulation (WMLES) for a model preswirl chamber showing mean circumferential velocity on the cavity mid-axial plane. $Re_\phi = 7.8 \times 10^5$, $\lambda = 0.37$. *Source:* Kusbeci and Chew (2018) / with permission of American Society of Mechanical Engineers (ASME).

6.3.3 Performance for More Complex Disc Cavity Flows

Kusbeci and Chew (2018) extended the study mentioned above to investigate mixing in the cavity interior for a modified cooling system with preswirled air supplied at two different temperatures. This was based on the preswirl chamber configuration shown in Figure 6.3 and previously described. Results with supply nozzles having alternating low and high air temperatures are shown in Figure 6.12. The unsteady RANS solution and WMLES assumed circumferential periodicity on a 30° sector containing two nozzles and five receiver holes. The rotational speed and total mass flow remained the same as for the single feed system at the higher flow rate in Figure 6.3, and mass flow was divided equally between the two supplies. This resulted in inlet swirl fractions ($v_{\phi,\text{in}}/\Omega r$) of ~1.3 and 2.2 for the cold and hot streams.

Comparison of instantaneous temperature contours on the mid-axial plane in the top row of Figure 6.12 clearly shows large-scale turbulence being resolved in the WMLES but not in the unsteady RANS solution. Some unsteadiness is shown in the RANS at high radii near the receiver holes. This is due to the relative motion of the non-axisymmetric rotor and stator. The time-averaged RANS solution (in the second row of the figure) is very similar to the instantaneous result. In contrast, the time-average WMLES results are very different from the instantaneous solution. The similarity between time-averaged results from the two models is reassuring, although the main hot streak is noticeably broader in the WMLES than the unsteady RANS solution. A quantitative measure of the predicted mixing is given by the comparison of maximum and minimum values of relative total temperatures on planes of constant radius. As expected, both models show mixing of the two streams as they flow radially outwards. The WMLES shows faster mixing, especially near the inlet jets. Both models predict the uncooled and cooled air supplies to be well mixed at the receiver hole inlets in this case. However, for a direct transfer preswirl system, with nozzles and receiver holes at the similar radii, differences could be significant.

More distinct differences between unsteady RANS solutions and LES have been found in other disc cavity flows. An important example is centrifugally buoyant flow in compressor disc cavities with axial throughflow. Sun *et al.* (2007) compared unsteady simulations with an RNG $k-\varepsilon/k-l_m$ model to LES with a Smagorinsky subgrid scale model, and experiments reported by Long *et al.* (2003) and Alexiou (2000). The disc cavity and results from the study are shown in Figure 6.13. The cavity is formed between two rotating discs, a heated cylindrical outer shroud and an inner shaft (which was stationary in this experiment). A forced axial throughflow is imposed in the gap between the disc cobs and the shaft. The main geometrical parameters are $a/b = 0.319$, $r_s/b = 0.283$

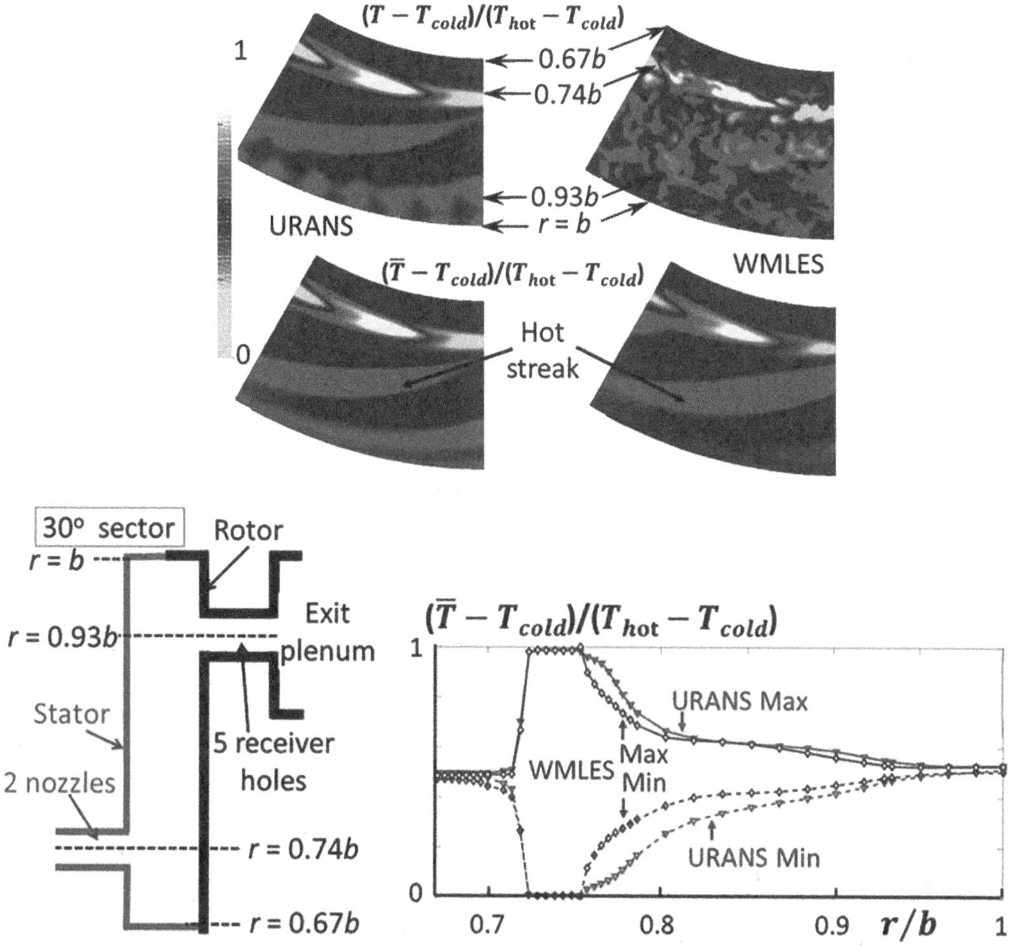

Figure 6.12 Comparison of Reynolds-averaged Navier–Stokes (RANS) and wall-modelled large eddy simulation (WMLES) for a model preswirl chamber showing computational fluid dynamics (CFD) domain, instantaneous and mean relative total temperature contours on the cavity mid-axial plane, and radial profiles of maximum and minimum mean temperatures on planes of constant radius. $Re_{\phi} = 8 \times 10^5$, $\lambda = 0.37$. *Source:* Kusbeci and Chew (2018) / with permission of American Society of Mechanical Engineers (ASME).

and $s/b = 0.195$. The flow condition is characterised by the rotational Reynolds number, inlet Rossby number ($Ro = v_{z,\text{in}}/\Omega a$, where $v_{z,\text{in}}$ is the mean axial velocity at inlet), and $\beta\Delta T$ (where β is the thermal expansion coefficient and ΔT is the shroud-to-inlet temperature difference), as given in the figure heading. Centrifugal buoyancy induces unsteady, 3D flows in the cavity which interact with the axial throughflow. The simulations were conducted in the rotating frame of reference with simulation times of around 200 disc revolutions. Following tests with domain sectors of different sizes, a 120° sector domain with circumferential periodicity was used for most simulations.

The instantaneous temperature and velocity contours in Figure 6.13 show plumes of hot and cold fluid generated at the shroud and from the axial throughflow in both RANS and LES. As expected, the LES shows finer flow structures. The unsteady nature of both solutions is clear from the predicted variation of area-averaged shroud heat flux (presented as a Nusselt number) with time. Both solutions also show the core mean circumferential velocity in the cavity to be less than disc speed. Thus there is some qualitative agreement between the two solutions although the heat

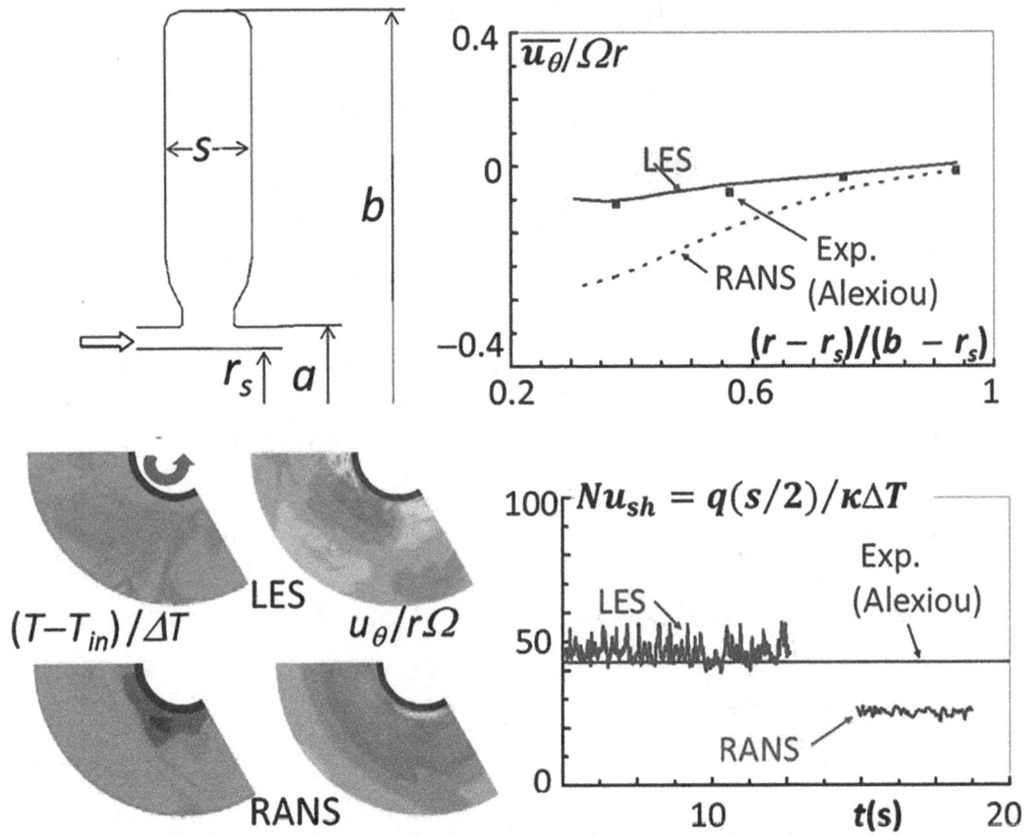

Figure 6.13 Comparison of Reynolds-averaged Navier–Stokes (RANS) and large eddy simulation (LES) for a model compressor disc cavity showing test configuration, radial variation of circumferential velocity on mid-axial plane, temporal variation of shroud Nusselt number, and instantaneous velocity and temperature contours on the mid-plane, $Re_\phi = 1.3 \times 10^6$, $Ro = 4$, $\beta\Delta T = 0.19$. *Source:* Sun *et al.* (2007) / with permission of American Society of Mechanical Engineers (ASME).

transfer and mean core velocity differ considerably. The measurements shown in the figure are at a slightly different condition than the CFD, but it is clear that the LES is closer to experiment than the RANS solution.

Further studies of buoyant disc cavity flows are available in the open literature and show some variable performance of RANS models. For example, Owen *et al.*'s (2007) unsteady RANS solutions for a cavity with a heated disc and axial throughflow show reasonable agreement with their measurements at some conditions but large differences at other conditions. Considerable progress has been made using LES methods in recent years and some of this work is described later in Chapter 8.

Another important class of flow for which RANS models have shown variable performance and for which LES has been shown to have advantages is turbine rim sealing. These flows are discussed further in Chapter 9. It may be observed that both the centrifugally buoyant flows in rotating cavities and rim seal flows display large-scale unsteadiness unrelated to any imposed frequency (such as the receiver hole passing frequency for the preswirl chamber example). Such phenomena may occur with lesser or greater overall impact in other test cases and may not be detected by steady flow measurements or CFD models assuming periodicity on small sectors or steady flow. This is an additional uncertainty to consider in turbulence modelling and CFD evaluations. Although the occurrence of inherent large-scale unsteadiness in RANS models has been shown to correlate with measured unsteadiness in many studies, this goes beyond the theoretical basis of the models. With no imposed flow unsteadiness, RANS models might have been expected to generate steady flow solutions with all unsteady effects modelled through the Reynolds stresses.

6.4 Direct Numerical Simulation (DNS)

Use of DNS, avoiding the modelling inaccuracies of RANS and LES, is conceptually very attractive and computing power has now evolved to the point where this is a useful research technique. Application at higher Reynolds numbers, as often occur in gas turbine engines, is prevented by the escalating computational demands, as discussed in the introduction to this chapter. However, DNS has now been applied for rotational Reynolds numbers up to 10^6 which is within the range of many laboratory experiments used in internal air systems research. Although the Navier–Stokes equations can be regarded as exact, DNS will be subject to numerical errors and uncertainties or approximations in boundary conditions. For example, computational limitations may restrict simulation times, domain extent and the degree to which mesh dependency can be tested. Transitional flows especially may be sensitive to vibration and inlet turbulence levels with experimental conditions unlikely to be exactly reproduced in DNS.

Examples of DNS for rotor–stator cavities with $Re_{\phi} = 10^6$ have been given in Figures 4.4 and 6.9, and Table 6.1. Figure 6.9 shows good agreement with measurements for mean velocity and Reynolds stresses. DNS improves on the LES Reynolds stress predictions. Further DNS results given by Gao and Chew show some wiggles in the core for the Reynolds stress component $\acute{v}_r\acute{v}_r$. These suggest the mesh may be slightly too coarse in this region, resulting in insufficient dissipation. The DNS was obtained using a spectral element method on a 30° sector domain. The circumferential direction was resolved using 512 Fourier planes, and $r - z$ planes were resolved with 10865 elements of polynomial order 5. This is equivalent to ~140 million mesh points in typical finite volume methods. The near-wall grid spacing in wall units is $\Delta n^+ \approx 1$. Mesh spacings in the radial and the circumferential directions are 10 to 20 and 3.5 to 30 wall units, respectively. Flow statistics were obtained by averaging over time and circumferential direction for five rotor revolutions, after an initial simulation of five revolutions to reach statistically steady state.

DNS for the simple rotor–stator configuration has clarified laminar-turbulent transition and provided benchmark solutions for the evaluation of other models. There is clear potential for using DNS to investigate other flows where neither RANS nor LES may be reliable. The nature of low Reynolds number effects in rotor–stator cavities with throughflow, such as the preswirl chamber and other examples discussed above, might well be elucidated by DNS studies. In principle, DNS can be used to improve RANS and LES models, as they can be interrogated to obtain full details of the flow. Considerable efforts have been made in this direction, but major advances have proved elusive. Many current attempts to improve RANS models focus on quite specific examples and machine learning techniques are increasingly being used to correlate models through comparison with LES or DNS.

Insight into centrifugally buoyant flows has been provided by DNS. In their review of buoyant flow in rotating cavities Owen and Long (2015) suggested that the transition from laminar to turbulent flow may occur at much higher Rayleigh numbers in centrifugal Rayleigh-Bénard convection than in the gravitational case. Further discussion and speculation about the nature of the disc and shroud heat transfer is given in many papers. DNS showing the development of turbulence at higher Rayleigh numbers is illustrated in Figure 4.7, and discussed in Chapter 4. Further results from LES for the closed cavity, higher *Ra* condition from Figure 4.7 are shown in Figure 6.14 along with results for a cavity with an imposed axial throughflow.

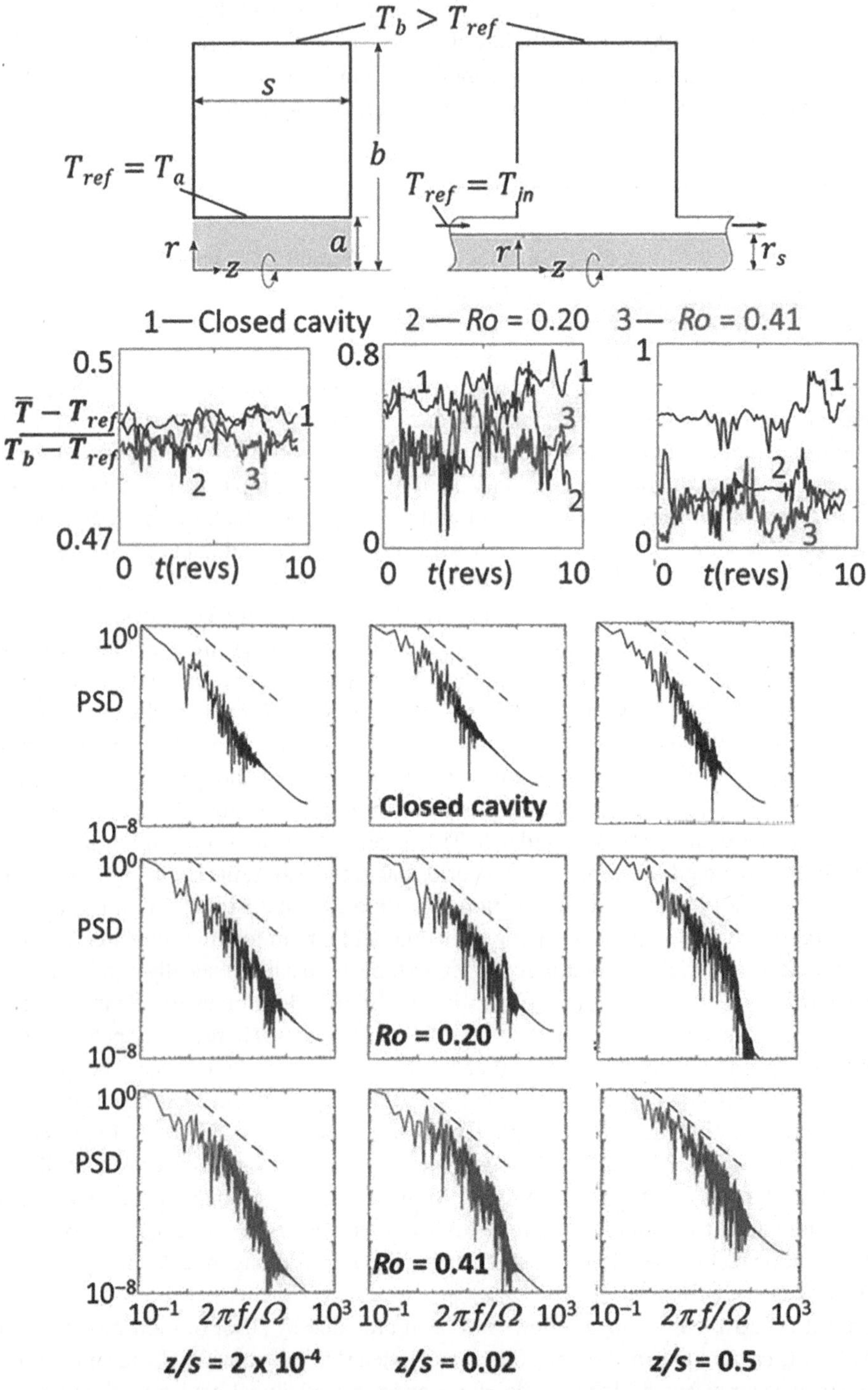

Figure 6.14 Centrifugally buoyant flow convection in a rotating annular cavity with and without axial throughflow showing temperature signals and associated power spectral densities at mid-cavity radius and 3 axial positions (from left to right), $Ra = 1.19 \times 10^9$, $Re_\phi = 1.13 \times 10^5$, $\beta \Delta T = 0.134$. *Source:* Pitz et al. (2019) / with permission of Elsevier.

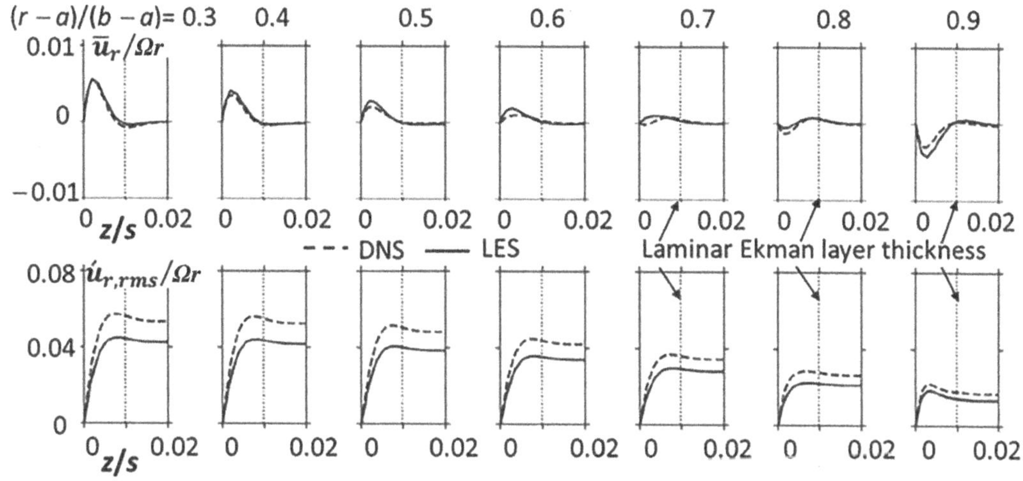

Figure 6.15 Centrifugally buoyant flow convection in a rotating annular cavity showing mean and fluctuating components of radial velocity in the disc boundary layer. $Ra = 1.19 \times 10^9$, $Re_\phi = 1.13 \times 10^5$, $\beta \Delta T = 0.134$. *Source:* Gao *et al.* (2020) / with permission of Elsevier.

The second row of Figure 6.14 shows temperature signals for the three flow conditions considered with columns corresponding to 3 axial positions on the cavity mid-radius. On the left-hand disc $z = 0$, and $z/s = 0.5$ is the mid-axial position. For each axial position, the power spectral densities of the signals are given in the third to fifth rows. The results for $z/s = 2 \times 10^{-4}$ (in the first column) are very close to the disc, well within the boundary layer, and show much reduced fluctuation levels compared to the other axial positions. However, the frequency content is similar to that just outside the boundary layer (at $z/s = 0.02$) and in the cavity interior. The closed cavity results show a lower level of unsteadiness than the cases with throughflow, with less high frequency content, but nevertheless indicate chaotic or turbulent flow characteristics. The dashed lines on the power spectral density curves show a −5/3 gradient representative of the inertial subrange in high Reynolds number turbulence.

Further examination of DNS for centrifugally buoyant flow shows that the instantaneous velocity profiles in the disc boundary layers agree well with the laminar Ekman layer solution described in Section 3.3.2. Recalling that the Ekman solution is a balance of Coriolis and viscous forces, the timescale for the layer to respond to changes in the core velocity will be of order of the disc rotation period. This is much faster than the low frequency unsteadiness shown in Figure 6.14. The higher frequency unsteadiness contains less energy and evidently has only a weak effect in the boundary layer. The role of the Coriolis force in producing a thin boundary layer with relatively high viscous shear is a factor here. As a further illustration, Figure 6.15 shows the mean and fluctuating radial velocity components in the disc boundary layer for the closed cavity case in Figure 6.14. DNS, which uses the Boussinesq approximation (described in Section 2.5.2), is compared with LES for the full equations assuming a perfect gas. Agreement with the Ekman layer thickness is apparent, and the agreement between the DNS and LES is considered reasonable. Other interesting points are that the mean velocities are an order of magnitude lower than the fluctuations and that the mean core radial velocity approximates well to zero in agreement with equation 3.3.7 describing the expected behaviour of steady, axisymmetric, rotationally-dominated flow.

6.5 Fluid–Solid Coupling

Engine disc and other component temperatures will strongly depend on the flow field through convective heat transfer. Hence, the energy equation describing heat conduction in the solid is coupled to the fluid energy equation describing convection and conduction in the fluid. Where buoyancy effects are significant, the flow field may also depend strongly on the component temperatures and two-way thermal coupling of the component temperature and fluid flow occurs. Fluid–solid coupling in forced convection through temperature dependency of fluid properties is comparatively weak. For example, Whitaker (1972) presents Nusselt number correlations with the dependency on fluid bulk-to-surface viscosity ratio $Nu \propto (\mu_{bulk}/\mu_0)^{0.14}$. However, flow coupling through thermal expansion of components affecting seal clearances, and hence fluid flow rates, can be strong. Running clearances must allow for relative movements of stationary and rotating components through an engine operating cycle and may depend on the degree of incursion into abradable surfaces during transient engine operation. Disc cavity pressures can also affect component movements and so combined aero-thermo-mechanical modelling is required for a full representation.

Many CFD codes include the facility for coupled or conjugate fluid–solid heat transfer calculations and capability for combined fluid–solid thermal and structural modelling either within the same code or coupling different codes is advancing rapidly. The main focus here is on thermal coupling in which any elastic deformation of the solid occurs much faster than the thermal adjustment. For steady state solutions using RANS models application of existing methods is relatively straightforward. For time-resolved, statistically steady CFD solutions, using unsteady RANS models, LES or DNS, the degree to which the solid temperature field needs to be resolved should be considered. For example, in transient simulations through an operating cycle the degree to which the fluid flow needs to be resolved in time should be considered.

As indicated by analytical solution of the 1D heat conduction equation for a semi-infinite solid, the time scale for changes in surface temperature is proportional to $\rho C_p \kappa / htc^2$, where ρ, C_p and κ are properties of the solid and htc is the convective heat transfer coefficient. For a thin plate or disc of thickness d a more relevant timescale may be $\rho C_p d/htc$. Both these timescales are typically much longer than engine rotation periods ($\propto 1/\Omega$), and timescales for flow to reach steady state (perhaps $\sim 10/\Omega$ or $100/\Omega$). Timescales for radial heat conduction in a disc ($\rho C_p r^2/\kappa$) can be longer still. Thus, as Ω changes in an engine operating cycle, rates of component temperature changes and thermal expansion are limited by convective heat transfer rate and heat conduction within the components. In contrast, elastic deformation due to changes in centrifugal force is very rapid and can be considered instantaneous, so it is limited by the rate of acceleration or deceleration of the engine. The separation of flow and structural thermal timescales allows averaged unsteady RANS, LES or DNS to be used in thermal modelling of components. A series of different steady or averaged flow solutions may then be used to account for the (relatively slow) changes of component temperatures and associated thermal distortion.

For illustration, some estimates of typical timescales for representative physical parameters are given in Table 6.2. These are obviously dependent on the assumptions made, but the wide range of timescales and distinct separation of flow and structure response times is clear and applies to many applications. Noting that the typical timestep for unsteady RANS calculations may be two orders of magnitude lower than the disc rotation period, and that LES and DNS must resolve turbulence timescales, the need for modelling and approximation in aero-thermo-mechanical modelling of an engine operating cycle is clear.

Table 6.2 Example of estimated flow and thermal response timescales.

Assumed properties and parameters						
Engine rpm	**Accel/decel time (s)**	ρ **(kg.m^{-3})**	C_p **(J.kg^{-1}.K^{-1})**	κ **(W.m^{-1}.K^{-1})**	**Structure length scale (m)**	**Convective** *htc* **(W.m^{-2}.K^{-1})**
10000	10	5000	500	20	0.005 to 0.1	1000
Estimated order of timescales (s)						
Disc rotation period	**Flow response**	**Structural centrifugal deformation**	**Structure surface thermal response**	**Structural bulk thermal response**	**Operating or flight cycle**	
10^{-2}	10^{-1}	10	50	10^3	10^4	

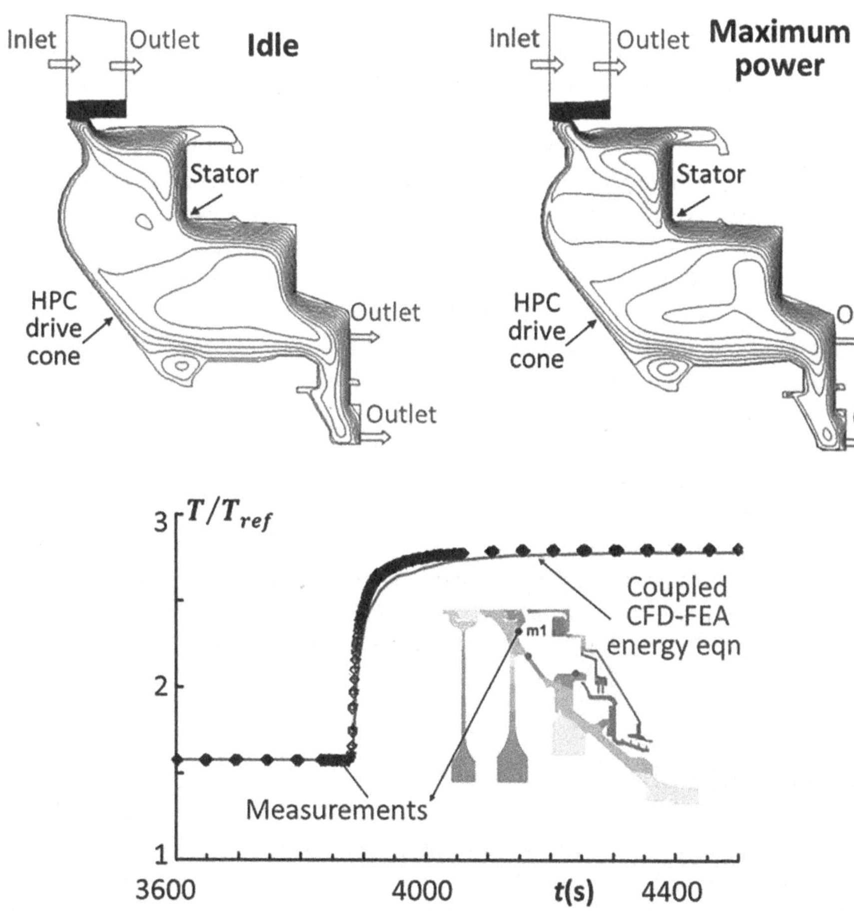

Figure 6.16 Coupled computational fluid dynamics-finite element (CFD-FE) prediction of compressor drive cone cavity flow and heat transfer during an acceleration from idle to maximum power. *Source:* Sun *et al.* (2010) / with permission of American Society of Mechanical Engineers (ASME).

In the method described by Sun *et al.* (2010), thermal coupling is achieved by an iterative procedure between finite element analysis (FEA) for the solid domain and CFD calculations, ensuring continuity of temperature and heat flux. The solid FEA model is treated as unsteady for a given flight cycle, while, for computational efficiency, steady CFD simulations are employed. To further enhance computational efficiency, a 'frozen flow' or 'energy equation only' coupling option was also developed, where only the energy equation is solved (for both solid and fluid domains) during the coupling process. This option takes advantage of the weak dependence of many flows on the thermal boundary conditions.

An example of a coupled CFD–FEA solution is illustrated in Figure 6.16. Here, CFD solutions of an aeroengine rear drive cone cavity provide boundary conditions for a critical area in an FEA model of a high pressure, axial compressor drum. Other boundary conditions in the FEA model are specified following industrial practice in applying heat transfer correlations. The model predicts metal temperatures as the engine accelerates from idle to maximum power, with contours of the stabilised temperatures shown in the inset. Two pre-defined, axisymmetric CFD models were supplied for this case, corresponding to idle and maximum power conditions. The CFD flow fields were fixed while the fluid energy equation was updated as the surface temperatures changed with time. Interpolation of results from the two CFD solutions were used during the (short) acceleration. Reasonably good agreement of prediction and engine temperature measurements was found, as shown by the results at a thermocouple location on the rotor.

The streamlines in Figure 6.16 show a relatively quiescent, rotating central flow core in the cavity with recirculation driven by the pumping effect of the rotating drive cone, and radial inflow in the boundary layer on the stator. The rotational Reynolds numbers at the two conditions are 4.3×10^6 and 2.3×10^7. The increase in Re_ϕ is reflected in thinner boundary layers at maximum power. The radial inflow also increases with engine speed resulting in the throughflow parameter λ increasing by a factor of about 1.6. The moderate changes in λ and the flow pattern give some justification for use of simple scaling of heat transfer coefficients with Reynolds number during the engine transient, as may be assumed in engineering calculations.

Similar levels of agreement with test data to that above have been shown for other test cases and methods have been extended to include coupled calculation of seal clearance and flow rates. Further examples are given by Amirante *et* al. (2012) and Ganine *et* al. (2021). Typically, a coupled thermal model of this type will require hundreds or thousands of updates to the steady CFD solution as modified surface temperatures and other conditions are supplied from the FEA model. Mesh adaptation or regeneration is required when component deformation is modelled. The computation time required for this type of coupled model may, therefore, be around one or two orders of magnitude greater than that for the stand-alone CFD model. The computational time and expense restricts application of fully coupled CFD-solid solutions but may be justified where increased accuracy may avoid the need for costly engine testing or where further investigation of specific issues is required.

6.6 Concluding Remark

While a century of experience has shown prediction of complex turbulent flows to be extremely difficult, CFD has now evolved to offer a powerful capability. Applications and accuracy are restricted by computing requirements, but CFD using RANS models of turbulence are now routinely applied in industry. Research exploiting LES and DNS is shedding light on flow and heat transfer problems that were not previously understood, and these methods will find increasing application in industry as computing power and software continue to develop. The progress made in recent decades and the capability of different modelling approaches for disc cavity flows has been illustrated and discussed above. For the foreseeable future efficient exploitation requires understanding of the strengths and limitations of the modelling approaches available and so some key points are summarised below.

RANS models can be sensitive to the choice of model coefficients, details of implementation, and other assumptions. This is particularly true in flows with boundary layer separation and complex flows where large-scale structures form. Menter (1994) states that 'the slow progress in engineering turbulence modelling, and the confusing picture it often presents, results to no small extent from theoretical concepts and a virtual denial of the empirical nature of the subject'. Menter also states that small changes in model constants (5% or 10%) can lead to significant improvement or deterioration in model performance. These general observations apply in rotating flows and it may also be observed that flow phenomena that are sensitive to model parameters, such as separation and transition over smooth surfaces, are often also found to be sensitive to conditions in physical experiments.

Comparisons with measurements have shown established eddy viscosity models (such as $k - \varepsilon$ and $k - \omega$ SST) to give good predictions for fully turbulent rotating disc cavity flows that are dominated by the radial flows in boundary layers on the rotor and stator. Some uncertainty remains regarding Reynolds number effects in flow regions with small differences between disc and freestream velocity. This and boundary condition uncertainties may account for sensitivities to the choice of eddy viscosity model that are observed in some cases. Standard wall functions work reasonably well for the rotor and stator disc boundary layers at high Reynolds numbers, but care may be needed in ensuring that the near-wall peak in radial velocity is resolved. For flows with strong radial gradients of swirl velocity (as can be generated in radial inflow of swirling flow), eddy viscosity models tend to overpredict turbulent diffusion. This has little overall effect in many cases but may give poor results in, for example, radial inflow between rotating discs at higher flow rates. In such cases, RSMs give better results than eddy viscosity models.

Large-scale, unsteady flow structures develop in some rotating flows (such as with centrifugal buoyancy and in turbine rim seals). Unsteady RANS models capture some of these effects, but comparisons with experimental data show mixed results. Large-eddy simulations generally show more consistent results than RANS models and their use is expected to expand as available computing power increases. Computational requirements also currently restrict model size and assumptions such as circumferential periodicity over a limited sector (rather than a full 360° domain) may preclude capture of some flow phenomena. As resolving the near-wall turbulence in LES is also computationally expensive, WMLES using a simple wall function approach has been proposed as a practical improvement over RANS modelling for engineering calculations. DNS offers the most accurate flow and heat transfer predictions and is a useful research tool but is far from being a practical proposition for high Reynolds number engineering applications. However, DNS has considerable potential for improving predictions of flows with laminar-turbulent transition.

References

Alexiou, A., 2000. Flow and heat transfer in gas turbine H.P. compressor internal air systems. D.Phil. thesis, School of Engineering, University of Sussex.

Amirante, D. and Hills, N. J., 2015. Large-eddy simulations of wall bounded turbulent flows using unstructured linear reconstruction techniques. ASME J. Turbomach., 137, 051006. https://doi.org/10.1115/1.4028549.

Amirante, D., Hills, N., and Barnes, C., 2012. A moving mesh algorithm for aero-thermo-mechanical modelling in turbomachinery. Int. J. Numer. Methods Fluids, 70, 1118–1138. https://doi.org/10.1002/fld.2734.

Bridel-Bertomeu T., Gicquel L. Y. and Staffelbach G., 2016. Wall modeled LES and its impact on rotor/stator cavity unsteady features. ASME Paper No GT2016-57244. https://doi.org/10.1115/GT2016-57244.

Bru Revert, A., Beard, P. F., and Chew, J. W., 2023. Measurement of inertial and acoustic waves in a turbine chute rim seal cavity. ASME J. Eng. Gas Turbines Power, 145, 061021. https://doi.org/10.1115/1.4056552.

Burden, S., Chew, J.W., Gao, F. and Marxen, O., 2022. Effects of turbine conditions on rim seal performance and prediction. ASME Turbo Expo 2022, GT2022-83194. https://doi.org/10.1115/GT2022-83194.

Cebeci, T. and Abbott, D. E., 1975. Boundary layers on a rotating disc. AIAA J., 13, 829–832. https://doi.org/10.2514/3.60447.

Chew, J.W., 1985. Prediction of flow in a rotating cavity with radial outflow using a mixing length turbulence model'. Proceedings of the 4th International Conference on Numerical Methods in Laminar and Turbulent Flow, Pineridge Press, pp. 318–329.

Chew, J. W., 1990. Prediction of rotating disc flow and heat transfer in gas turbine engines. Proceedings of the 3rd International Symposium on Transport Phenomena and Dynamics of Rotating Machinery. Hemisphere Publiching Corporation.

Chew, J.W. and Hills, N.J., 2009. Computational fluid dynamics and virtual aeroengine modelling. Invited paper. Proceedings of the IMechE Vol. 223 Part C: J. Mechanical Engineering Science, 50th Anniversary special edition. https://doi.org/10.1243/09544062JMES1597.

Chew, J. W. and Vaughan, C. M., 1988. Numerical predictions for the flow Induced by an enclosed rotating disc. Proceedings of the ASME 1988 International Gas Turbine and Aeroengine Congress and Exposition. Paper 88-GT-127. https://doi.org/10.1115/88-GT-127.

Da Soghe, R., Bianchini, C., and D'Errico, J., 2018. Numerical characterization of flow and heat transfer in preswirl systems. ASME J. Eng. Gas Turbines Power, 140, 071901. https://doi.org/10.1115/1.4038618.

Daily, J. W., Ernst, W. D. and Asbedian, V. V., 1964. Enclosed rotating disks with superposed throughflow, Report 64, M.I.T, Department of Civil Engineering. https://hdl.handle.net/1721.1/142983.

Daily, J. W. and Nece, R. E., 1960. Chamber dimension effects on induced flow and frictional resistance of enclosed rotating disks. ASME J. Basic Eng., 82, 217–230. https://doi.org/10.1115/1.3662532.

Davidson, P., Kaneda, Y., Moffatt, K., and Sreenivasan, K. (Eds.), 2011. A voyage through turbulence, Cambridge: Cambridge University Press.

Farthing, P. R., Chew, J. W., and Owen, J. M., 1991. The use of de-swirl nozzles to reduce the pressure drop in a rotating cavity with a radial inflow. ASME J. Turbomach., 113(1), 106–114. https://doi.org/10.1115/1.2927727.

Ganine, V., Chew, J. W., Hills, N. J., Mohamed, S. N., and Miller, M. M., 2021. Transient aero-thermo-mechanical multidimensional analysis of a high pressure turbine assembly through a square cycle. ASME J. Eng. Gas Turbines Power, 143, 081008. https://doi.org/10.1115/1.4049498.

Gao, F. and Chew, J. W., 2021. Evaluation and application of advanced CFD models for rotating disc flows. J. Mech. Eng. Sci. (Part C, Proc. IMechE), 235, 6847–6864. https://doi.org/10.1177/09544062211013850.

Gao, F., Ma, W., Zambonini, G., Boudet, J., Ottavy, X., Lu, L., and Shao, L., 2015. Large-eddy simulation of 3-D corner separation in a linear compressor cascade. Phys. Fluids, 27, 085105. https://doi.org/10.1063/1.4928246.

Gao, F., Pitz, D. B., and Chew, J. W., 2020. Numerical investigation of buoyancy-induced flow in a sealed rapidly rotating disc cavity. Int. J. Heat Mass Transfer, 147, 118860. https://doi.org/10.1016/j.ijheatmasstransfer.2019.118860.

Hart, K.J. and Turner, A. B., 1994. Influence of radial inflow on rotor-stator cavity pressure distributions. Proceedings of the ASME 1994 International Gas Turbine and Aeroengine Congress and Exposition. Volume 1: Turbomachinery, paper 94-GT-106. https://doi.org/10.1115/94-GT-106.

Hirsch, C., 2007. Numerical computation of internal and external flows: The fundamentals of computational fluid dynamics, vol. 1, Butterworth-Heinemann.

Itoh, M., Yamada, Y., Imao, S., and Gonda, M., 1992. Experiments on turbulent flow due to an enclosed rotating disk. Exp. Therm. Fluid Sci., 5, 359–368. https://doi.org/10.1016/0894-1777(92)90081-F.

Javiya, U., Chew, J., Hills, N., Zhou, L., Wilson, M., and Lock, G., 2012. CFD analysis of flow and heat transfer in a direct transfer pre-swirl system. ASME J. Turbomach., 134, 031017. https://doi.org/10.1115/1.4003229.

Kakade V. U., Lock G. D., Wilson M., Owen J. M. and Mayhew J. E., 2009. Accurate heat transfer measurements using thermochromic liquid crystals. Part 1: Calibration and characteristics of crystals. Int. J. Heat Fluid Flow, 30, 939–949. https://doi.org/10.1016/j.ijheatfluidflow.2009.04.007.

Koosinlin, M. L., Launder, B. E., and Sharma, B. I., 1974. Prediction of momentum, heat and mass transfer in swirling, turbulent boundary layers. ASME J. Heat Transfer, 96, 204–209. https://doi.org/10.1115/1.3450165.

Kumar, B. G. V., Chew, J. W., and Hills, N. J., 2013. Rotating flow and heat transfer in cylindrical cavities with radial inflow. ASME J. Eng. Gas Turbines Power, 135, 032502. https://doi.org/10.1115/1.4007826.

Kusbeci, M.E., and Chew, J.W., 2018, Assessment of wall-modeled LES for pre-swirl cooling systems. Proceedings of the ASME Turbo Expo ASME, Paper No GT2018-75112,. https://doi.org/10.1115/GT2018-75112.

Launder, B. E. and Spalding, D. B., 1974. The numerical computation of turbulent flows. Comput. Methods Appl. Mech. Eng., 3, 269–289. https://doi.org/10.1016/0045-7825(74)90029-2.

Long, C. A., Alexiou, A., and Smout, P. D., 2003. Heat transfer in H.P. compressor gas turbine Internal air systems: Measurements from the peripheral shroud of a rotating cavity with axial throughflow. HEFAT2003, 2nd International Conference on Heat Transfer, Fluid Mechanics, and Thermodynamics.

Menter, F. R., 1994. Two-equation eddy-viscosity turbulence models for engineering applications. AIAA J., 32, 1598–1605. https://doi.org/10.2514/3.12149.

Millward, J. A. and Robinson, P. H., 1989. Experimental investigation into the effects of rotating and static bolts on both windage heating and local heat transfer coefficients in a rotor/stator cavity. ASME paper 89-GT-196. https://doi.org/10.1115/89-GT-196.

Owen, J. M., Abrahamsson, H., and Linblad, K., 2007. Buoyancy-induced flow in open rotating cavities. ASME J. Eng. Gas Turbines Power, 129, 893–900. https://doi.org/10.1115/1.2719260.

Owen, J. M. and Long, C. A., 2015. Review of buoyancy-induced flow in rotating cavities. ASME J. Turbomach., 137, 111001. https://doi.org/10.1115/1.4031039.

Palermo, D. M., Bru-Revert, A., Gao, F., Amirante, D., Chew, J. and Beard, P., 2021. Computational and experimental assessment of rim sealing flows in axial turbine chute rim seals. AIAA 2021-3464. AIAA Propulsion and Energy 2021 Forum. https://doi.org/10.2514/6.2021-3464.

Piomelli, U., Moin, P., and Ferziger, J. H., 1988. Model consistency in large eddy simulation of turbulent channel flows. Phys. Fluids, 31, 1884. https://doi.org/10.1063/1.866635.

Pitz, D. B., Chew, J. W., and Marxen, O., 2019. Effect of an axial throughflow on buoyancy-induced flow in a rotating cavity. Int. J. Heat Fluid Flow, 80, 108468. https://doi.org/10.1016/j.ijheatfluidflow.2019.108468.

Poncet, S., Da Soghe, R., Facchini, B., 12010. RANS modeling of flow in rotating cavity system. European Conference on Computational Fluid Dynamics ECCOMAS CFD, Lisbon, Portugal. https://hal.science/hal-00679124v1.

Pope, S., 2000. Turbulent flows. Cambridge: Cambridge University Press. https://doi.org/10.1017/CBO9780511840531.

Richardson, L. F., 1920. The supply of energy from and to atmospheric eddies. Proc. R. Soc. London, Ser. A, 97, 354–373. https://doi.org/10.1098/rspa.1920.0039.

Richardson, L. F., 1922. Weather prediction by numerical process, Cambridge University Press.

Roy, R. P., Xu, G., and Feng, J., 2001. A study of convective heat transfer in a model rotor–stator disk cavity. ASME. J. Turbomach., 123, 621–632. https://doi.org/10.1115/1.1371776.

Séverac, E., Poncet, S., Serre, E., and Chauve, M. P., 2007. Large eddy simulation and measurements of turbulent enclosed rotor-stator flows. Phys. Fluids, 2007(19), 085113. https://doi.org/10.1063/1.2759530.

Smagorinsky, J., 1963. General circulation experiments with the primitive equations: I. The basic experiment. Monthly Weather Rev., 91, 99–164. https://doi.org/10.1175/1520-0493(1963)091<0099:GCEWTP>2.3.CO;2.

Spalart, P. R., 2000. Strategies for turbulence modelling and simulations. Int. J. Heat Fluid Flow, 21, 252–263. https://doi.org/10.1016/S0142-727X(00)00007-2.

Sun, Z., Chew, J. W., Hills, N. J., Volkov, K. N., and Barnes, C. J., 2010. Efficient finite element analysis/computational fluid dynamics thermal coupling for engineering applications. ASME. J. Turbomach., 132, 031016. https://doi.org/10.1115/1.3147105.

Sun, Z., Lindblad, K., Chew, J. W., and Young, C., 2007. LES and RANS investigations into buoyancy-affected convection in a rotating cavity with a central axial throughflow. ASME J. Eng. Gas Turbines Power, 129, 318–325. https://doi.org/10.1115/1.2364192.

Virr, G. P., Chew, J. W., and Coupland, J., 1994. Application of computational fluid dynamics to turbine disk cavities. ASME J. Turbomach., 116, 701–708. https://doi.org/10.1115/1.2929463.

Volchkov, E. P., Semenov, S. V., and Terekov, V., 1991. Heat transfer and shear stress at the end wall of a vortex chamber. Exp. Therm Fluid Sci., 4, 546–557. https://doi.org/10.1016/0894-1777(91)90033-N.

Whitaker, S., 1972. Forced convection heat transfer correlations for flow in pipes, past flat plates, single cylinders, single spheres, and for flow in packed beds and tube bundles. AlChE J., 18, 361–371. https://doi.org/10.1002/aic.690180219.

Wilcox, D. C., 1988. Reassessment of the scale-determining equation for advanced turbulence models. AIAA J., 26, 1299–1310. https://doi.org/10.2514/3.10041.

Will, B. C., 2011. Theoretical, numerical and experimental investigation of the flow in rotor-stator cavities with application to a centrifugal pump. dissertations. Duisburg, Essen, Universität Duisburg-Essen. bibliographie.ub.rub.de.

Part II

Examples and Applications

7

Rotor–Stator Disc Cavities

This chapter considers a methodical classification of rotor–stator disc cavities. The flow and heat transfer mechanisms involved and the predictive capability of theoretical and computational methods are illustrated. As discussed in Chapter 1, and illustrated in Figures 1.1, 1.2 and 1.3, a wide variety of rotor–stator configurations are found in rotating machines. There is a large body of publications in this area, and the coverage in this chapter is not exhaustive. The aims are to present a range of examples that show the extent and limitations of current knowledge and to inform engineers in system design and analysis of their configurations.

7.1 Flow in Plane Disc Cavities

7.1.1 Closed Cavities

Considering flow in a closed rotor–stator cavity with plane discs Daily and Nece (1960) identified four flow regimes, depending on the cavity width-to-radius ratio s/b and the rotational Reynolds number Re_ϕ. Delineation of the regimes is shown in Figure 7.1. This includes the empirical correlations for the moment coefficient in each regime and the equations for the regime boundaries determined by matching the correlations in adjacent regions. Daily and Nece's experiments were in the gap-to-radius ratio range $0.0127 \leq s/b \leq 0.217$ and Reynolds number range $\sim 10^3 \leq Re_\phi \leq \sim 10^7$, for a cavity with rotating inner shaft and stationary outer cylinder with radius ratio $a/b = 0.1$.

A comparison of Daily and Nece's regime boundaries with more theoretical estimates is shown in Figure 7.2. Daily and Nece's moment correlation for regime I assumes simple shear between the rotor and stator with no significant radial flow. Assuming, somewhat arbitrarily, that this assumption fails when the spacing is greater than 3× the laminar boundary layer length scale $\sqrt{\nu/\Omega}$ gives the theoretical boundary between regimes I and II, with scaling $Re_\phi^{-1/2}$. Similarly, assuming turbulent boundary layer scaling from the $1/7th$ power-law integral solutions in Chapter 5 gives the boundary between regimes III and IV (merged and separate turbulent boundary layers) scaling with $Re_\phi^{-0.2}$. The multiplying factor 0.5 corresponds to transition between regimes when the free disc boundary layer thickness at the outer radius is approximately equal to the cavity spacing. The boundary $Re_\phi = 3.5 \times 10^5$ between regimes II and IV (laminar and turbulent separate boundary layers) corresponds to the absolute instability prediction in Table 4.1 giving onset of turbulence at the outer radius for a forced vortex rotating $\sim 0.45\Omega$ adjacent to the rotor. It was noted in Section 4.3 that Daily and Nece's results suggested earlier transition than this for the enclosed disc. The remaining boundary I to III (laminar to turbulent merged boundary layers) is estimated using

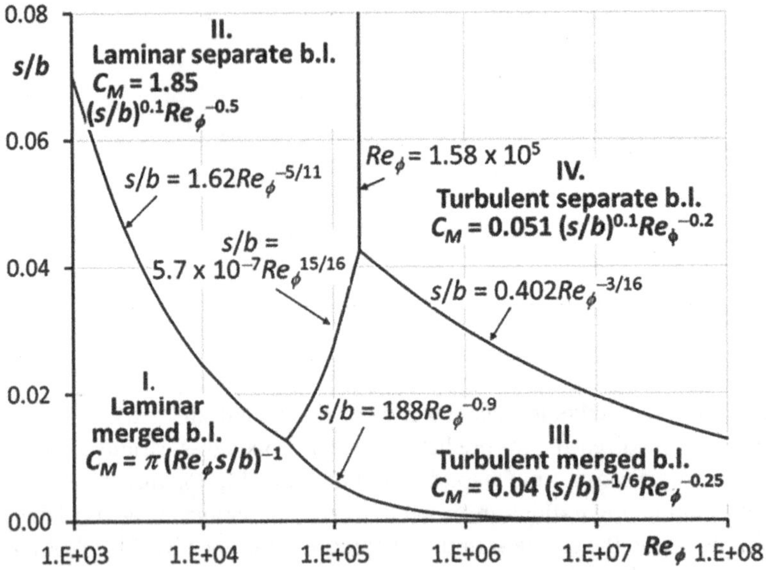

Figure 7.1 Flow regimes and moment correlations. *Source:* Adapted from Daily and Nece (1960).

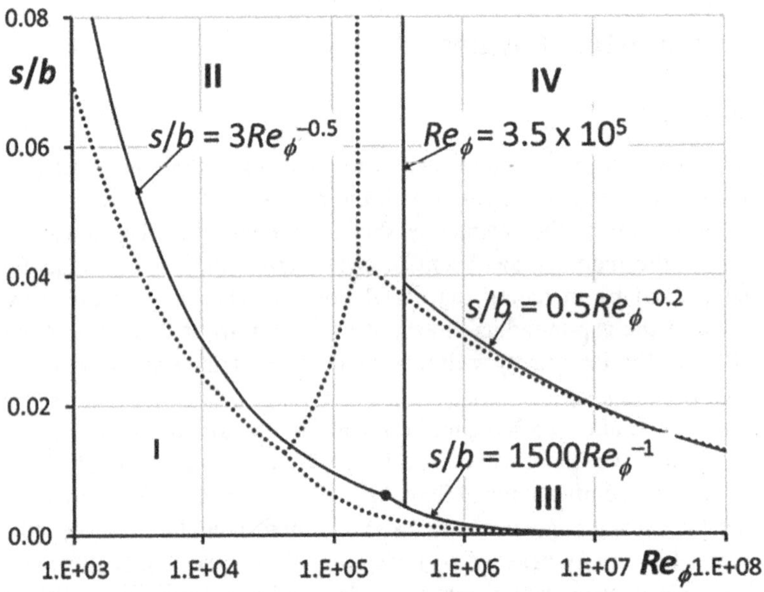

Figure 7.2 Comparison of regime boundaries estimated from theory (full lines) with those of Daily and Nece (broken lines).

Lundbladh and Johansson's (1991) estimate of the minimum Reynolds number for plane Couette flow to support turbulence. Considering that Daily and Nece's torque measurements showed a gradual transition between regimes, the general level of agreement of their regime boundaries with the theoretical estimates is good.

The flow structure for regimes II and IV is of the Batchelor-type, as described in Section 3.1.2.3 and Figure 3.5 for flow between infinite discs. Radial outflow and inflow occur in the rotor and stator boundary layers which are separated by a rotating core with a weak axial flow from the stator to the rotor. In the closed cavity, a boundary layer on the shroud channels recirculation of the rotor pumping flow. This is illustrated in Figure 4.3, with a discussion of laminar-turbulent transition in Section 4.3. Hart and Turner's (1994) experimental results give further details of the variation of core velocity with radius, rotor-to-stator spacing and Reynolds number. Core velocities were estimated from pressure measurements using equation 2.1.15. All cases show lower core rotation at inner radii, consistent with transitional flow. For example, at the highest Re_ϕ of 1.2×10^6, the core swirl fraction drops from ~0.4 at $r/b = 0.6$ to ~0.3 at $r/b = 0.25$. At these two radii, $\rho\Omega_0 r^2/\mu = 4.3 \times 10^5$ and 0.75×10^5 in reasonable correspondence with observations on laminar to turbulent transition in Section 4.3.

As the core velocities in regime IV determine the pressure distribution in the cavity, they are important in considering thrust loads on the rotor and stator. Estimates of core swirl fraction in regime IV from several experimental studies are given in Figure 7.3. Daily and Nece's results are from velocity measurements. It should be noted that their pressure measurements indicate core velocities about 5% lower than those presented. Applying this adjustment would bring these results much closer to the lower values deduced from other studies. These include results from Schultz–Grunow (as reported by Daily and Nece), velocity measurements from Poncet *et al.* (2005a), Hart and Turner's results discussed above and high Reynolds number results deduced from Schröder *et al.*'s (2020, 2021) pressure measurements.

As discussed in Chapters 3 and 5, flows with separated rotor and stator disc boundary layers succumb to analysis by integral methods, and a detailed treatment of regime IV has been described in Section 5.2.5. The enclosed rotating disc in regime IV has also frequently been used as a test case in CFD. Most standard RANS models give reasonable predictions for flow and disc moments, with the flow assumed to be axisymmetric and steady. The overall angular momentum balance given by equation 2.1.7 can also be used to obtain quick estimates of rotor moment and vortex strength. This requires assumptions regarding friction factors on the stationary and rotating surfaces. Examples are given by Haaser *et al.* (1988). Here, a forced vortex is assumed in the core flow with rotor and stator disc moments estimated from equations 5.2.9 and 5.2.13, and the moment

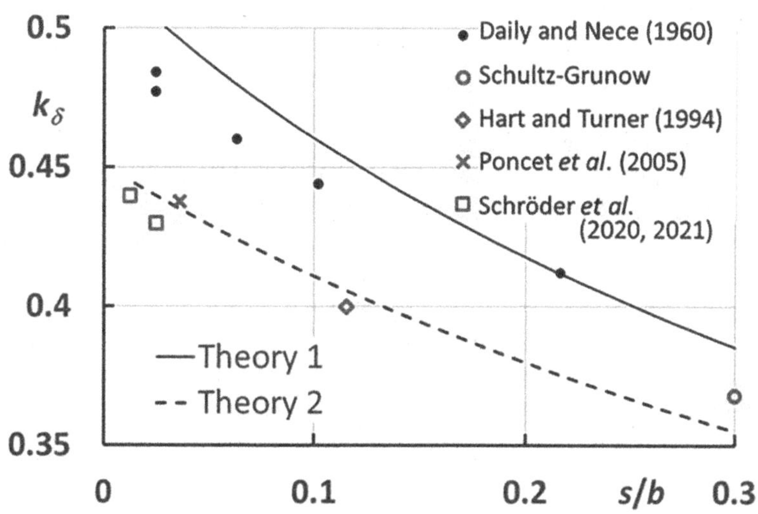

Figure 7.3 Core swirl fraction for an enclosed rotating disc. Theory 1 uses equation 7.1.1. Theory 2 reduces s/b by 0.12.

on the stationary shroud estimated from the second of equation 5.2.13. Balancing moments on the stator and rotor to find an equilibrium (steady flow) condition then gives the following relation between core swirl fraction $k_\delta = v_\phi/\Omega r$ and aspect ratio s/b:

$$\frac{s}{b} = \frac{2.12B_M(1-k_\delta)^{1.6}}{k_\delta{}^{1.8}} - 0.216 \tag{7.1.1}$$

where B_M is given in equation 5.2.10. For the case $n = 7$, a convenient curve fit is $B_M = 0.148k_\delta^2 + 0.0617k_\delta + 0.0738$. So, for a given swirl fraction, a corresponding s/b value can be calculated. Equation 7.1.1 is shown as 'Theory 1' in Figure 7.3. It appears to overestimate data from most sources, perhaps due to underestimating the shroud moment. Impingement of the radial flow in the rotor disc boundary layer (as shown in Figure 4.3) will tend to increase shroud moment. To account for this and achieve a better match to more recent data, 0.12 is subtracted from s/b calculated from equation (7.1.1), to give the curve 'Theory 2' in the figure. Using this result with equation 5.2.9 gives rotor moments in generally good agreement with Daily and Nece's regime IV results. The maximum difference of 10% occurs at $s/b = 0.025$ as the interface with regime III is approached in Figure 7.1.

The merged boundary layer regimes I and III have received less attention from researchers. Both regimes might be expected to succumb to analysis and RANS CFD modelling, but there are some challenges. The interaction of the two boundary layers may introduce additional sensitivity to turbulence modelling compared to regime IV. For plane laminar Couette flow, there is no theoretical upper limit to the Reynolds number above which the flow must be turbulent. Transition therefore depends on the strength of finite disturbances. Similar considerations are likely to apply here. The dependency of C_M on $Re_\phi^{-1/4}$ given by Daily and Nece for regime III is reflected in Kurokawa and Sakuma's (1988) analysis (which covers both regimes III and IV).

Some modification to the Reynolds number scaling in Figure 7.1 might be expected at very high Reynolds numbers, although this has not been thoroughly investigated. For the free disc, as discussed in Section 5.2.1, scaling deduced from the 1/7*th* power law may become inaccurate for Re_ϕ above ~10^7. Similar effects might be expected for the enclosed disc. Schröder *et al.*'s (2020, 2021) results for $2 \times 10^7 \leq Re_\phi \leq 2.5 \times 10^8$ from a pressurised enclosed disc rig filled with carbon dioxide show some interesting effects. At the widest gap ratio test $s/b = 0.0375$, the flow is clearly in regime IV according to Figure 7.1. At the low gap ratio, tested $s/b = 0.0125$ transition from regime III to IV occurs at $Re_\phi = 1.1 \times 10^8$ according to Figure 7.1. Measured moments for both cases are close to Daily and Nece's correlations at $Re_\phi = 2.5 \times 10^7$ but are clearly higher than the correlations at $Re_\phi = 2.5 \times 10^8$. The effect is less pronounced for the narrower cavity, but this could be associated with the change of flow regime. Pressure distributions for both clearances at $Re_\phi = 2.5 \times 10^8$ indicate core rotation consistent with regime IV, as shown in Figure 7.3. For the smaller clearance, core speed increases as Re_ϕ reduces, consistent with transition to regime III. The results appear to confirm changes to the scaling at Reynolds numbers of order 10^8, although the authors caution that their measurements may be affected by the roughness of the stator surface and balance holes drilled in the disc rim. A preliminary, unpublished CFD study of an enclosed rotating disc by Hall (2019) using the transitional $k - \omega$ SST model with $s/b = 0.15$ also suggests that a reduced power-law exponent better fits CFD results for the highest Re_ϕ considered of ~6×10^7.

7.1.2 Simple Radial Outflow

With radial outflow, the flow structure in cavities wide enough for separate rotor and stator boundary layers is illustrated by the schematic in Figure 5.7 and the streamlines obtained from CFD in Figure 6.1. The model presented for turbulent flow in Section 5.2.5 assumes an inner source region where the supplied flow is entrained into the disc boundary layer. Outboard of this region, boundary layers on the rotor and stator (with radial outflow and inflow) are separated by a rotating core. The core swirl fraction increases with radius tending towards values expected for closed cavities. With high throughflow rates, the source region may fill the cavity.

For the simplest case of incompressible flow, dimensional analysis, as described in Section 2.3,2, shows that nondimensional flow parameters such as core swirl, disc moment and pressure distribution will depend on Re_ϕ, $C_W = \dot{m}/\mu b$ and inlet swirl fraction $v_{\phi,in}/\Omega a$ (in addition to geometrical parameters). When the radius ratio of the cavity a/b is low and $v_{\phi,in}/\Omega a \sim 1$ or lower, the effect of inlet swirl will be weak and largely confined to low radii. As described in Chapter 5, integral boundary layer treatments assuming $1/7th$ power laws indicate that for turbulent, rotationally-dominated conditions, the dependency on Re_ϕ and C_W may be captured through the mass flow parameter $\lambda = C_W/Re_\phi^{0.8}$. Figure 6.1 shows the dependency of the moment coefficient on lambda as obtained in early experimental and CFD studies and previously discussed. Other windage studies, for example, by Coren *et al.* (2009), show similar trends with variation in results most evident at higher values of λ, as in Figure 6.1.

Further experimental and CFD studies include those of Hart and Turner (1994) and Poncet *et al.* (2005a). These papers give details of core velocity and pressure distributions. Hart and Turner deduced core swirl fraction from radial pressure distributions, while Poncet *et al.* measured velocity directly using laser-Doppler anemometry. Poncet *et al.* identify the transition from Stewartson to Batchelor-type flow (marking the edge of the source region) slightly inboard of the point at which von Kármán's free disc solution indicates entrainment of the supplied flow. Their experiments and CFD had an axial inlet, and their results are consistent with the flow patterns shown in Figure 6.1. They give a correlation for core swirl fraction that can be written in terms of the local flow parameter $\lambda_r = \lambda(r/b)^{-2.6}$. There is broad agreement in trends between the two studies, although a detailed comparison reveals some differences. For example, Hart and Turner's results for $Re_\phi = 10^6$, $\lambda = 0.05$, $s/b = 0.115$ indicate lower swirl fractions (by around 0.05) in the outer part of the cavity than Poncet *et al.*'s correlation. This may be associated with gap ratio effects as Poncet *et al.* considered $s/b = 0.012$ and 0.036. Scaling the correlation with the ratio of core swirl fractions for the zero throughflow condition would improve agreement, but comparing Hart and Turner's results for $s/b = 0.025$ and 0.115 suggests a more complex effect of s/b. Some differences between Poncet *et al.*'s measurements and CFD predictions with an RSM turbulence model should also be noted. Hart and Turner show some good agreement of core velocity measurements with CFD using a mixing length model of turbulence, but with some discrepancies in the source region and close to the outer shroud and flow outlet.

Incorporating scaling with aspect ratio from closed cavity results into Poncet *et al.*'s correlation for core swirl velocity gives

$$k_\delta = (k_{\delta,\lambda=0}/0.438).max\left\{2(0.63 - 0.939\lambda_r)^{5/7} - 1|0.032 + 0.32e^{-5.68\lambda_r}\right\} \tag{7.1.2}$$

with $k_{\delta,\lambda=0}$ denoting the closed cavity value. Together with Figure 7.3, this equation allows quick estimates of the effect of radial outflow for the enclosed rotor. Cavity pressure distributions calculated using Poncet *et al.*'s correlation and equation 2.1.15 are compared with Schröder *et al.*'s (2021) measurements at $Re_\phi = 2.5 \times 10^8$ with $s/b = 0.0375$ in Figure 7.4. As for the core velocity, in Figure 7.3, the two studies give similar results for the closed cavity ($\lambda = 0$). The effect of throughflow in the two studies is also in reasonable agreement. Differences could be due to variations in inlet geometry and some loss in accuracy in using λ to scale results at very high Re_ϕ.

With further approximations regarding friction factor, equation 7.1.2 can be used to estimate circumferential shear stress on the rotating disc, and hence moment coefficient. For example, it might be supposed that the shear stress is given (as a function of k_δ) by the integral solution for

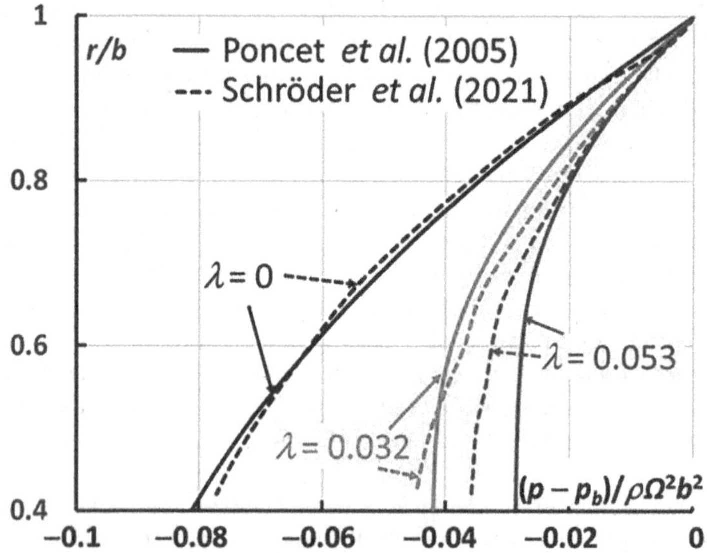

Figure 7.4 Pressure distribution in a rotor–stator cavity with outflow for $s/b = 0.0375$, $\lambda = 0$, 0.032, and 053.

forced vortex flow above a rotating disc in equation 5.2.9. Applying this result and integrating over the disc area gives an estimate for the moment coefficient, which depends on λ and is proportional to $Re_\phi^{-0.2}$. With numerical integration, the following curve fit was obtained for λ in the range 0–0.3:

$$\frac{C_M}{C_{M0}} = 1 + 8.09\,\lambda - 28.6\lambda^2 + 35.1\lambda^3 \tag{7.1.3}$$

where C_{M0} is the moment coefficient for $\lambda = 0$. This may be obtained from the regime IV correlation in Figure 7.1. Equation 7.1.3 agrees reasonably well with the results in Figure 6.1 and could be convenient for quick estimates.

Hart and Turner's pressure measurements for $Re_\phi = 10^6$ with $s/b = 0.115$ are nondimensionalised and compared with estimates obtained from equation 7.1.2 in Figure 7.5. These data were obtained from the graphs published by Hart and Turner, neglecting the outermost measurement point, estimating rig operating conditions from the information given and (arbitrarily) matching to the estimates at the innermost measurement point. Although the data points for $\lambda = 0$ are displaced from the estimates, the slope of the curve matches the data well in the main part of the cavity. Differences can be attributed to the effects of transition at lower radii. Very good agreement of the data with estimates is found at higher values of λ.

As described in Section 6.2, most RANS turbulence models give reasonable flow predictions for rotor–stator cavities with separated boundary layers. Further illustration of this, with relatively high outflow, is given by Moghaddam *et al.* (2013) who compared CFD with measurements for $\lambda = 0.09$, $Re_\phi = 8.1 \times 10^5$ and $\lambda = 0.21$, $Re_\phi = 2.7 \times 10^6$. Generally good agreement with measured moment coefficients was found using a number of turbulence models, although a Reynolds stress model gave noticeably differing results. The moment coefficients for the two cases were ~2 and 2.3 times the closed cavity values given by Daily and Nece's correlation for regime IV. These results are within the range of flow rates sensitive to inlet conditions in Figure 6.1 and are higher than the values given in the figure and equation 7.1.3. While RANS models are clearly useful, some uncertainty remains regarding inlet and outlet conditions, laminar-turbulent transition and narrow cavities with merged boundary layers.

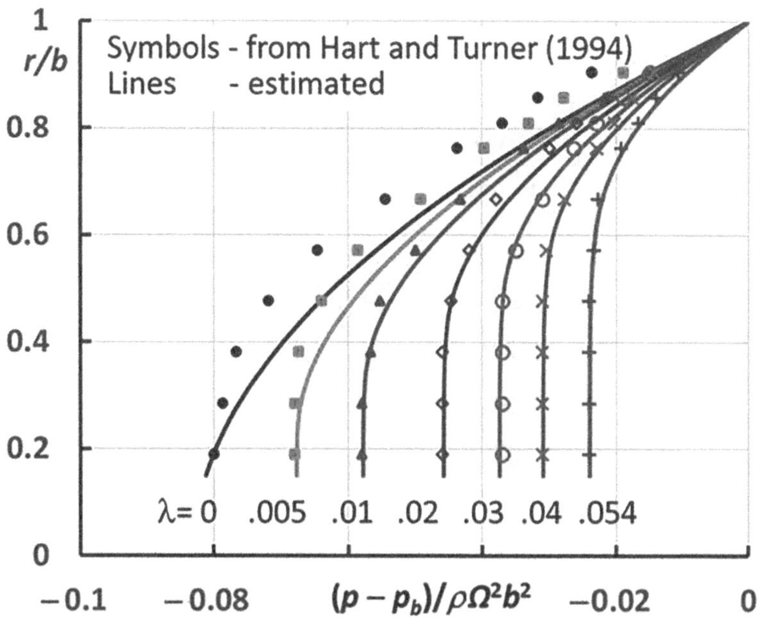

Figure 7.5 Pressure distribution in a rotor–stator cavity with outflow from Hart and Turner's (1994) measurements for $Re_\phi = 10^6$, $s/b = 0.115$, and estimated from equation 7.1.2. *Source:* Adapted from from Hart et al. (1994).

7.1.3 Simple Radial Inflow

For radial inflow, the expected rotationally-dominated flow in cavities wide enough for separate rotor and stator boundary layers is illustrated in Figure 7.6. It is assumed here that inlet swirl is non-negative and less than the disc speed. The fluid circumferential velocity is then lower than the disc speed in the outer part of the cavity, and flow is pumped radially outwards in the boundary layer on the rotor. Further inboard, a point is reached where the fluid tangential velocity equals that of the disc, and a stagnation line is assumed at this radius. Below this radius, the circumferential velocity exceeds that of the disc, and radial inflow occurs on the rotor. Radial inflow occurs on the stationary disc at all radii, with the flow rate varying to preserve the net mass flow rate as the flow in the rotor boundary layer varies. With high throughflow rates, the source region may fill the cavity. As for the radial outflow case, for incompressible flow, core swirl, disc moment and pressure distribution will depend on Re_ϕ, $C_W = \dot{m}/\mu b$ and inlet swirl fraction ($v_{\phi,in}/\Omega b$ in this case), in addition to geometrical parameters. Inlet swirl is expected to have a significant influence. The mass flow parameter $\lambda = C_W/Re_\phi^{0.8}$ is again expected to be a useful correlating parameter.

As for radial outflow, Hart and Turner (1994) conducted a series of experiments with Re_ϕ held constant and flow rate adjusted to give a range of values of λ. Inlet swirl ratios of 0 and 1 were achieved by passing the inflow through holes in either the stator or rotor. Their results for core swirl velocity (obtained from pressure measurements) at $Re_\phi = 10^6$ and $s/b = 0.115$ are reproduced in Figure 7.7. The effects of the flow rate are evident and similar for both swirl ratios. As noted by Hart and Turner, the effect of inlet swirl is most significant at higher radii. The data points at $r/b \sim 0.9$ in Figure 7.7 show a higher range of variation for the swirl fraction of 1 than for 0 inlet swirl.

The dotted lines in Figure 7.7 are fourth-order polynomial fits to the data for each case with inflow. The dashed lines join points on these curves corresponding to the given values of the local flow parameter λ_r. If the local swirl fraction was solely a function of λ_r, then the dashed lines would

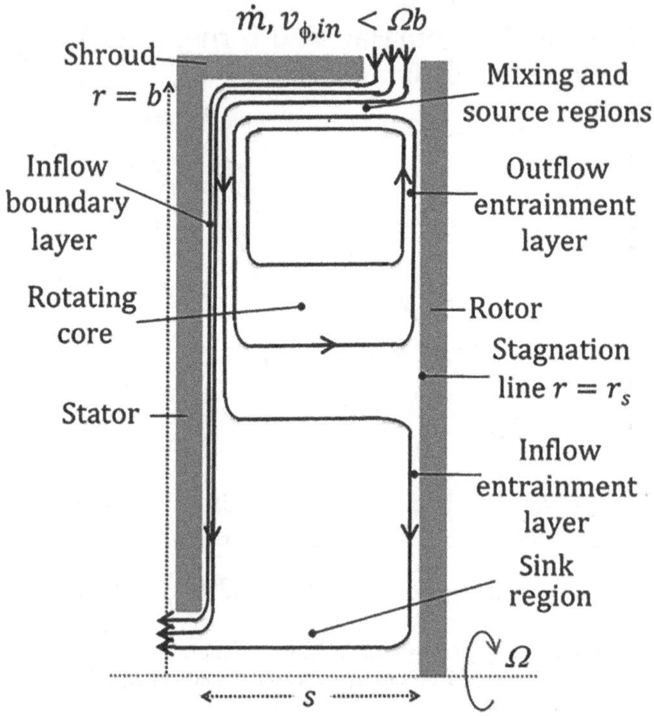

Figure 7.6 Schematic of flow in a rotor–stator cavity with inflow.

be horizontal. Apart from the lowest flow rate, with $\lambda = 0.005$, the variation of swirl fraction at each value of λ_r is modest suggesting, as for radial outflow, that this is a useful correlating parameter. The results for $\lambda = 0.005$ and 0 may be affected by laminar-turbulent transition at lower radii, as discussed for the closed cavity case. Note also that the core swirl fraction for $\lambda = 0.005$ is well below 1, so there will be no inflow region on the rotor for this case.

Poncet *et al.* (2005b) presented pressure and velocity measurements for inflow with $v_{\phi,in}/\Omega b$ stated to be from 0.43 to 0.5, $Re_\phi = 10^6$ to 4.15×10^6, and $s/b = 0.024$ to 0.048. Core swirl velocity fractions deduced from pressure measurements agreed well with the velocity measurements and were correlated as a function of λ_r. With adjustment to take some account of cavity aspect ratio s/b (as was done for radial outflow), their correlation can be written as

$$k_\delta = (k_{\delta,\lambda=0}/0.438)\left[2(0.63 + 0.939\lambda_r)^{5/7} - 1\right]. \tag{7.1.4}$$

Hart and Turner's results from Figure 7.7, for $\lambda \geq 0.01$, are compared with equation 7.1.4 in Figure 7.8. There is clearly a good degree of consistency between the two studies. As seen in the inset highlighting values at low λ_r, the case without inlet swirl shows significant scatter in this region. The case with an inlet swirl fraction of 1 shows less scatter, as also shown by Poncet *et al.*'s measurements. Low values of λ_r include the results from near the inlet and display a tendency of increasing swirl, with inlet swirl fraction, as expected. The intermediate inlet swirl condition represented by the adjusted correlation from Poncet *et al.* falls between the two cases from Hart and Turner.

Equation 7.1.4 or the curve fits to Hart and Turner's data given in Table 7.1 and Figure 7.7 might be used for quick estimates of core swirl velocity based directly on experimental data. The correlations in Table 7.1 could be adjusted to allow for variation in s/b, as has been done in equation 7.1.4. Pressure distributions can then be estimated using equation 2.1.15. Friction factors

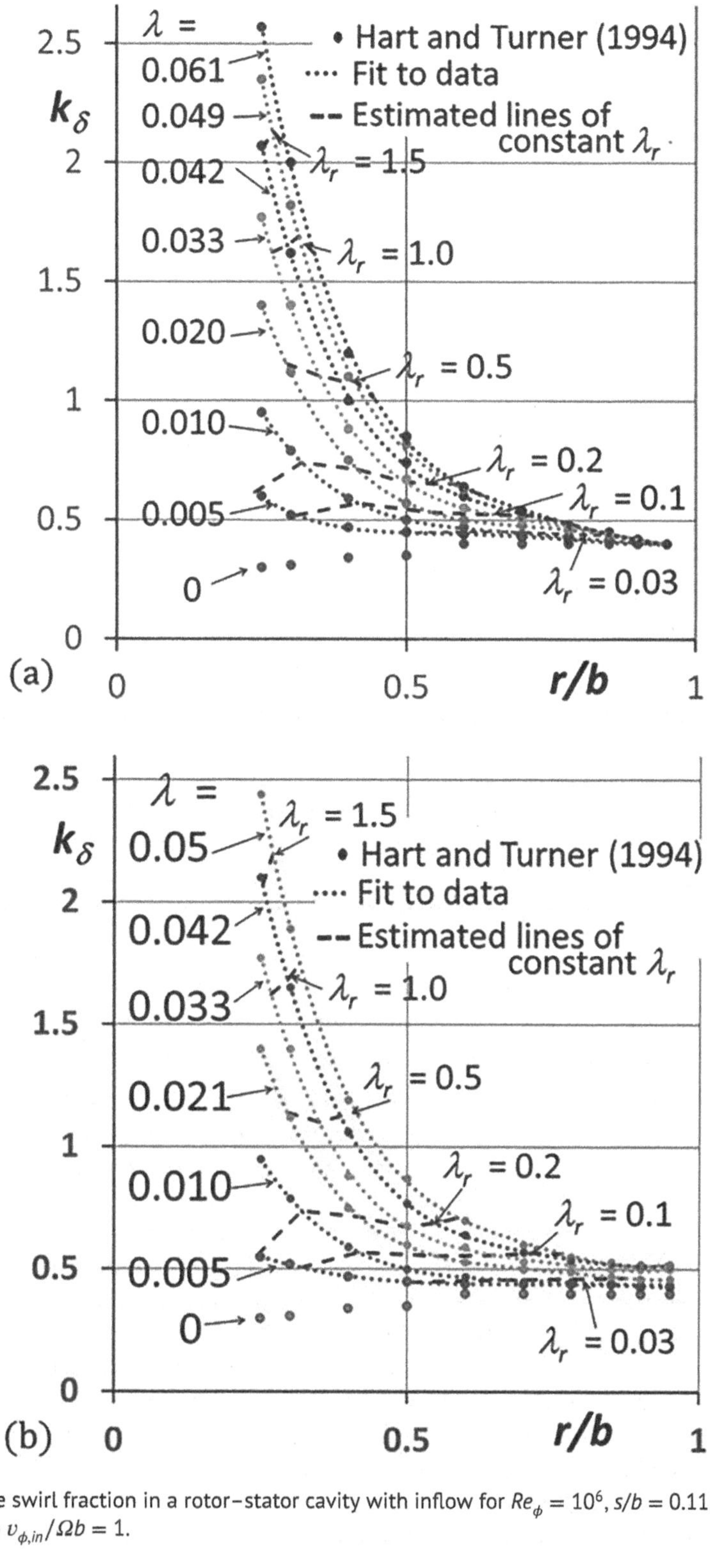

Figure 7.7 Core swirl fraction in a rotor–stator cavity with inflow for $Re_\phi = 10^6$, $s/b = 0.115$ with $v_{\phi,in}/\Omega b = 0$, (b) $v_{\phi,in}/\Omega b = 1$.

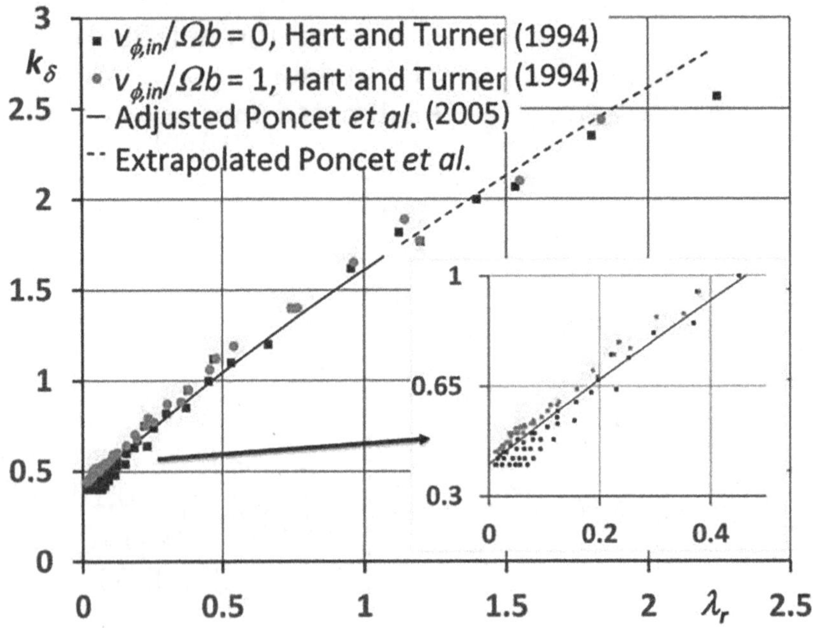

Figure 7.8 Core swirl fraction in a rotor–stator cavity with inflow plotted against local flow parameter.

Table 7.1 Curve fits to Hart and Turner's (1994) core velocity data in a rotor–stator cavity with inflow for $Re_\phi = 10^6$, $s/b = 0.115$.

$v_{\phi,fl}/\Omega r = A(r/b)^4 + B(r/b)^3 + C(r/b)^2 + D(r/b) + E$											
$v_{\phi,in}/\Omega b = 0$						$v_{\phi,in}/\Omega b = 1$					
λ	A	B	C	D	E	λ	A	B	C	D	E
0.0048	6.1291	−16.567	16.283	−6.9922	1.56	**0.0051**	0.3408	−1.9608	3.1516	−1.9945	0.8823
0.010	7.6183	−22.814	25.389	−12.643	2.851	**0.010**	7.0382	−21.314	24.166	−12.248	2.8073
0.020	8.5853	−29.062	35.974	−19.597	4.4734	**0.021**	12.609	−38.23	43.301	−21.939	4.7288
0.033	13.797	−43.805	51.732	−27.342	6.0067	**0.033**	17.644	−52.716	59.175	−29.871	6.3003
0.042	18.597	−57.036	65.456	−33.885	7.2727	**0.042**	16.842	−51.868	60.415	−31.893	7.0424
0.049	23.314	−69.749	78.429	−40.017	8.4522	**0.050**	24.519	−72.116	80.251	−40.706	8.6292
0.061	21.394	−66.237	77.309	−40.996	8.9429						

must be assumed to obtain estimates of disc moments from the core swirl distribution. As discussed in Section 5.2.3, equation 5.2.9 from the extended von Kármán solution is unreliable for $k_\delta > 1$. Equation 5.2.11 is a plausible alternative and matches the friction factor from equation 5.2.9 well for $k_\delta < 1$, as shown in Figure 5.5. Note, though, that this has not been tested against measurement for radial inflow in which the core swirl fraction can vary strongly with radius.

Radial inflow has received less attention than outflow in CFD studies, but, as noted above, Hart and Turner found reasonable agreement with measurements using a mixing length model of turbulence. Thus, similar RANS model performance to that for radial outflow can be expected. Again, some uncertainty remains regarding inlet and outlet conditions, laminar-turbulent transition and narrow cavities with merged boundary layers.

7.1.4 Open Cavities and Rotating Flow Modes

As discussed in Section 3.1.2.3, whether a rotor–stator cavity is open or closed at the outer radius has a profound effect on the flow. With an open boundary, flow is drawn in from a quiescent environment to satisfy the rotor disc pumping. If the axial spacing between the rotor and stator (s/b) is sufficient, the flow on the rotor will be close to that of a free disc. At lower spacings, other effects may be expected.

Maroti *et al.* (1960) observed unsteady flow phenomena for a rotating disc adjacent to stationary walls extending beyond the rotor diameter. Separate inflow and outflow regions rotated in the same direction as the disc but at a lower speed. The number of these regions depended on the spacing between the rotor and the stator and the rotor speed. For smaller stators, no unsteady effects were reported. The paper clearly illustrated the importance of outer boundary conditions and is a precursor to more recent identification of unsteady flow phenomena. The discussions from contemporary researchers in the field published alongside Maroti *et al.*'s paper provide some possible explanations for the unsteady flow.

Bayley and Conway (1964) measured pressure distributions and stator torque in an open rotor–stator cavity with no flow and with inflow for gap ratios of 0.004–0.06 for rotational Reynolds numbers up to 4×10^6. The pressure measurements show strong dependence on cavity width and on inflow rate. The authors report that they found no time-dependent effects or evidence of instability. However, unsteady RANS CFD modelling of this case reported by Chew (2009) suggests that large-scale rotating flow structures may occur in the outer part of the cavity. This links the open cavity to research on the interaction of the main gas path and disc cavity flows in turbomachinery and in turbine disc cavity rim sealing. Gaps between rotating and stationary parts at the cavity periphery are unavoidable in practice and must be sufficient to accommodate relative movements of the components during transient operations. Such cavities may be partially open in that they allow a degree of interaction of the two flows. This often leads to unsteady large-scale flow structures unrelated to imposed frequencies (such as that due to turbine blade passing) and will be discussed further in Chapter 9. As described by Bru Revert *et al.* (2023), these flow structures can be interpreted as rotating inertial waves and may interact with rotating acoustic waves. Considering potential vibration in centrifugal compressors for high-density fluids, Unglaube and Brillert (2022) have also identified acoustic waves in a rotor–stator disc cavity. In this case, coupling with structural vibration modes was investigated.

7.1.5 Transient Flow

Flow behaviour during a change in operating conditions, or following a component failure, is often of interest and may require consideration of how disc cavity flows respond to changes in boundary conditions. This imposed unsteadiness is considered here, as distinct from the intrinsic unsteadiness at steady operating conditions discussed in Section 7.1.4 and in later sections. However, it should be noted that rotating flow features can occur and might interact with imposed unsteadiness. Examples of modelling of transient air systems are given by Alexiou and Mathioudakis (2008), Gallar *et al.* (2009) and Calcagni *et al.* (2009).

Estimates of the flow response time in cavities without rotation are often obtained from the flow residence time (i.e. the time taken to replace the mass in the cavity). For a simple rotor–stator cavity, the residence time corresponds to the order of $\sim(s/b)Re_\phi^{0.2}/2\lambda$ disc revolutions. However, the response of the core vortex that controls the pressure distribution in the cavity (or spin-up time) should also be considered. For a closed rotor–stator cavity or at low flow rates, this timescale will dominate. Spin-up time may be estimated from an overall unsteady angular momentum balance, as in equation 2.1.7. Assuming that net surface torque scales as for free and enclosed discs with separate turbulent boundary layers, the number of disc revolutions needed for flow to adjust to new steady boundary conditions may be of order $\sim 5(s/b)Re_{\phi}^{0.2}$. For $s/b = 0.1$ and Re_{ϕ} up to 10^7, this indicates flow adjustment within around 10 revolutions. With these estimates, the spin-up and residence times will be similar when λ is $\sim$0.1, which is representative of many engine flows. For a rotor speed of 10,000 rpm, the spin-up time is of order 0.1 s. This corresponds to the flow time in Table 6.2, which compares this to other representative timescales associated with a 10 s acceleration. From CFD solutions for a simple rotor–stator cavity with outflow May and Chew (2010) found that spin-up effects led to a doubling of the response time scale compared to a non-rotating cavity. The viability of an elementary model based on overall mass, angular momentum and energy balances for the cavity was also demonstrated.

As was discussed in Section 6.5, a series of steady flow solutions may often be used in thermal analysis through engine accelerations or decelerations. More rapid responses may have to be considered in failure modes analysis or for very fast adjustment of engine conditions. For these cases, pressure waves at the speed of sound will transmit disturbances through the engine air system, changing the boundary conditions to the disc cavities. The associated timescales are (again) much faster than those for component temperature changes and should be considered along with those for flow residence and spin-up.

Some vortex flows are subject to hysteresis, where the steady or periodic flow depends on the initial conditions. An example of rotating cavities with radial inflow is given in the next chapter. While such cases might be constructed for rotor–stator cavities, they would be unusual and so are not considered here.

7.2 Flow in More Complex Geometries

As shown by the examples in Figures 1.1, 1.2 and 1.3, engine discs are rarely fully planar. Relatively thin rotating compressor and turbine discs have thicker cobs at inner radii to mitigate centrifugal mechanical force. Other features connect the discs to shafts, accommodate seals, channel cooling flows or are associated with fixing blades and other components. Stationary components also often differ considerably from the planar ideal. In addition, some cavities may be traversed by pipes to supply oil to bearings. A further illustration of engine geometries is given in Figure 7.9. This shows an axisymmetric finite element model of a high-pressure (HP) turbine disc and surrounding components. Component colours or shades indicate their rotational speed. In this case, the low-pressure (LP) and intermediate-pressure (IP) shafts rotate in opposite directions to the HP components, and a bearing supporting the HP shaft is assumed to rotate at half the HP speed.

The subsections below discuss some observed effects in experimental and CFD studies and indicate areas of uncertainty and modelling approaches.

7.2.1 Axisymmetric Geometries or Approximations

There are many examples where flow structures in more complex, but predominantly axisymmetric, geometries are similar to those in ideal plane disc geometries. The effect of disc radius ratio a/b has been discussed for the free disc in Section 5.2.1. The correction factors in equation 5.2.7 could be used to estimate the effects on disc moment, disc pumping and heat transfer in rotor–stator cavities. The effect of cone angle, for half angles above about 30°, has also been included in the boundary integral solutions with the scaling given in equations 5.1.15 and 5.1.21. The coupled rotor–stator integral boundary layer model given in Section 5.2.5 also incorporates these effects and might be extended to accommodate changes in cone angle. However, with increasing complexity, the integral methods become less reliable and CFD may be preferred.

Figures 6.4 and 6.16 illustrate steady, axisymmetric RANS model solutions for higher radius ratio cavities with step changes in cone angles for the rotor and stator. As shown by the streamlines for radial inflow in Figure 6.16, the flow in rotationally dominated conditions retains the rotating core/boundary layer structure of the plane disc cavities. For outflow, in Figure 6.4, the accuracy of the mean flow prediction is confirmed by comparison with mean pressure measurements. As discussed in Section 6.2.5, the pressure distribution in this case suggests the core flow rotates at different fractions of disc speed in the inner and outer parts of the cavity and further work identified flow unsteadiness in the cavity. The weaker inner vortex may be associated with a higher ratio of stator area to rotor area in the inner region. The unsteadiness is linked to ingestion through the rim seal, but the maximum amplitude from fast response pressure transducers on the stator occurred close to the corner at $r/b \sim 0.94$. Whether or not the flow unsteadiness would occur without ingestion is unknown. It does appear that cavity geometry affects unsteady flows with potential impact on ingestion through the rim seal.

Figures 1.1, 1.2 and 7.9 feature drive cone cavities with the inner boundary given by the drive cone/shaft connecting the high-pressure compressor and turbine. Long *et al.* (2003) compared velocity measurements with 2D and 3D CFD for such a cavity and found similar recirculation and levels of core swirl velocity to those expected in a classic rotor–stator disc cavity. However, some uncertainty regarding the inlet condition (taking air from the HP compressor outlet) remained and the study focused on the flow in the front part of the cavity (near the compressor) where the cone angles are relatively high. The flow in the rear of the cavity, where the shaft radius is almost constant, was not investigated experimentally but is expected to be dominated by the turbine disc pumping. Further CFD, for the drive cone cavity in Figure 7.9, is given by Javiya *et al.* (2015), with streamlines from their 2D solution and circumferential velocity predictions from 2D

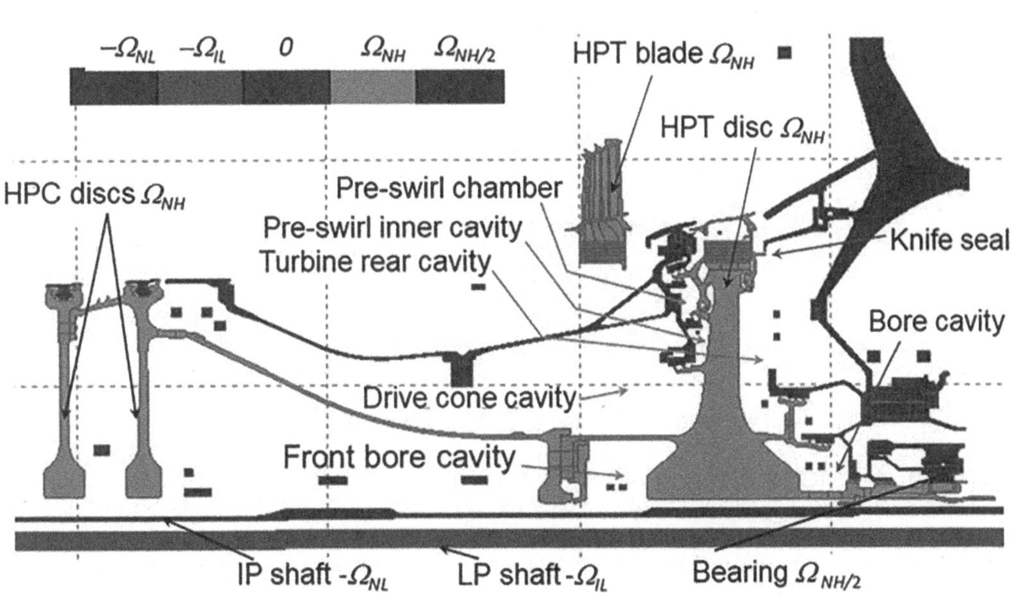

Figure 7.9 Finite element model for a high-pressure turbine disc and surrounding components. *Source:* Javiya *et al.* (2015) / with permission of SAGE Publications.

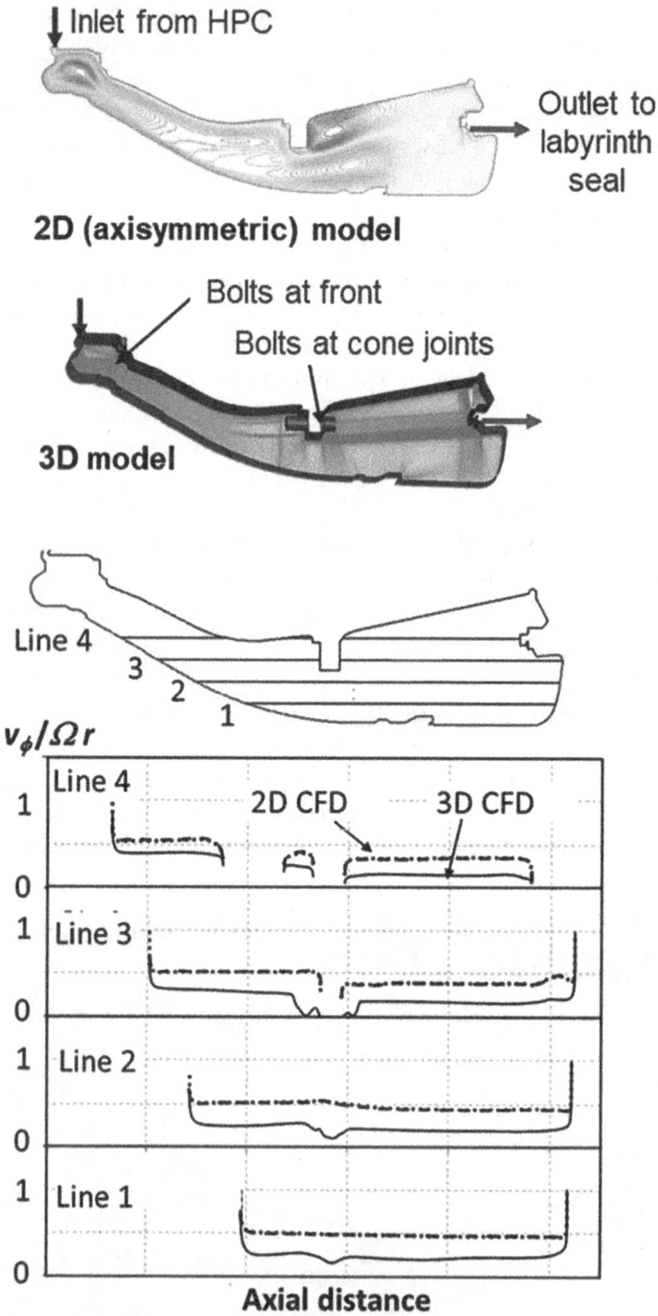

Figure 7.10 Axisymmetric streamline predictions, mesh for a 3D CFD model, and circumferential velocity predictions for a drive cone cavity, $\lambda = 0.1$. *Source:* Javiya *et al.* (2015) / with permission of SAGE Publications.

and 3D CFD shown in Figure 7.10. The CFD uses a RANS model and assumes steady flow. The 3D model includes bolts on the stator.

The axisymmetric streamlines in Figure 7.10 show the expected disc pumping with outflow boundary layers on the rotating cone and turbine disc. At low radii, the flow core rotation is axially uniform, but the stator flange divides the flow into two regions at higher radii. On the compressor side, flow is similar to that expected for a conical rotor–stator cavity, as shown in Figure 5.7 for a radial outflow case. On the turbine side, recirculation is driven by the rotating disc, and axially uniform core rotation is still evident, but the recirculation pattern is more complicated. Without experimental verification or more advanced CFD modelling, there is some uncertainty in predictions of flow near the outer stator and inner rotor surface in this region. The bolts clearly increase stator drag and reduce the flow core rotation.

The turbine rear cavity in Figure 7.9 provides a further interesting example. Javiya *et al.*'s CFD models for this are shown in Figure 7.11. Again, 2D and 3D steady RANS solutions were obtained. In this case, a row of inlet nozzles that provided inlet swirl was approximated by an axisymmetric slot. The flow exits to the main annulus at the outer radius and through an inner seal. The clearance for the rim seal was set sufficiently small to avoid any hot gas ingestion from the main annulus. In practice, any ingestion would be expected to be confined to the outer cavity (between the rim seal and the knife seal). The 3D model included some bolts on the rotor in the inner part of the cavity. Similar flows are predicted by 2D and 3D CFD in the outer part of the cavity, but there are clear differences in the inner part of the cavity. In particular, the inlet stream follows different paths. Further investigations showed that, depending on starting conditions, the 2D model could give a different solution with the flow pattern closer to that of the 3D model. Thus, it appears that in certain circumstances more than one solution is possible, at least when steady flow is assumed. In practice, sensitivity to exact inlet conditions and flow unsteadiness could well occur.

The circumferential velocities given in Figure 7.11 on two radial lines show some difference between the two solutions. Lines of constant angular momentum rv_ϕ are included for reference. The 3D solution allows more mixing between the outer and inner parts of the cavity and this is reflected in the velocity profiles. Angular momentum is expected to be uniform in well-mixed, inviscid flow regions. The 3D solution is closer to this, with the 2D solution suggesting a split between inner and outer regions at $r/b \sim 0.78$. It may also be noted that Rayleigh's condition, in equation 3.1.4, indicates instability if rv_ϕ reduces as radius increases for a vortex with no radial or axial velocity. This could suggest that this case could be subject to some instability and unsteadiness, but this was not fully investigated. While the flow solutions show effects of disc pumping and rotationally-driven inflow on the stator, the interior flow differs from that in the classic rotor–stator cavity. This is attributed to the relatively high flow rate ($\lambda = 0.2$) and inlet angular momentum, and is consistent with observations of sensitivity to inlet conditions at higher flow rates noted for simple radial outflow in Section 7.1.2.

Flow boundary conditions at inlets and outlets are often a source of asymmetry in otherwise axisymmetric configurations. In many cases, these can be approximated in CFD by axisymmetric models. A frequent practice is to replace circumferential rows of holes or nozzles with axisymmetric slots of the same area. This allows both inlet mass flow (or total pressure and temperature) and momentum fluxes to be set correctly. Uncertainty in accuracy and possible numerical difficulties are inevitable if recirculating flow crosses the inlet or outlet planes, as is likely for low flow rates across larger inlet or outlet areas. Allowance for drag from boltheads may also be made in axisymmetric models, as described in the following section.

7.2.2 Effects of Boltheads and Other Features

Engine structures often require bolts on the stationary or rotating components. Examples are shown in Figure 1.2. Unless these are recessed or covered, they will usually result in increased rotor and stator moments. The additional drag on a bolthead or other 3D protrusion directly

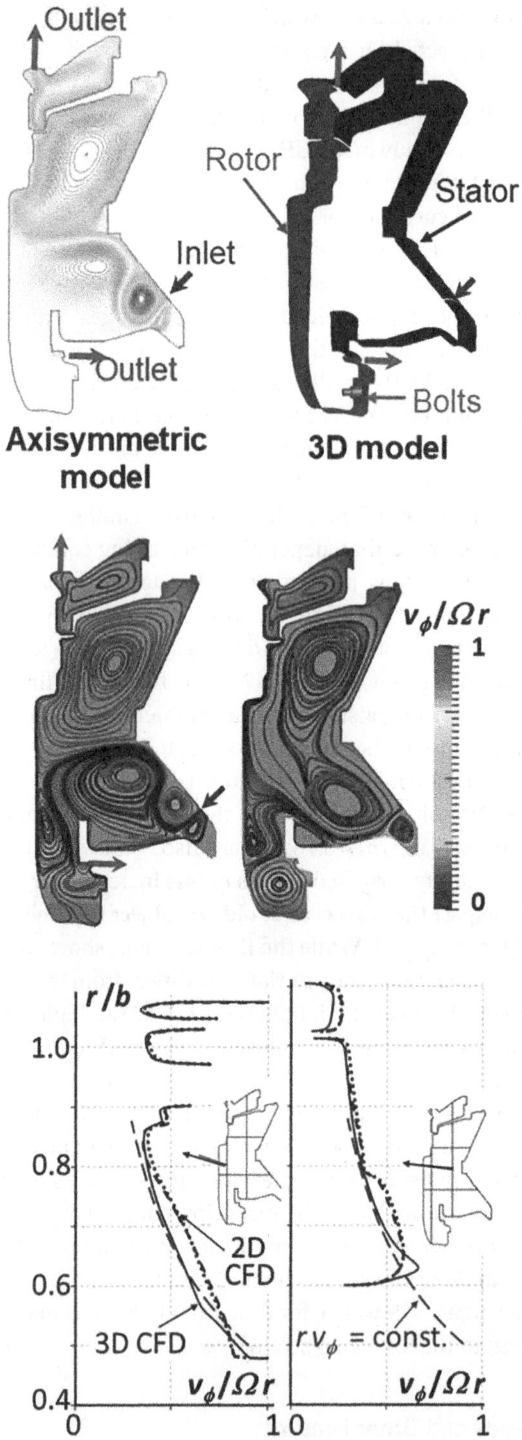

Figure 7.11 Axisymmetric streamline predictions, CFD domain for a 3D model, path lines, and circumferential velocity predictions for a turbine rear cavity, $\lambda = 0.2$. *Source:* Javiya *et al.* (2015) / with permission of SAGE Publications.

changes the moment on the rotor or stator to which it is attached and indirectly changes the moment on other surfaces through its effect on fluid swirl velocity. Provided either the rotor or stator surface remains axisymmetric, the effect is limited by the fluid swirl matching the speed of the bolthead. Thus, for a wide cavity with prominent boltheads on the stator, the rotor moment may be similar to that for a free disc. Such effects can be included in angular momentum balances for the fluid and simplified models by relating bolt drag to the core swirl velocity, assuming a drag coefficient. Unless the protrusion is streamlined, the drag coefficient is expected to be of order 1. Wake interference effects in circumferential rows of bolts are likely to be less than those in non-rotating flows owing to radial flow and the Taylor–Proudman effect suppressing axial gradients, as discussed in Chapter 3. A further point to note is that the mass flow parameter λ loses some significance in combining the effects of flow rate and viscosity, as drag is no longer solely due to viscous effects.

A number of experimental and CFD studies have examined the effect of protrusions on windage in rotor–stator cavities. The early experiments of Zimmermann *et al.* (1986) and Millward and Robinson (1989) illustrate the effects well. Zimmermann *et al.* tested different bolt shapes on the rotor. It was found that bolts on the disc increase the torque substantially, by a factor of 2.5. Flat bolts on the rotor of the same circumferential projection area as staged bolts gave a further increase in moment of around 20%. A partial shroud around the bolts had little effect on windage, compared to the uncovered bolts. Fully covered bolts only increased the torque above that of a plane disc by ~20%, of similar magnitude to the increase in the rotor surface area over that for a plane disc. Millward and Robinson varied the number of bolts, bolt diameter, pitch and cross-sectional area attached to either the rotating disc or the stationary casing. They noted that the windage due to bolts located towards the outer radius of the disc was most significant. Bolts located towards the inner radius had little effect.

Some of Zimmermann *et al.*'s measurements are reproduced and compared with simplified model estimates in Figure 7.12. There were 18 protrusions with equal circumferential spacing on the disc at $r_b/b = 0.75$, and the projected face area to the circumferential flow of each protrusion was $A = 0.00952b^2$. The present plane disc estimates use Daily and Nece's correlation for the cavity without throughflow and equation 7.1.3 to account for throughflow. Without throughflow, Zimmermann *et al.*'s measurements suggest higher plane disc moments than given by Daily and Nece's correlation. This could be associated with differences in geometries, including a higher radius shaft and inlet, and a slightly higher gap ratio s/b. The throughflow rate $C_W = 26000$ corresponds to λ in the range ~0.08–0.24 for the experimental conditions. For the plane disc, the measurements with throughflow are close to the estimate. The results for the covered bolts, staged bolts and cylindrical protuberances show the trends noted above. Estimates for the bolts and cylinders assuming a drag coefficient $C_D = 1$, overestimate the disc moment in these cases. The estimates extend the angular momentum balance used to obtain equation 7.1.1 to include bolt windage and throughflow. Somewhat questionably but conveniently, a forced vortex $k_\delta\Omega$ is again assumed in the core flow. Rotor and stator disc moments are estimated from equations 5.2.9 and 5.2.13, and the moment on the stationary shroud is estimated from the second of equations 5.2.13. The aspect ratio s/b is adjusted by 0.12 as for 'Theory 2' in Section 7.1, and the bolt drag is taken to be $NC_D\rho A[(1-k_\delta)\Omega r_b]^2$, where N and r_b are the number and radial location of the bolts. The angular momentum balance gives the following relation:

$$2\lambda k_\delta = N(r_bA/b^3)C_D(1-k_\delta)^2 Re_\phi^{0.2} + B_M(1-k_\delta)^{1.6} - 0.102k_\delta^{1.8} - 0.471(s/b+0.12)k_\delta^{1.8} \quad (7.2.1)$$

This neglects departures from full disc geometry and the shaft and angular momentum of the inflow (which are at low radii), and assumes outflow at the outer radius, but could easily be adapted to

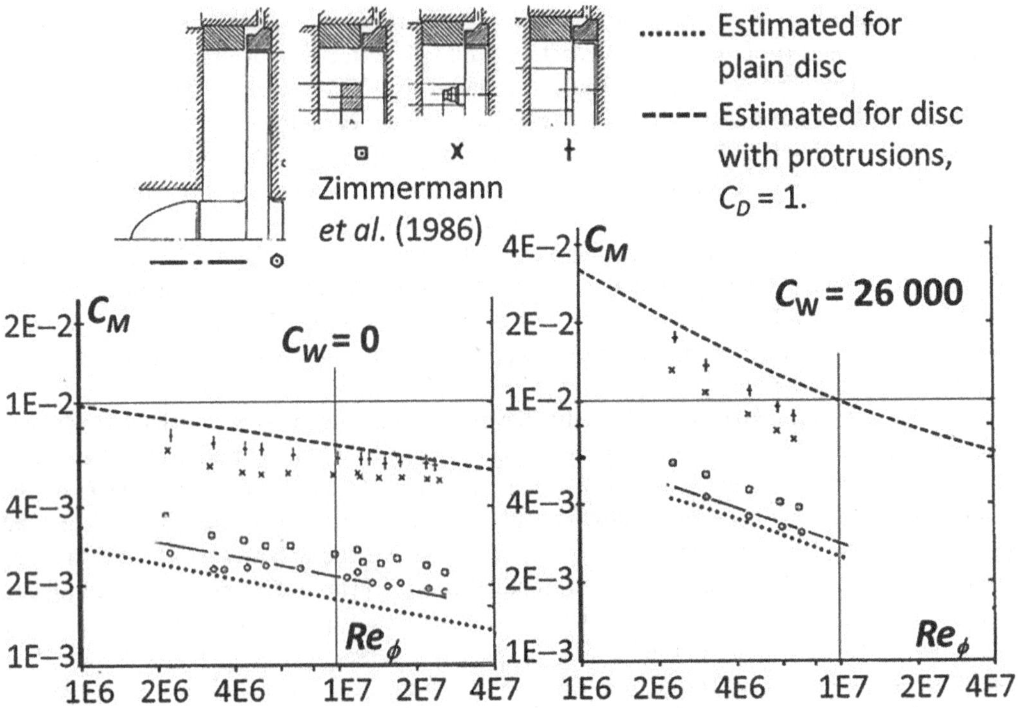

Figure 7.12 Moment coefficients for a rotor–stator cavity with protrusions on the rotor $s/b = 0.2375$, $r_b/b = 0.75$, $A = 0.00952b^2$. *Source:* Adapted from Zimmermann *et al.* (1986).

other conditions. As described in the text following equation 7.1.1, B_M is a function of k_δ only. Equation 7.2.1 can be solved iteratively for k_δ, which then allows moments to be evaluated.

Further comparisons of moment coefficients from the above equation with measurements are shown in Figure 7.13. The experimental data here is from Luo *et al.* (2014), who studied different protrusion shapes and orientations. In all cases, there is a row of 18 protrusions with equal circumferential spacing at a radius $r_b/b = 0.889$. The length D in the figure was $0.058b$, and the protrusion height was $0.049b$. These give a face area $A = 0.002843b^2$ for the orientations A, B, C and F in the figure. For each configuration, measurements were taken at the 3 values of C_W indicated. Lines connect data points taken at $Re_\phi = 1.8 \times 10^6$, 2.1×10^6 and 2.5×10^6. The current estimates overlaid on the experimental results are for $Re_\phi = 2 \times 10^6$ and use the same methods as above. At low flow rates, the plane disc measurements are close to estimates from equation 7.1.3. Departure at higher flow rates may be associated with impingement of the axial inflow on the rotor. The experiments show consistent trends with protrusion shape at all conditions. Hexagonal bolts (A,D) give moments above those for a circular cylinder (C) and below those for a square cross-section (B,E). There is some sensitivity to the orientation of the hexagonal and square sections, and a more streamlined section (F) shows significantly lower moment. Estimates from equation 7.2.1 are given for a range of drag coefficients, $C_D = 0.25$–1.25. Comparing with the measurements at $Re_\phi = 2.1 \times 10^6$ suggests that C_D values are 0.75–1.25 for the hexagonal bolts, around 1.25 for the square sections and around 0.75 for the circular section.

Additional experiments on bolt windage are reported by Long *et al.* (2012), Coren (2007) and Miles (2011). Correlations derived by these authors and in publications discussed above can be used directly in estimating moments in new applications, offering an alternative or check on estimates

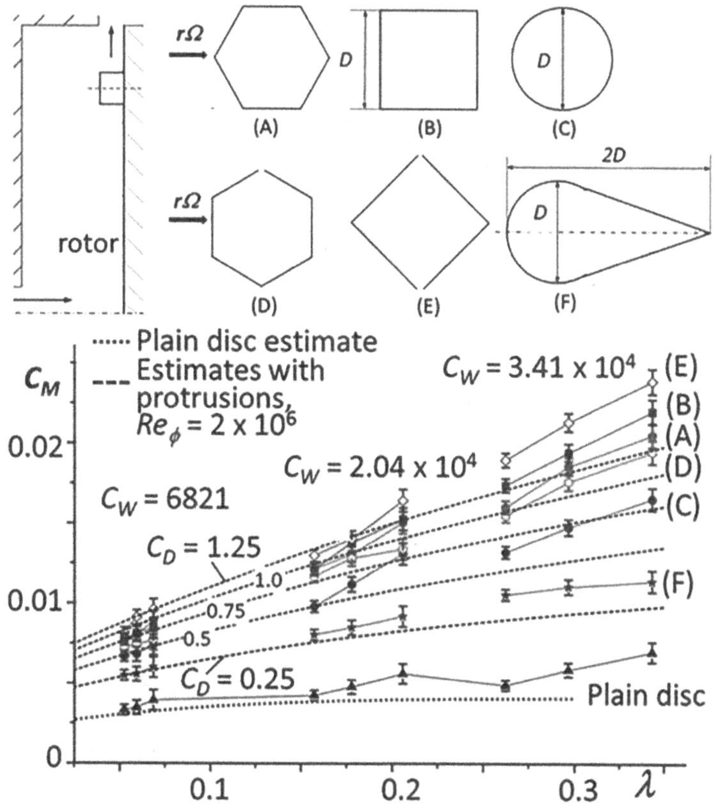

Figure 7.13 Moment coefficients for a rotor–stator cavity with protrusions on the rotor, $s/b = 0.12, r_b/b = 0.889, A = 0.002843b^2$. *Source:* Adapted from Luo et al. (2014).

using the elementary analysis indicated above. However, the correlations are valid only for a range of nondimensional parameters and specific geometries. The form of equation 7.2.1 suggests that the nondimensional group $NRe_\phi^{0.2} r_b A/b^3$ may be useful in developing more general correlations. CFD offers a more general predictive capability and can provide further insight.

Noor Mohamed *et al.* (2015) considered CFD modelling of a configuration studied by Coren (2007) and Miles (2011) with hexagonal boltheads on the rotor. Two steady flow CFD models were used. A 3D sector model included the bolthead geometry and assumed circumferential periodicity for a sector associated with one bolthead. An axisymmetric model added terms to the momentum and energy equations to account for aerodynamic drag in the circumferential direction and the associated windage heating. These terms were estimated assuming bolthead drag at any point within the porous region representing the boltheads to be equal to the number of bolts times a drag coefficient $C_{D,\text{CFD}}$ times the dynamic head associated with the relative circumferential velocity at that point. (This drag coefficient is not directly comparable to that considered above, as the core is no longer constrained to be a forced vortex.) Both models used the standard $k - \varepsilon$ turbulence model with wall functions. The CFD domain for the axisymmetric model, representing Coren and Miles' experiment, and some results are shown in Figure 7.14. Only bolts at the outer radial location $r_b/b = 0.89$ were considered in this study. The protrusion diameter for this case is $0.071b$ with height $0.049b$, giving a face area $A = 0.00348b^2$.

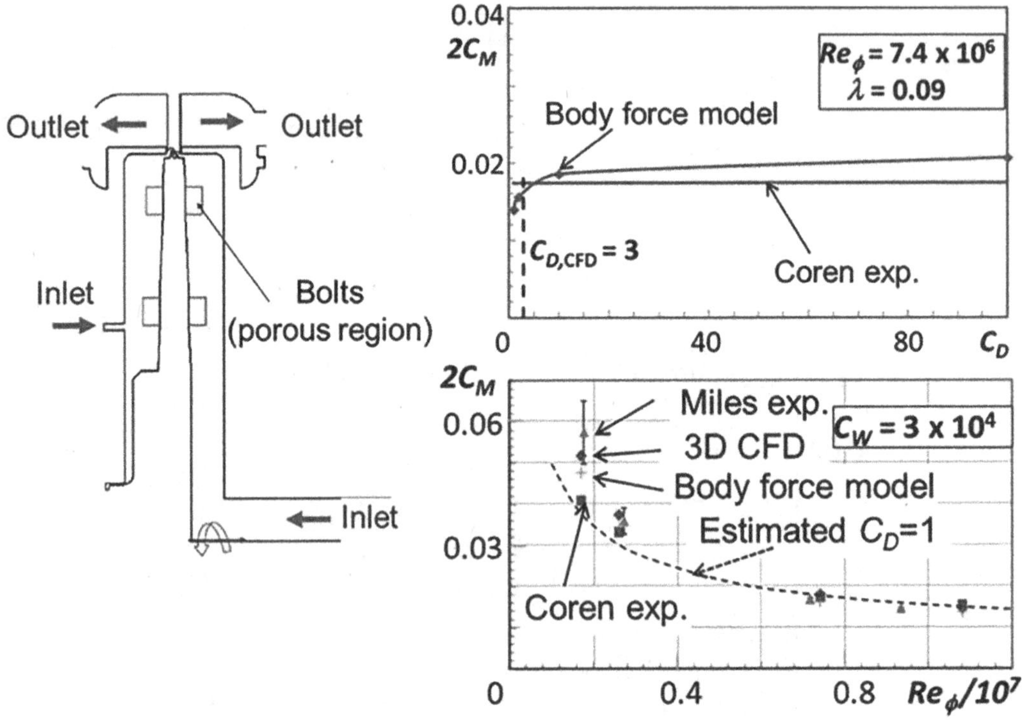

Figure 7.14 Steady CFD modelling of boltheads on a rotor disc, with 18 equally spaced bolts at the outer radial position $r_b/b = 0.89, s/b = 0.1, A = 0.00348b^2$. *Source:* Adapted from Noor Mohamed *et al.* (2015).

As shown in the bottom right graph of Figure 7.14, both steady flow models and estimates from equation 7.2.1 give reasonable predictions of disc moment compared to measurements. Here, $2C_M$ gives moment on both sides of the disc and, for comparison, von Kármán's solution, equation 5.2.2, gives $2C_M = 0.008$ and 0.0058 for $Re_\phi = 2 \times 10^6$ and 10^7, respectively. The measurements of Coren and Miles show significant differences at the lowest Re_ϕ considered, with the CFD values lying between the two measurements. The scatter may be associated with lower levels of torque at this condition. The top right graph in Figure 7.14 illustrates variations in the axisymmetric model moment predictions with different values of drag coefficient. The highlighted value $C_{D,\text{CFD}} = 3$ was adopted for the results shown for $C_W = 3 \times 10^4$ and gave good results over a wider range of conditions. Notwithstanding the good agreement found with torque measurements, the CFD models showed some differences with measurements of swirl velocity and adiabatic disc temperature.

Moghaddam *et al.* (2013) and Pett *et al.* (2007) also applied steady 3D CFD to compare with Coren's experiments. Moghaddam *et al.* considered rotor mounted protrusions and showed good agreement with experimentally derived moment coefficients. CFD solutions were obtained varying the number of bolts from 0 to 60. Increasing the number of bolts increased the moment coefficient up to a plateau at ~52 bolts. Pett *et al.* carried out CFD investigations into stator mounted protrusions. Moment coefficients from CFD were within 15% and 20% of measurements for the plane disc and static protrusion analyses, respectively.

Time-dependent, 3D CFD solutions for Coren's configuration with rotor-mounted protrusions were presented by Noor Mohamed *et al.* (2017a,b). These revealed an intrinsic unsteady flow that explained the discrepancies with velocity and disc temperature measurements observed in steady 3D CFD solutions. Examples are given in Figure 7.15. The contours of instantaneous core swirl velocity and adiabatic disc temperature show a three-lobed large-scale flow structure.

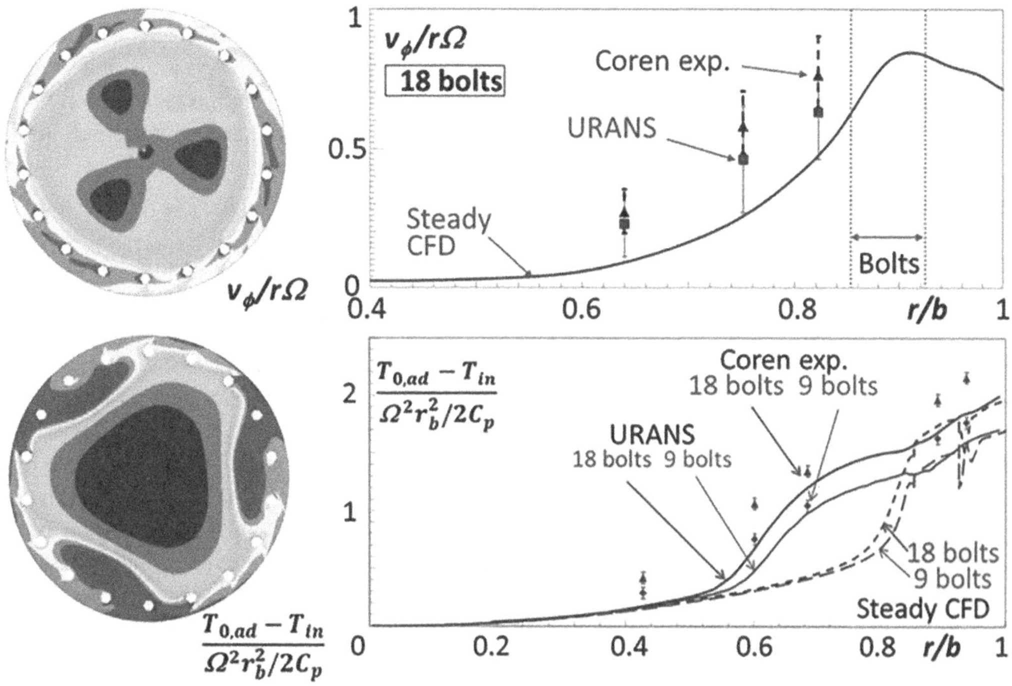

Figure 7.15 Unsteady CFD modelling of boltheads on a rotor disc, with 18 and 9 equally spaced bolts at the outer radial position $r_b/b = 0.89$, $s/b = 0.1$, $Re_\phi = 2.1 \times 10^6$, $C_W = 3 \times 10^4$. *Source:* Adapted from Noor Mohamed *et al.* (2017b).

This rotates at about half disc speed. Compared to steady CFD solutions, the mixing within the cavity is enhanced and much better agreement with core velocity and surface temperature measurements is achieved. The level of velocity fluctuations is also similar to that reported in Coren's experiments.

Zimmermann *et al.* and Long *et al.* have reported rotor moment measurements for configurations with cylindrical holes or pockets in the rotating disc, at the same radial location as their discs with protrusions reported above. Zimmermann *et al.* had 18 equally spaced 'double holes' with total depth of $0.0825b$. The first 12% of the depth had a diameter of $0.1125b$, and the remainder was a threaded hole with an outer diameter of $0.055b$. Increases in the moment coefficient of around 50% above plane disc case were observed. Long *et al.* had up to 18 holes with a diameter of $0.0444b$ and a depth of $0.02667b$. Any increase in moment was within experimental uncertainty.

7.3 Preswirl Systems

7.3.1 System Description

High-pressure turbine blades are the hottest rotating components in gas turbine engines, and a considerable portion of engine cooling air supplies blade internal cooling channels and film cooling holes. Air from the high-pressure compressor is used for this purpose. Increases in air temperature and losses in total pressure must be limited in the transfer to the turbine. Unless the cooling air itself is to be cooled (which is unusual), the transfer from compressor to turbine would ideally be isentropic. Frictional torque on both stators and rotors is parasitic, generating entropy. Other stator and rotor torque with associated shaft work (i.e. windage) is not always detrimental. For example, as air is ducted through rotating pipes or channels in rotating components, work is done according to the Euler work equation 2.1.28 as the angular momentum rv_ϕ changes with radius. For steady, isentropic flow through rotating passages, rotary stagnation pressure and rotary stagnation temperature are constant, although the work done by or on the rotor will change the absolute total pressure and temperature.

In practical cooling air delivery systems, cooling air is likely to flow through rotor–stator cavities before entering rotating passages leading to the blade. As the flow enters the rotating passages work on the air may be required to increase its angular momentum. This will also increase its total temperature. If the circumferential velocity v_ϕ of the air exceeds that of the rotor, the air will do work on the rotor as it enters the rotating passages and its temperature drops. Thus, at the expense of a drop in pressure, using angled nozzles in the stator to impart swirl on the air can avoid the increase in temperature associated with rotor work and, for $v_\phi/\Omega r > 1$, can reduce coolant temperature. This is referred to as preswirl and one possible arrangement is shown in Figure 7.16. The space between the stator containing the nozzles and the rotor containing the cooling air receiver holes is called the preswirl chamber. If the preswirl fraction in the chamber $v_\phi/\Omega r$ is ~1, the system allows the flow to enter the receiver holes relatively smoothly, avoiding pressure losses and entropy generation associated with strong crossflow at the entry to the holes. As indicated in the figure, seal leakage may flow into and/or out of the chamber, depending on the arrangement. This seal leakage can have a significant effect on performance. The example shown is a relatively high radius preswirl system. In other configurations, transfer to the rotor occurs at lower radii, with the choice depending on pressure and temperature requirements for blade cooling, seal clearances and other factors.

7.3.2 System Performance Parameters

A common approach to characterising preswirl system performance is to use an effective nozzle velocity ratio, as introduced by Meierhofer and Franklin (1981). Considering isentropic flow at constant radius, the relative total temperature at nozzle exit, state 1, is related to the absolute total temperature at nozzle inlet, state 0, by the relation

$$T_{t,0} - T_{t,rel,1} = \frac{V_{isen,1}^2 - U_{isen,1}^2}{2C_p}. \tag{7.3.1}$$

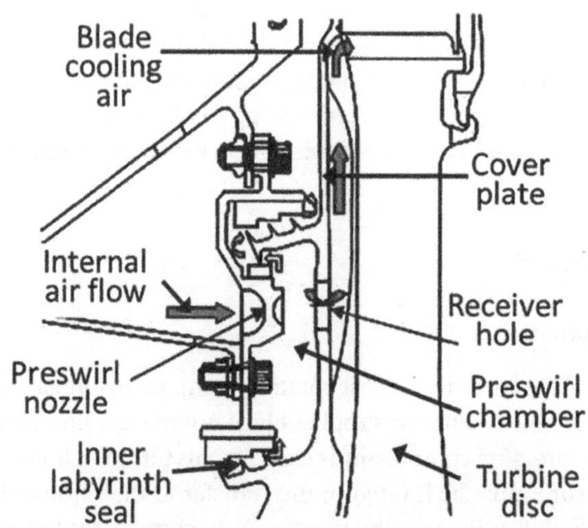

Figure 7.16 Preswirl system components.

where V_{isen} and U_{isen} are ideal fluid velocity magnitudes in the stationary and rotating reference frames. For a known rotor speed, assuming the flow aligns with the nozzle exit angle to the circumferential direction (α) gives a relationship between these two velocities. Meierhofer and Franklin used this and measurements of temperature at the inlet and in the receiver holes to derive an effective nozzle exit velocity, as

$$\frac{V_{eff}}{\Omega r} = \frac{C_p(T_{t,0} - T_{t,rel,rh})_{meas}}{\Omega^2 r^2 \cos(\alpha)} + \frac{1}{2\cos(\alpha)} \tag{7.3.2}$$

where changes in radius are neglected. The effective velocity ratio is V_{eff}/V_{isen} and the effective preswirl ratio is $(cos\alpha)V_{eff}/\Omega r$. In design, assuming a value of the nozzle velocity ratio allows the temperature in the receiver holes to be estimated.

The isentropic velocity is usually calculated using the cavity static pressure. The velocity ratio does not then account for pressure losses as the flow enters the receiver holes. Other workers, for example, Dittmann *et al.* (2002), have used discharge coefficients to characterise the nozzle and receiver hole performance. Geis *et al.* (2004) introduced a cooling effectiveness relating the achieved temperature drop to the 'polytropic nozzle' theory.

Kong *et al.* (2022) considered preswirl nozzles and receiver holes at different radii. They highlighted the importance of minimising entropy production and examined various sources of entropy and pressure losses. Overall system performance was characterised as an effectiveness given by the ratio of achieved temperature drop $T_{t,0} - T_{t,rel,\,rh}$ to a maximum temperature drop.

An overall isentropic efficiency for the preswirl system comprising nozzles, chamber and receiver holes may be defined considering rotary stagnation temperature T^* and rotary stagnation pressure p^*, as introduced in Section 2.2.3. From the energy balance equation 2.1.27, the moment on the stator associated with turning of the flow in the nozzles will produce a drop in rotary stagnation temperature. For an isentropic process from state 0 to state *rh* in the receiver holes, the drop of rotary stagnation temperature $T_0^* - T_{rh,isen}^* = T_0^*[1 - (p_{rh}^*/p_0^*)^{(\gamma-1)/\gamma}]$. In a real process, the full temperature drop is not achieved, so a suitable efficiency is

$$\eta_{ps} = \frac{\text{actual temperature drop}}{\text{ideal temperature drop}} = \frac{T_0^* - T_{rh}^*}{T_0^* - T_{rh,isen}^*} = \frac{1 - T_{rh}^*/T_0^*}{1 - \left(p_{rh,isen}^*/p_0^*\right)^{(\gamma-1)/\gamma}}. \tag{7.3.3}$$

This expression corresponds directly to that for turbine efficiency in a stationary reference frame with rotary stagnation properties replacing absolute stagnation properties. Assuming negligible fluid velocity upstream of the nozzle $p_0^* = p_{t,0} = p_0$ and $T_0^* = T_{t,0} = T_0$. If $T_{t,rel,\,rh}$ is measured T_{rh}^* may be calculated from equation 2.2.11a.

For the ideal process $T_{rh,isen}^* = T_{1,isen}^*$ and $p_{rh,isen}^* = p_{1,isen}^*$, and these may be calculated from the nozzle inlet conditions and chamber pressure $p_{1,isen}$ (at the nozzle exit) using isentropic flow relations and the nozzle angle. The relationship of $p_{1,isen}$ to the receiver hole pressure p_{rh} may then be determined. If the relative velocity in the receiver holes is negligible, then the isentropic process from nozzle inlet conditions to within the receiver holes at static pressure p_{rh} is described by the following equations:

$$\left(\frac{T_{1,isen}}{T_{t,0}}\right) = \left(\frac{p_{1,isen}}{p_{t,0}}\right)^{\frac{\gamma-1}{\gamma}}, \quad V_{isen,1}^2 = 2C_p\,(T_{t,0} - T_{1,isen}), \quad v_{\phi,isen,1} = V_{isen,1}\cos\alpha,$$

$$T_{1,isen}^* = T_{t,0} - v_{\phi,isen,1}\frac{\Omega r_1}{C_p}, \quad p_{1,isen}^* = p_{1,isen}\left(\frac{T_{1,isen}^*}{T_{1,isen}}\right)^{\gamma/(\gamma-1)},$$

$$T_{rh,isen} = T_{1,isen}^* + \frac{\Omega^2 r_{rh}^2}{2C_p}, \quad p_{rh} = p_{1,isen}^*\left(\frac{T_{rh,isen}}{T_{1,isen}^*}\right)^{\gamma/(\gamma-1)}. \tag{7.3.4}$$

Using these equations $p_{1,isen}$ may be adjusted to match p_{rh} to that of the real process, and η_{ps} can then be evaluated.

Considering the flow through the nozzles alone, equations 2.1.26 and 2.1.27 show that the drop in rothalpy ($= C_pT^*$) equates to the increase in $\Omega r v_\phi$ produced. Thus, without including the receiver holes and for $v_{\phi,0} = 0$, the ratio of actual to isentropic change in rotary stagnation temperature, and hence η_{ps} equates to the effective velocity ratio V_{eff}/V_{isen}. However, pressure losses as the flow enters the receiver hole are likely to be of the order of the relative dynamic head and can significantly reduce system efficiency. Noting also that $T_{t,0} - T_{t,rel,rh} = T_0^* - T_{rh}^* + \Omega^2 r^2/2C_p$, it is clear that both the effective velocity ratio and Kong *et al.*'s effectiveness are closely related to η_{ps}.

7.3.3 Elementary Modelling

A schematic representation of an elementary model for a preswirl chamber is shown in the top right of Figure 7.17. In this case, the model assumes inflow at an inner seal and outflow at an outer seal. A fully mixed assumption is used for the circumferential velocity and rotary stagnation temperature in the chamber. Alternative models might assume a forced or free vortex in the chamber, and the approximations made can be significant for chambers with lower radius ratio, a/b. The nozzle exit flows may be related to the feed pressure and temperature, chamber pressure and nozzle angle using the isentropic flow relations and assumed nozzle efficiency or effective velocity ratio. The receiver hole flow is related to the chamber pressure and temperature, and receiver hole discharge coefficient $C_{d,rh,}$. If the effective preswirl ratio is not close to unity, $C_{d,rh}$ will be affected by the crossflow at the inlet, introducing a further dependency on chamber circumferential

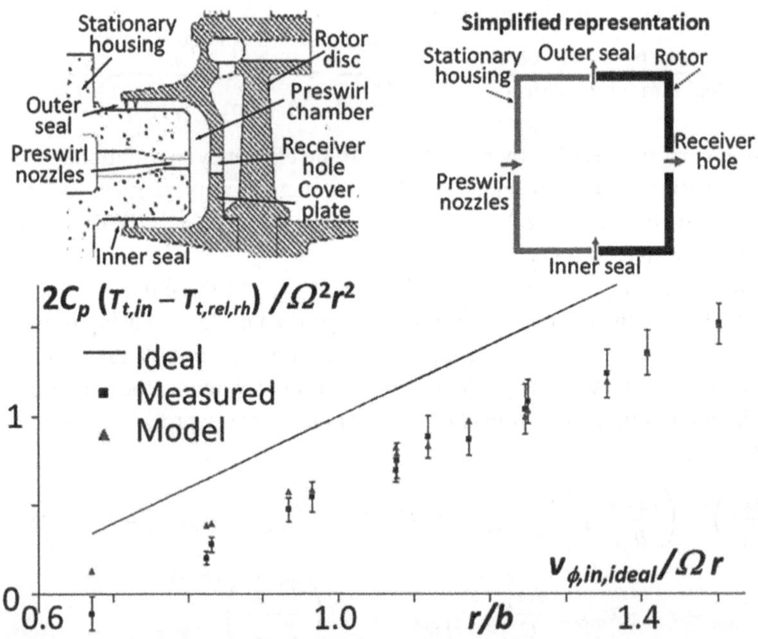

Figure 7.17 Preswirl system performance, 18 nozzles, 72 receiver holes $s/r_{ps} = 0.036$, 9600 rpm. *Source:* Adapted from Chew *et al.* (2003).

velocity. In air system modelling, seal flows will depend on chamber and external conditions and would require further assumptions regarding discharge coefficients and seal clearances. If the nozzle velocity ratio is assumed or chosen to include the effects of surface drag in the chamber and heat transfer is negligible, no further modelling of the chamber flow is required in the ful l air system model.

Here, the preswirl system alone, and the chamber flow in particular, are considered in more detail. Simplified, steady flow mass, angular momentum and energy balances for the chamber give

$$\sum_{outlets} \dot{m} + \sum_{outlets} \dot{m} = 0$$

$$\sum_{inlets} \dot{m} r v_{\phi,in,eff} + \sum_{outlets} \dot{m} r v_{\phi,imix} = -M_{rotor} - M_{stator}$$

$$\sum_{outlets} \dot{m} C_p T^*_{mix} + \sum_{inlets} \dot{m} C_p T^*_{in} = -Q + \Omega M_{stator}. \tag{7.3.5}$$

where positive values of flow rate $\dot{m}$, moment M and heat transfer Q will increase mass, angular momentum and energy in the chamber, as in Section 2.1. The subscript *in* denotes values at inlets to the chamber. These correspond to the nozzle and seal exit flows. For axisymmetric rotors and stators, the moments in these equations arise from frictional drag only, as the nozzles are not included in the chamber. As discussed above for other rotor–stator cavities, friction factors from integral solutions or other correlations may be used in estimating moments. Then $v_{\phi,mix}$ and T^*_{mix} may be calculated using measured or assumed mass flow rates and inlet conditions. The rotor moment needed to change $v_{\phi,mix}$ to disc speed as the flow enters the receiver holes is not included in the above equations. This moment does not change the rotary stagnation temperature, so $T^*_{rh} = T^*_{mix}$. For an ideal chamber with zero moments, heat flux and seal flows, the temperature in the receiver holes $T^*_{rh} = T^*_{in}$. The difference between the absolute total temperature at the inlet and the relative total temperature in the receiver holes is then given by

$$\frac{2C_p(T_{t,in} - T_{t,rel,rh,ideal})}{\Omega^2 r_{rh}^2} = \frac{2r_{in} v_{\phi,in,eff}}{\Omega\, r_{rh}^2} - 1 \tag{7.3.6}$$

The inlet swirl $v_{\phi,in,eff}$ is the effective preswirl, as previously estimated using the nozzle velocity ratio and nozzle angle. Neglecting heat transfer in the nozzle, $T_{t,in}$ is equal to the nozzle inlet total temperature. This equation can be adapted to account for seal flows using inlet mass flow weighted averages of rv_ϕ and $T^*(= T_t - \Omega r\, v_\phi / C_p)$. Equation 7.3.6 extends equation 7.3.1 to allow for pressure loss in the nozzles and changes in radius and agrees with Kong *et al.*'s (2022) analysis.

Introducing average inlet angular momentum $(rv_{\phi,in})_{av}$ and temperature $T^*_{in.av}$ obtained from nozzle and seal flow inlet mass flow weighted averages, equation (7.3.5) gives

$$\frac{(rv_\phi)_{mix}}{\Omega b^2} = \frac{(rv_{\phi,in})_{av}}{\Omega b^2} + \frac{(C_{M,rotor} + C_{M,stator})Re_\phi^{0.2}}{2\lambda}$$

$$\frac{C_p T^*_{rh}}{\Omega^2 b^2} = \frac{C_p T^*_{in,av}}{\Omega^2 b^2} + \left(\frac{2Q}{\rho \Omega^3 b^5} - C_{M,stator}\right) \frac{Re_\phi^{0.2}}{2\lambda}. \tag{7.3.7}$$

where the mass flow parameter λ is based on the total inflow. The first two terms in the angular momentum equation above are expected to be of order unity. The third term will depend on the cavity swirl. Assuming friction factors proportional to $Re_\phi^{-0.2}$, the moment coefficients will depend on the squares of relative and absolute mixed out swirl velocity. For full stator and rotor discs, at least one of the moments is expected (from, for example, equations 5.2.7 and 5.2.13) to be of order $0.05/\lambda$.

For the high radius ratio (a/b) chambers considered by Meierhofer and Franklin and the example in Figure 7.16, the reduced surface areas may lead to significantly lower values. Nevertheless, for the experimental range of λ from 0.13 to 0.66 reported by Chew *et al.* (2003), the frictional moments could have a significant effect on cavity swirl. Similarly, the stator moment in the energy equation may significantly increase coolant delivery temperature. From the Reynolds analogy, heat transfer can be expected to be of similar magnitude or higher than the windage heating when the Eckert number ($Ec = \Omega^2 b^2 / C_p \Delta T_{ref}$ with the temperature difference ΔT_{ref} driving convective heat transfer) is order unity or less.

7.3.4 Examples

Measurements for an engine-representative preswirl system and a simplified flow analysis were presented by Chew *et al.* (2003). The test section geometry, representation of the elementary model for the preswirl chamber and some results are shown in Figure 7.17. In this case, the model neglects radius changes and uses friction coefficients derived from von Kármán's free disc solution. Seal flows are significant in this example.

The ideal temperature drop given by equation 7.3.6 assuming isentropic nozzle flow is compared with values from measurements and the elementary model of the cavity for a range of operating conditions in Figure 7.17. The model accounts for pressure losses in the nozzle by multiplying the ideal velocity by the measured discharge coefficient, effectively using the measured mass flow, and this accounts for most of the difference from the ideal temperature drop. Measurements with the cavity width s increased from 0.036 to 0.092 times the radius at the centre of the nozzles r_{ps} (increasing the rotor area) did not show a clear trend in temperature drop. Typically, nozzle discharge coefficients were in the range 0.8–0.9, consistent with Dittmann *et al.*'s measurements. More approximate estimates of discharge coefficients for the receiver holes were in the range 0.25–0.8, again consistent with Dittmann *et al.* Low values are associated with high crossflow conditions at the receiver holes. The model agrees reasonably well with measurements but overpredicts the temperature drop at low preswirl ratios $v_{\phi,in,rv}/\Omega r$. This suggests that the stator moment may be underestimated or the nozzle velocity coefficient (estimated as the measured discharge coefficient) may be overestimated at these conditions.

A further series of experimental studies of high radius preswirl systems were undertaken at Karlsruhe Institute of Technology with publications, for example, by Geis *et al.* (2004), Bricaud *et al.* (2007) and Bricaud (2017). Comparisons with elementary modelling and CFD are discussed by Cagan et al. (2009), Chew *et al.* (2005), Ciampoli *et al.* (2008) and Javiya *et al.* (2013). Preswirl nozzles, formed as straight drilled angled holes, performed significantly worse than more aerodynamically designed nozzles and presented modelling challenges with flow separation being sensitive to CFD modelling. Otherwise, results appear consistent with the elementary modelling. However, with a low number of receiver holes or a reduced cavity width $s/r_{rh} = 0.023$, strong unsteady flow behaviour was observed at some experimental conditions. This was captured by CFD solutions for a high overswirl case (with an isentropic preswirl ratio of 2).

A preswirl system with nozzles at low radius and receiver holes at higher radius, as shown in Figure 6.3, was studied experimentally at the University of Bath. The emphasis in these studies was on flow structure in the cavity and heat transfer on the rotating disc. As illustrated in Figures 6.3 and 6.11, conventional CFD captures much of the flow physics. RANS models with other model simplifications, such as mixing planes and slot representations of the nozzles or receiver holes, allow simplified modelling but may not capture full details of mixing or strong unsteady effects, such as those illustrated in Section 6.3.2. Other studies have confirmed and augmented the findings noted above. Amongst these, the combined experimental and computational studies of Kong *et al.* (2022) show good agreement between CFD and measurements. The effect of the radial location of the nozzles compared to the receiver holes was investigated with CFD by these authors.

In practice, nozzle design may be a compromise between aerodynamic performance and manufacturing cost and simplicity. For aerodynamically designed nozzles, CFD predictions of the discharge coefficient are generally close to measurements. Using CFD, Ciampoli *et al.* (2007) produced an optimised drilled nozzle design, assuming a nozzle angle of 15° and a pressure ratio of 1.45. The predicted velocity coefficient was 0.88 (with a discharge coefficient of 0.83). Again, for 15° nozzles and using CFD, Javiya *et al.* (2011) compared three different nozzle designs. They confirmed that straight drilled nozzles performed relatively poorly. Their aerodynamically designed vaned and drilled nozzles gave discharge coefficients around 0.92 and 0.88, and velocity coefficients around 0.95 and 0.94, respectively. Further studies by Liu *et al.* (2017) confirmed these trends and explored the vaned nozzle design further. Effective velocity ratios from calculations including the preswirl chamber are up to 0.93.

Receiver hole discharge coefficients vary strongly with preswirl ratio. As described by Dittmann *et al.* (2002), crossflow associated with overswirl or underswirl is associated with pressure losses and low discharge coefficients. A method for estimating the effects of orifice length-to-diameter ratio, corner radiusing and crossflow was proposed by McGreehan and Schotsch (1988) and could be applied in receiver hole design. The preswirl system studies referenced above have all used axial receiver holes. Angled holes matched to the preswirl ratio could reduce losses for overswirl or underswirl conditions. Liu *et al.* (2018) have proposed the use of vaned receiver holes to reduce pressure losses and improve system performance, supporting this with CFD analysis. In their example, the stationary nozzles produce a preswirl ratio of 1, and the rotating vanes produce overswirl, which then reduces as the air flows radially outward in a rotating cavity formed between a cover plate and the disc. This was compared to a similar system using conventional receiver holes and to a high radius system with conventional holes. Work done by the rotor increased the total pressure of the air flowing through the rotor vanes, allowing a lower pressure for expansion through the stationary nozzles and hence a greater drop in rotary stagnation temperature overall than with the axial holes (with pressure losses due to underswirl). Practical benefits of a vaned rotor system would depend on other system factors such as available pressure, achievable seal clearances and trade-offs in performance, weight and manufacturing cost.

7.4 Heat Transfer

As described in Chapters 2 and 5, dimensional analysis and the Reynolds analogy along with integral methods for turbulent flow give a useful indication of how heat transfer in rotor–stator systems will scale with operating conditions. Considering rotor–stator cavities with throughflow as in Figures 2.3 and 7.6, the dimensional analysis in equation 2.3.17 shows that for a given geometry with fixed thermal boundary conditions involving just two arbitrary temperatures Nusselt number Nu is a function of C_W, Re_ϕ. $Ec.Pr$, γ, $\beta(T_{t,in} - T_0)$ and $v_{\phi,in}/\Omega b$. For air Pr and γ can usually be assumed constant, and the coefficient of thermal expansion β is approximated by $1/T_{t,in}$. In addition, for rotationally-dominated flow with separate disc boundary layers, the 1/7*th* power-law integral equations allow C_W, and Re_ϕ to be replaced by λ. The analysis then gives

$$NuRe_\phi^{-0.8} = f(\lambda, v_{\phi,in}/\Omega b, Ec, (T_{t,in} - T_0)/T_{t,in}) \tag{7.4.1}$$

For many laboratory experiments, Eckert number and buoyancy effects are negligible and just two independent parameters λ and $v_{\phi,in}/\Omega b$ remain, as for the flow field. However, Eckert number effects are significant at high-speed engine conditions and the requirement for similarity of thermal boundary conditions is a severe restriction. Elementary models used above to calculate flow swirl and disc moments can be extended to include an energy balance for flow temperature, as in equation 7.3.5. The Reynolds analogy, described in Section 2.5.3 and applied to laminar and turbulent boundary layers in Chapters 3 and 5, may be used in estimating surface heat transfer.

In industry the use of numerical solutions for heat conduction in engine components is commonplace. This is illustrated by the finite element thermal model in Figure 1.1. As described by Dixon *et al.* (2004), such solutions provide complete temperature distributions through an engine operation cycle but are reliant on modelling of coolant flows throughout the engine. Convective heat transfer is traditionally modelled using heat transfer correlations from relatively simple configurations, such as the free disc. Reynolds number scaling may be explicit in the Nusselt number correlations used. Further modelling may be needed to obtain meaningful (usually local) estimates of cooling air temperatures and swirl and to account for windage heating. Unless carefully matched against engine temperature measurements such models are subject to considerable uncertainty. Methods of matching engine test data and detailed modelling of convection in engine models have received little attention in the open literature. Some insight into the challenges is given by Saunders *et al.* (2007).

Examples given in Chapter 6, such as those in Figures 6.2, 6.3 and 6.16, confirm the potential of CFD for heat transfer prediction in rotor–stator cavities while also illustrating the difficulties and uncertainties associated with sensitivity to thermal boundary conditions, inlet conditions and transition to turbulence. CFD is increasingly used to inform traditional thermal models or to couple directly with component models, replacing empirical correlations. Further to the coupled CFD/FEA calculations discussed in Section 6.5, evaluations of coupled CFD/FEA modelling for temperature prediction are given by Javiya *et al.* (2013, 2015). The earlier publication considered the Karlsruhe preswirl rig discussed in Section 7.3.3. The second publication considered the engine geometry shown in Figure 7.9.

Figure 7.18 illustrates Javiya *et al.*'s coupled solution for the Karlsruhe preswirl chamber and rotating disc. The fluid domain included the inner and outer cavities and an axisymmetric slot representing the preswirl nozzles. Seal leakages were set to zero to match experimental conditions, and the stator surface was assumed adiabatic. Temperature and heat flux distributions on the rotor surfaces were obtained by matching the fluid and solid domain solutions. The solid domain represented the rotor with the temperature specified on the disc's back face from measurements. Windage heating of the outer cylindrical surface from a labyrinth seal was estimated using an elementary model with an assumed friction factor. This is similar to rotor–stator cavity models described above, and the approach was verified by comparison with CFD. The main CFD solution used the $k - \varepsilon$ turbulence model with wall functions. The level of agreement with disc temperature measurements obtained was considered satisfactory. As shown by the example in Figure 7.18, some differences were observed at lower radii. These may be due to neglect of stator heat transfer.

Section 7.2.1 described some of the CFD solutions used in Javiya *et al.*'s thermal model for the engine high-pressure turbine disc in Figure 7.9. Figure 7.19 shows component temperatures obtained from traditional and coupled CFD-FEA solutions. The contour plots are at a stabilised condition in the transient test cycle, shown by the graphs of HP rotor speed and turbine disc rim

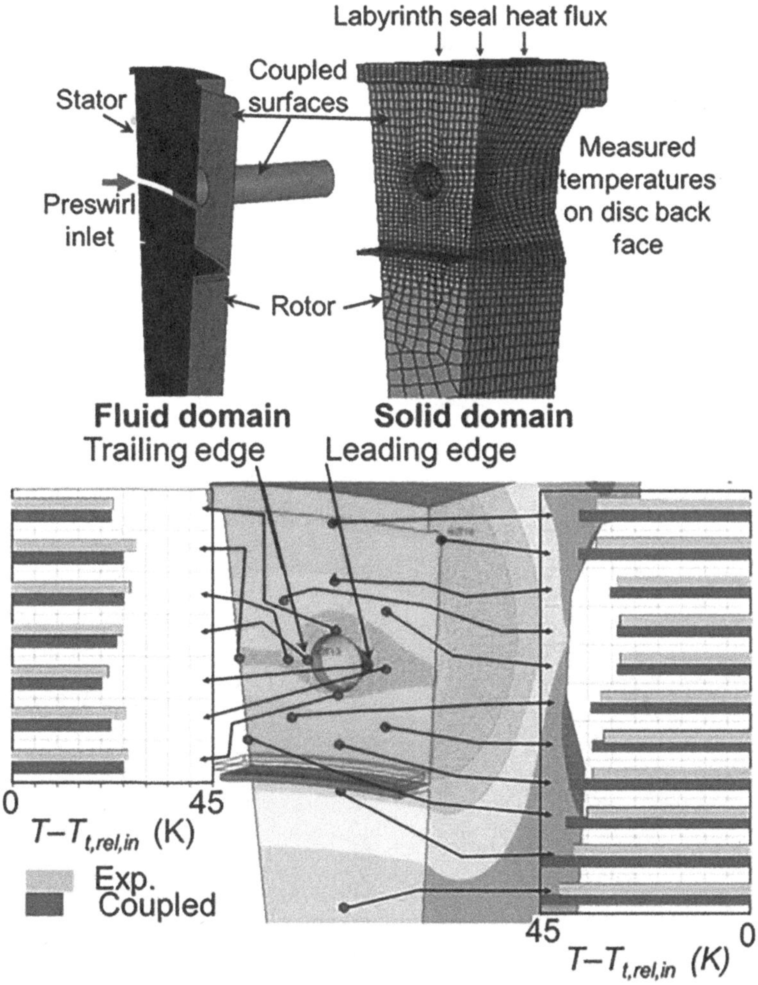

Figure 7.18 Coupled CFD–FEA disc temperature prediction for a preswirl system, rotation is clockwise $Re_\phi \sim 3 \times 10^6$, $\lambda \sim 0.2$. *Source:* Javiya *et al.* (2013) / with permission of American Society of Mechanical Engineers (ASME).

temperature against time. The finite element model assumed axisymmetric components with adjustments made to account for 3D features. For coupled solutions, convective heat transfer to the HP turbine disc was obtained from steady-state CFD solutions using the $k - \varepsilon$ turbulence model with wall functions. Other convective boundary conditions were, as provided, in an industry standard model using correlations adjusted to match FEA predictions with measured temperatures. A series of steady CFD solutions for each of the five cavities surrounding the disc used surface temperatures from the solid model. Iterations were performed to match surface temperature and heat flux at each point in the transient calculation for the solid component temperatures. Other boundary conditions for CFD models, including rotor speed, flow rates and pressures, were taken from a 1D network model of the engine's internal air system based on engine pressure measurements, seal flow correlations and best understanding. As these varied during the test cycle, three different models were defined for each cavity and used at different engine powers, as indicated on the speed–time diagram. Solutions using axisymmetric CFD approximations were obtained using two different CFD codes and a solution using 3D CFD models was obtained for the first half of the test period.

Considering comparisons with thermocouple measurements in some detail, Javiya *et al.* concluded that the accuracy of CFD-based modelling is comparable to that of a traditional model that had been adjusted to match engine data. Some differences between the results from 2D and 3D CFD modelling were noted. Regions where the 2D model, in particular, gave lower temperatures than the baseline, engine-matched model are indicated in Figure 7.19. On the front side of the disc, the effect of the boltheads in the drive cone cavity, as discussed in Section 7.2.1 and illustrated in

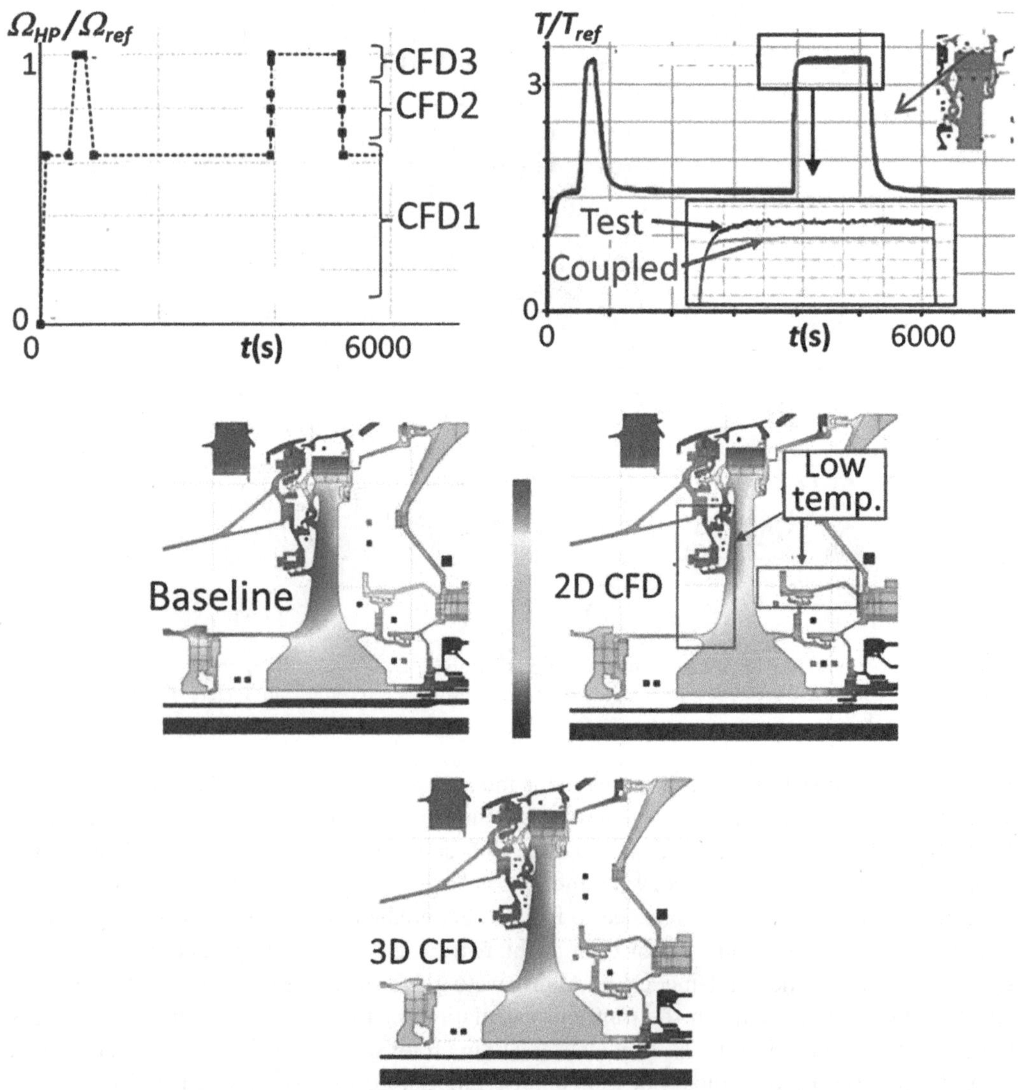

Figure 7.19 Coupled CFD-FEA disc temperature prediction for an aeroengine HP turbine disc. *Source:* Javiya *et al.* (2015) / with permission of SAGE Publications. Normalised temperature contours are at a stabilised low-power condition.

Figure 7.10, may be significant. The 2D CFD did not include any modelling of these features. There were no temperature measurements on the rear cavity drive arm and so the CFD predictions in this area might be preferred to the baseline model.

7.5 Concluding Remark

Understanding of rotor–stator disc cavity flow in turbomachinery has advanced greatly since Daily and Nece's seminal contribution in 1960 identified four flow regimes in an enclosed disc cavity. Many practical examples relate to regime IV, with recirculation driven by turbulent boundary layers on the rotating and stationary discs separated by a rotating core flow. Investigations of plane disc cavities have included the effects of radial inflow and outflow and have shown consistent results with a generally good degree of correlation between independent studies. Further investigations have confirmed the relevance of the plane disc flow mechanisms and modelling methods identified, to more complex cavity geometries, and show the effects of features such as boltheads, holes and highly swirled inflow. Methods available for engineering estimates of flow pressure, temperature and circumferential velocity include experimental correlations, elementary angular momentum and energy balances, integral boundary layer methods, axisymmetric, 3D, steady and unsteady RANS CFD and LES. The examples given above illustrate the capabilities of each method and indicate limitations and uncertainties.

Disc temperature prediction through an engine operating cycle remains challenging due to the interdependency of heat transfer and surface temperature distributions within disc cavities and other uncertainties. The latter include cavity inlet flow rates and air temperatures which are themselves coupled to disc temperatures through dependence on seal clearances and heat transfer in upstream elements of the internal air system. CFD has been shown to provide useful estimates of disc heat transfer in many cases and is having impact in this area. Full exploitation relies on integrating CFD with other engine system analysis. Development of user-friendly, versatile, efficient methods and systems for use with intensive computer modelling is a key issue.

Significant uncertainty can arise when inflow and outflow to the disc cavities are not well defined. This was illustrated by early studies of open cavities showing rotating inflow and outflow regions in the outer part of the cavity. It is also associated with a large body of more recent research on turbine disc cavity rim sealing and rotating flow structures or waves in the rim seals and cavities.

References

Alexiou, A. and Mathioudakis, K., 2008. Secondary air system component modelling for engine performance simulations. Proceedings of ASME Turbo Expo 2008, Paper GT2008-50771. https://doi.org/10.1115/GT2008-50771.

Bayley, F. J. and Conway, L., 1964. Fluid friction and leakage between a stationary and a rotating disc. J. Mech. Eng. Sci., 6, 164–172. https://doi.org/10.1243/JMES_JOUR_1964_006_026_02.

Bricaud, C., 2017. Charakterisierung der nerluste in vordralldüsensystemen: Wege zu erhöhter kühleffektivität in gasturbinen. Doctoral Thesis, Karlsruhe Institute of Technology. https://publikationen.bibliothek.kit.edu/1000043053.

Bricaud, C., Geis, T., Dullenkopf, K., and Bauer, H., 2007. Measurement and analysis of aerodynamic and thermodynamic losses in pre-swirl system arrangements. Proceedings of the ASME Turbo Expo 2007. https://doi.org/10.1115/GT2007-27191.

Bru Revert, A., Beard, P. F., and Chew, J. W., 2023. Measurement of inertial and acoustic waves in a turbine chute rim seal cavity. ASME J. Eng. Gas Turbines Power, 145, 061021. https://doi.org/10.1115/1.4056552.

Cagan, M., Benim, A. C., and Gunes, D., 2009. Computational analysis of gas turbine pre-swirl system operation characteristics. WSEAS Trans. Fluid Mech., 4, 117–126.

Calcagni, C., Gallar, L., and Pachidis, V., 2009. Development of a one dimensional dynamic gas turbine secondary air system model – Part 2: Assembly and validation of a complete network. Proceedings of ASME Turbo Expo 2009, Paper GT2009-60051. https://doi.org/10.1115/GT2009-60051.

Chew, J. W., 2009. Developments in turbomachinery internal air systems. Proc. Inst. Mech. Eng., Part C: J. Mech. Eng. Sci., 223, 189–234. https://doi.org/10.1243/09544062JMES1140.

Chew, J. W., Ciampoli, F., Hills, N. J., and Scanlon, T., 2005. Pre-swirled cooling air delivery system performance. ASME Paper No. GT2005-68323. https://doi.org/10.1115/GT2005-68323.

Chew, J. W., Hills, N. J., Khalatov, S., Scanlon, T., and Turner A. B., 2003. Measurement and analysis of flow in a pre-swirled cooling air delivery system. Proceedings of ASME Turbo Expo, GT2003-38084. https://doi.org/10.1115/GT2003-38084.

Ciampoli, F., Chew, J. W., Shahpar, S., and Willocq, E., 2007. Automatic optimization of preswirl nozzle design. ASME J. Eng. Gas Turbines Power, 129, 387–393. https://doi.org/10.1115/1.2364194.

Ciampoli, F., Hills, N. J., Chew, J. W., and Scanlon, T., 2008. Unsteady numerical simulation of the flow in a direct transfer preswirl system, ASME Paper No. GT2008-51198. https://doi.org/10.1115/GT2008-51198.

Coren, D., 2007. Windage due to Protrusions in Rotor-Stator Systems. D. Phil Thesis, Thermo-Fluid Mechanics Research Centre, University of Sussex.

Coren, D., Childs, P. R. N., and Long, C. A., 2009. Windage sources in smooth-walled rotating disc systems. Proc. Inst. Mech. Eng., Part C: J. Mech. Eng. Sci., 223, 873–888. https://doi.org/10.1243/09544062JMES1260.

Daily, J. W. and Nece, R. E., 1960. Chamber dimension effects on induced flow and frictional resistance of enclosed rotating disks. ASME J. Basic Eng., 217–230. https://doi.org/10.1115/1.3662532.

Dittmann, M., Geis, T., Schramm, V., Kim, S., and Wittig, S., 2002. Discharge coefficients of a preswirl system in secondary air systems. ASME J. Turbomach., 124, 119–124. https://doi.org/10.1115/1.1413474.

Dixon, J. A., Verdicchio, J. A., Benito, D., Karl, A., and Tham, K. M., 2004. Recent developments in gas turbine component temperature prediction methods, using computational fluid dynamics and optimization tools, in conjunction with more conventional finite element analysis techniques. Proc. Inst. Mech. Eng., Part A: J. Power Energy, 218, 241–155. https://doi.org/10.1243/0957650041200641.

Gallar, L., Calcagni, C., Pachidis, V., and Pilidis, P., 2009. Development of a one dimensional dynamic gas turbine secondary air system model – Part 1: Tool components development and validation. Proceedings of ASME Turbo Expo 2009, Paper GT2009-60058. https://doi.org/10.1115/GT2009-60058.

Geis, T., Dittmann, M., and Dullenkopf, K., 2004. Cooling air temperature reduction in a direct transfer preswirl system. ASME J. Eng. Gas Turbines Power, 126, 809–815. https://doi.org/10.1115/1.1765124.

Haaser, F., Jack, J., and McGreehan, W., 1988. Windage rise and flowpath gas ingestion in turbine rim cavities. ASME J. Eng. Gas Turbines Power, 110, 78–85. https://doi.org/10.1115/1.3240090.

Hall, S., 2019. MEng project dissertation, University of Surrey.

Hart, K. J. and Turner, A. B., 1994. Influence of radial inflow on rotor-stator cavity pressure distributions. Proceedings of the ASME 1994 International Gas Turbine and Aeroengine Congress and Exposition. Volume 1: Turbomachinery, Paper 94-GT-106. https://doi.org/10.1115/94-GT-106.

Javiya, U., Chew, J., Hills, N., and Scanlon, T., 2011. A comparative study of cascade vanes and drilled nozzle design for pre-swirl. ASME Paper No. GT2011-46006. https://doi.org/10.1115/GT2011-46006.

Javiya, U., Chew, J., Hills, N., and Scanlon, T., 2015. Coupled FE–CFD thermal analysis for a cooled turbine disk. Proc. Inst. Mech. Eng., Part C: J. Mech. Eng. Sci., 229, 3417–3432. https://doi.org/10.1177/0954406215572430.

Javiya, U., Chew, J., Hills, N., Dullenkopf, K., and Scanlon, T., 2013. Evaluation of computational fluid dynamics and coupled fluid-solid modeling for a direct transfer preswirl system. ASME J. Eng. Gas Turbines Power, 135, 051501. https://doi.org/10.1115/1.4007752.

Kong, X., Huang, T., Liu, Y., Chen, H., and Lu, H., 2022. Effects of pre-swirl radius on cooling performance of a rotor-stator pre-swirl system in gas turbine engines. Case Studies. Therm. Eng., 37, 102250. ISSN:2214-157X. https://doi.org/10.1016/j.csite.2022.102250.

Kurokawa, J. and Sakuma, M., 1988. Flow in a narrow gap along an enclosed rotating disk with through-flow. JSME Int. J., Ser. B, 31, 243–251. https://doi.org/10.1299/jsmeb1988.31.2_243.

Liu, G., Liu, Y., Kong, X., and Feng, Q., 2017. A new design of vane-shaped hole pre-swirl nozzle. Proc. Inst. Mech. Eng., Part A: J. Power Energy, 231, 14–24. https://doi.org/10.1177/0957650916673462.

Liu, Y., Liu, G., Kong, X., and Wang, Y., 2018. Design and numerical analysis of a vane shaped receiver hole in a cover-plate preswirl system. ASME J. Eng. Gas Turbines Power, 141, 041001. https://doi.org/10.1115/1.4041628.

Long, C. A., Miles, A. L., and Coren, D. D., 2012. Windage measurements in a rotor stator cavity with rotor mounted protrusions and bolts. Proceedings of the ASME Turbo Expo 2012. https://doi.org/10.1115/GT2012-69385.

Long, C. A., Turner, A. B., Kais, G., Tham, K. M., and Verdicchio, J. A., 2003. Measurement and CFD prediction of the flow within an HP compressor drive cone. J. Turbomach., 125, 165–172. https://doi.org/10.1115/1.1516195.

Lundbladh, A. and Johansson, A., 1991. Direct simulation of turbulent spots in plane Couette flow. J. Fluid Mech., 229, 499–516. https://doi.org/10.1017/S0022112091003130.

Luo, X., Zhang, X., Tao, Z., Xu, G., and Wang, Q., 2014. Windage measurements in a rotor-stator system with superimposed cooling and rotor-mounted protrusions. ASME J. Eng. Gas Turbines Power, 136(4), 042505. https://doi.org/10.1115/1.4026086.

Maroti, L. A., Deak, G., and Kreith, F., 1960. Flow phenomena of partially enclosed rotating disks. ASME. J. Basic Eng., 82, 539–549. https://doi.org/10.1115/1.3662650.

May, D. and Chew, J. W., 2010. Response of a disk cavity flow to gas turbine engine transients. Proceedings of the ASME Turbo Expo 2012, Paper GT2010-22824, 1113–1122. https://doi.org/10.1115/GT2010-22824.

McGreehan, W. F. and Schotsch, M. J., 1988. Flow characteristics of long orifices with rotation and corner radiusing. ASME J. Turbomach., 110, 213–217. https://doi.org/10.1115/1.3262183.

Meierhofer, B. and Franklin, C. J., 1981. An investigation of a preswirled cooling airflow to a turbine disc by measuring the air temperature in the rotating channels. ASME Paper 81-GT-132.

Miles, A., 2011. An experimental study of windage due to rotating and static bolts in an enclosed rotor-stator system. D. Phil Thesis, Thermo-Fluid Mechanics Research Centre, University of Sussex.

Millward, J. A. and Robinson, P. H., 1989. Experimental investigation into the effects of rotating and static bolts on both windage heating and local heat transfer coefficients in a rotor-stator cavity. ASME Gas Turbine and Aero-engine Congress and Exposition, Paper, 89-GT-196. https://doi.org/10.1115/89-GT-196.

Moghaddam, E. R., Long, C., and Sayma, A., 2013, 2013. A numerical investigation of moment coefficient and flow structure in a rotor–stator cavity with rotor-mounted bolts. Proc. Inst. Mech. Eng., Part A: J. Power Energy, 227(3), 306–327. https://doi.org/10.1177/0957650912473543.

Noor Mohamed, S., Chew, J. W., and Hills, N. J., 2015. Simplified protrusion drag and heat transfer modelling of bolts on a rotating disc. Proceedings of ASME Turbo Expo, 56734, V05CT15A025. https://doi.org/10.1115/GT2015-43501.

Noor Mohamed, S., Chew, J. W., and Hills, N. J., 2017a. Effect of bolts on flow and heat transfer in a rotor–stator disk cavity. ASME J. Eng. Gas Turbines Power, 139, 051901. https://doi.org/10.1115/1.4035144.

Noor Mohamed, S., Chew, J., and Hills, N., 2017b. Flow and windage due to bolts on a rotating disc. Proc. Inst. Mech. Eng., Part C: J. Mech. Eng. Sci., 231, 2925–2941. https://doi.org/10.1177/0954406216642260.

Pett, A., Coren, D., and Childs, P., 2007. Model validation for a shrouded rotor-stator system with superposed cooling and static protuberances. Proceedings of ASME Turbo Expo, Paper GT2007-27744. https://doi.org/10.1115/GT2007-27744.

Poncet, S., Schiestel, R., and Chauve, M., 2005a. Centrifugal flow in a rotor-stator cavity. ASME J. Fluids Eng., 127, 787–794. https://doi.org/10.1115/1.1949645.

Poncet, S., Chauve, M. P., and Le Gal, P., 2005b. Turbulent rotating disk flow with inward throughflow. J. Fluid Mech., 522, 253–262. https://doi.org/10.1017/S0022112004002046.

Saunders, K., Alizadeh, S., Lewis, L. V., and Provins, J., 2007. The use of CFD to generate heat transfer boundary conditions for a rotor-stator cavity in a compressor drum thermal model. ASME Turbo Expo Paper GT2007-28333. https://doi.org/10.1115/GT2007-28333.

Schröder, T. R., Dohmen, H.-J., Brillert, D., and Benra, F.-K., 2020. Impact of leakage inlet swirl angle in a rotor–stator cavity on flow pattern, radial pressure distribution and frictional torque in a wide circumferential Reynolds number range. Int. J. Turbomach. Propul. Power, 5, 7. https://doi.org/10.3390/ijtpp5020007.

Schröder, T. R., Schuster, S., and Brillert, D., 2021. Experimental investigation of centrifugal flow in rotor–stator cavities at high Reynolds numbers >10^8. Int. J. Turbomach. Propul. Power, 6, 13. https://doi.org/10.3390/ijtpp6020013.

Unglaube, T. and Brillert, D., 2022. Acoustoelastic modes in rotor-cavity systems: An overview on frequency shift effects supported with measurements. Int. J. Turbomach. Propul. Power, 7, 15. https://doi.org/10.3390/ijtpp7020015.

Zimmermann, H., Firsching, A., Dibelius, G. H., and Ziemann, M., 1986. Friction losses and flow distribution for rotating discs with shielded and protruding bolts. ASME J. Eng. Gas Turbines Power, 108, 547–552. https://doi.org/10.1115/1.3239945.

8

Corotating Disc Cavities

This chapter considers cavities in which the two discs and shroud rotate at the same speed. Such cavities are common in multistage axial compressors and turbines, as shown in Figures 1.1 and 1.2. As in the previous chapter, the aims are to present a range of examples that show the extent and limitations of current knowledge and to inform engineers in system design and analysis of their configurations. A distinguishing feature of rotating cavities from rotor–stator cavities (and other systems with differentially rotating discs) is the absence of recirculating flow when there is no throughflow. In a closed, rotating cavity fluid with constant density and viscosity rotates as a solid body or forced vortex, as in equation 3.1.2 with $B = 0$. As described in Sections 8.1 and 8.2, radial throughflow results in flows that have much in common with those for rotor–stator cavities. In rotor–stator systems, the variation in centrifugal force due to changes in fluid density is usually small compared to the variation due to changes in circumferential velocity. This is not always the case in rotating cavities with limited throughflow in which centrifugal buoyancy may dominate, as discussed in Sections 8.3 and 8.4.

8.1 Radial Outflow

The flow structure in a cavity with radial outflow is shown for laminar flow in Figure 3.9 and for turbulent flow in Figures 5.9 and 8.1. In a sufficiently wide, axisymmetric cavity, the disc boundary layers first entrain the supplied flow at lower radii and constant mass flow layers develop at higher radii. As the rotation of the core flow between the layers increases (with radius), the flow is described by the linear Ekman layer solutions given in Sections 3.3.2 and 5.3.1. Provided ingress of flow at the outlet on the cavity outer boundary is suppressed, the influence of flow exit conditions may be limited to a small outer region, and the bulk flow is primarily radially outwards at all other radii.

Outflow in rotating cavities has not been so widely investigated experimentally as rotor–stator cavities, but measurements were obtained on several experimental rigs at Sussex University for Re_ϕ up to 1.3×10^6, as described by Owen *et al.* (1985). These measurements, the comparison in Figure 5.10, and further work including comparisons with CFD and heat transfer measurements give confidence in integral solutions and RANS CFD for turbulent flow. An example of the agreement obtained is shown in Figure 8.1. Here, core velocity measurements are available from $r/b = 0.5$–0.8. Overall agreement with the integral solution and CFD is good. The measurements are coincident with the CFD results for lower radii and with the integral solution at higher radii.

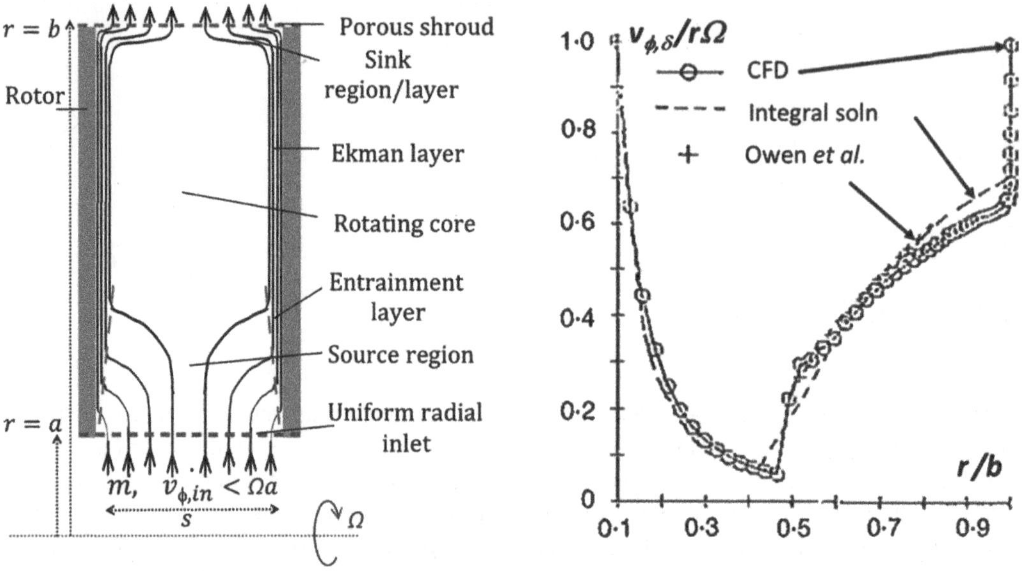

Figure 8.1 Core circumferential velocity in a rotating cavity with turbulent radial outflow for $s/b = 0.13$, $Re_\phi = 10^6$ and $C_W = 2500$. CFD uses a mixing length model.

8.1.1 Integral Solutions for the Flow Field

Dimensional analysis for incompressible flow is similar to that for rotor–stator systems in Section 7.1.2. Nondimensional flow parameters describing core swirl, disc moment and pressure distribution will depend on $Re_\phi, C_W = \dot{m}/\mu b$ and inlet swirl fraction $v_{\phi,in}/\Omega a$ (in addition to geometrical parameters). For turbulent flows, integral boundary layer treatments assuming $1/7th$ power laws again indicate that for rotationally dominated conditions, the dependency on Re_ϕ and C_W may be captured through the mass flow parameter $\lambda = C_W/Re_\phi^{0.8}$. The integral momentum equation 5.1.14 has been shown to give useful predictions for turbulent flow above.

Further solutions of the integral equations are presented in Figure 8.2. In this model, the effects of the outlet are neglected. The integral equations were solved numerically, starting from specified inlet conditions at $r = a$ and marching radially outwards. It is assumed that the inlet radial and axial velocity components are small compared to disc speed and that the disc boundary layers start from a negligible thickness at $r = a$. The principal independent parameters are then inlet swirl fraction and the local mass flow parameter λ_r, as used to correlate core velocities for rotor–stator systems in Sections 7.1.2 and 7.1.3. The values of both $v_{\phi,in}/\Omega a = v_{\phi,\delta}/\Omega a$ and λ_r at $r = a$ are required to start the integral solution and interpret Figure 8.2. The outer limit of the cavity corresponds to $\lambda_r = \lambda$. Note that λ and λ_r are defined here using the full cavity mass flow and that Figure 8.2 includes the linear turbulent Ekman layer solution given by equation 5.3.13.

The results in Figure 8.2 are presented in a general form to allow application to cavities with different radius ratios, a/b. This figure may be compared with Figure 5.10, giving results for $a/b = 0.1$, $v_{\phi,in}/\Omega a = 1$ and comparing with measurements. Here, results are given for $v_{\phi,in}/\Omega a = 0$ or 1, and show the effects of both flow rate and radius ratio for $(a/b)/(\lambda/2)^{5/13} = 0.5$, 1, 2, 3, 4 or 5. Each line gives the solution for specific inlet values of $v_{\phi,in}/\Omega a$ and λ_r. Taking, for example, $\lambda = 0.05$ as representative of the measurements in Figure 5.10, values of $a/b = 0.1$, 0.3, 0.6 and 0.9 correspond to $(a/b)/(\lambda/2)^{5/13} = 0.41$, 1.24, 2.48, 3.72 and $r = b$ when $(r/b)/(\lambda/2)^{5/13} = 4.13$. For the experimental case of $a/b = 0.1$, the flow becomes fully entrained in the disc boundary layers close

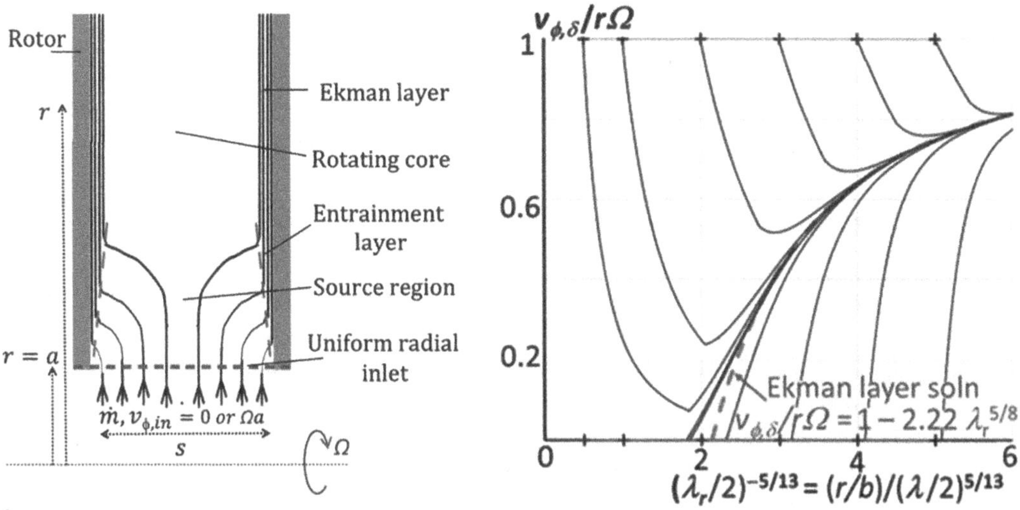

Figure 8.2 Integral method solutions for core circumferential velocity in a rotating cavity with turbulent radial outflow for inlet swirl fractions of 0 and 1. + indicates values at $r = a$.

to $(r/b)/(\lambda/2)^{5/13} = 1.83$, which corresponds to an estimate from von Kármán's free disc solution for flow entrainment. For $a/b = 0.6$ or lower, the solutions indicate convergence to the Ekman layer solution at the outer radius. For $a/b = 0.9$, the solution is strongly dependent on the inlet swirl throughout the cavity.

All cases in Figure 8.2 show free vortex behaviour, $rv_{\phi,\delta} = constant$, in the source region at lower radii. The core velocity then adjusts to the linear Ekman solution at higher radii as the constant mass flow Ekman layers develop. As for rotor–stator flows, some sensitivity to inlet conditions is expected, and although turbulent flow is expected in most engineering applications, laminar-turbulent transition may occur in areas with lower local boundary layer Reynolds numbers. This was discussed in Section 4.4.

Disc moments can be estimated from the angular momentum increase between the inlet and outlet flows. Assuming all the flow is entrained into the disc boundary layers and using equations 5.1.7 and 5.1.11 for $n = 7$, the angular momentum outflow in each boundary layer at radius r (outside the source region) is given by

$$\frac{2 \times AM_{outflow}}{\rho \Omega^2 b^5} = \lambda Re_\phi^{-0.2} \left(\frac{r}{b}\right)^2 \left[1 - 0.833\left(1 - \frac{v_{\phi,\delta}}{\Omega r}\right)\right]. \tag{8.1.1}$$

Applying this result at $r = b$ and accounting for inflow, the moment coefficient on one disc face is

$$C_M = \lambda Re_\phi^{-0.2} \left[1 - 0.833\left(1 - \frac{v_{\phi,\delta,out}}{\Omega r}\right) - \frac{v_{\phi,in}}{\Omega a}\left(\frac{a}{b}\right)^2\right]. \tag{8.1.2}$$

If the Ekman layer solution holds at the outer radius, this gives

$$C_M = \lambda Re_\phi^{-0.2} \left[1 - 1.85\lambda^{5/8} - \frac{v_{\phi,in}}{\Omega a}\left(\frac{a}{b}\right)^2\right]. \tag{8.1.3}$$

Moment on the shroud may be estimated using the integral solution for $v_{\phi,\delta}$, at $r = b$ and with assumptions for the friction factor, as was described in Sections 5.2.5 and 7.1.1 for a rotor–stator system. Also, as described in Section 7.1, the core velocity can be used with equation 2.1.15 to calculate pressure loading on the discs, and boundary layer thickness from integral solutions can be used to estimate when transition to a merged boundary layer regime is expected. While numerical integration to obtain solutions such as those in Figure 8.2 is relatively straightforward; the results in Chapter 5 may also be useful for quick estimates. Depending on the flow region indicated in Figure 8.2, equation 5.3.13 for Ekman layers, equation 5.2.14 for free vortex flow above a rotating disc or equations 5.2.2 and 5.2.3 for the free disc solution may be relevant.

8.1.2 CFD Solutions for the Flow Field

Axisymmetric, steady CFD models have confirmed the flow field described above and extended investigations to consider departures from the plane disc cavities studied experimentally. A laminar flow simulation is shown in Figure 3.9. Several early CFD studies of turbulent flow used mixing length models. These include studies by Chew (1985a, 1990) and Lapworth and Chew (1992), with an example in Figure 8.1. Morse (1988, 1991) used a tuned low Reynolds number $k - \varepsilon$ model to extend studies into transitional flows and to predict high Reynolds number conditions. He showed very good agreement with measurements. As discussed in Section 6.2.3.2, Virr *et al.* (1994) evaluated the $k - \varepsilon$ model with wall functions or, alternatively, with a mixing length treatment near the wall. This showed that wall functions could be used for higher Reynolds number conditions. Further evaluations against measurements include heat transfer results and are illustrated in the next subsection.

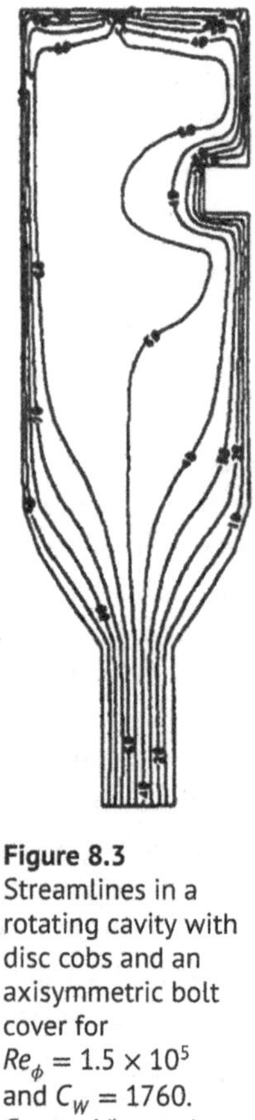

Figure 8.3 Streamlines in a rotating cavity with disc cobs and an axisymmetric bolt cover for $Re_\phi = 1.5 \times 10^5$ and $C_W = 1760$. *Source:* Virr *et al.* (1994) / with permission of American Society of Mechanical Engineers (ASME).

Examples of CFD flow predictions for more complex, but still axisymmetric, geometries are given in Figures 8.3 and 8.4. In Figure 8.3, the discs have cobs, and one disc has a bolt cover. Neither of these features changes the overall flow structure. The shaped cobs will affect the entraining flow boundary layers slightly, but variations in inlet conditions may well have stronger effects. The integral method might be extended to account for the cobs using a cone angle as described in Chapter 5. The bolt cover deflects the streamlines and induces a slight asymmetry to the flow but does not induce significant flow separation. This can be attributed to the strong circumferential velocity component, not shown directly by the streamlines, and the Taylor–Proudman effect suppressing axial gradients in the inviscid core, as in equation 3.3.7.

Figure 8.4 shows streaklines from CFD calculations for several disc cavities in a three shaft aeroengine. These use a $k - \varepsilon$ model with wall functions and model a maximum power condition. The flow in each cavity was calculated from separate CFD solutions using seal flows based on clearance estimates and leakage correlations. The two central cavities, bounded by corotating, low-pressure turbine discs, have imposed radial outflows. In both these cases, the source flow regions extend over much of the cavity. The disc pumping is clearly apparent for the left central cavity but less so for the right cavity, which has a high velocity inlet jet. Noting that the figure also includes rotor–stator cavities and the IPT-LPT cavity with differentially corotating discs, this is an early example of CFD providing estimates of flow and pressure field that would have been costly to obtain experimentally.

The effects of boltheads and other 3D geometrical features in rotating cavities have not been investigated to the extent they have for rotor–stator flows. Similar effects to those described in

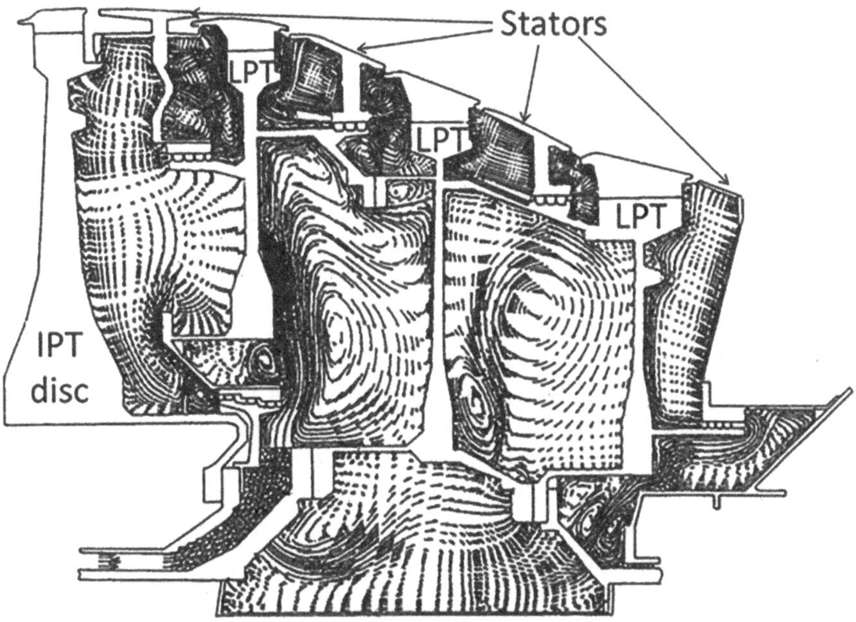

Figure 8.4 Streaklines in typical turbine disc cavities at maximum power condition. *Source:* Virr *et al.* (1994) / with permission of American Society of Mechanical Engineers (ASME).

Section 7.2.2 can be expected with some loss of significance of the mass flow parameter λ in scaling results. Elementary angular momentum balances, similar to that given by equation 7.2.1, may again be used to estimate effects. In this case, there is no stator moment, and the extra moment due to protrusions will be limited to that required to bring the cavity flow up to rotor speed.

Depending on outlet geometry, external flow may be ingested at the cavity outlet. Owen *et al.* observed this in some experiments with larger outlet holes in the rotating cylindrical shroud to the cavity. For one experimental rig with a central axial inlet, coupling of ingestion with highly unsteady oscillation of the inlet jet was observed. Unlike rotor–stator systems, which must accommodate relative movement of the components, it is straightforward to avoid ingestion by reducing the size of the outlet holes. A fully open system would allow sufficient ingestion to satisfy disc pumping.

8.1.3 Heat Transfer

The dimensional analysis and elementary modelling for heat transfer in rotor–stator systems discussed in the opening paragraph of Section 7.4 generally apply to rotating cavities with radial throughflow. For rotating cavities with outflow, as in Figures 8.1–8.3, the flow is principally or wholly in the radial direction, rather than recirculating. This simplifies heat transfer modelling as a local fluid temperature may be more easily estimated within the thermal model. For example, an integral energy equation can be efficiently solved numerically by marching in the flow direction. If the mass flow in a disc boundary layer on the rotor is assumed constant and *Pr* is assumed to be close to 1 in the viscous terms, the energy balance for the boundary layer gives the equation below:

$$\frac{d(\dot{m}C_p\overline{T^*})}{ds} = 2\pi r q_0 = 2\pi r\, htc\left(T_0^* - \overline{T^*}\right) \tag{8.1.4}$$

Here, $\dot{m}$ is the mass flow in the boundary layer, $\overline{T^*}$ is a bulk fluid rotary stagnation temperature, s is streamwise distance in the $r - z$ plane, q_0 is surface heat flux, *htc* is a heat transfer coefficient and T_0^* is surface rotary stagnation temperature. If estimates of *htc* and the fluid temperature at the start of boundary layer are available, equation 8.1.4 can provide the boundary condition for a heat conduction solution in the rotor. Estimates of *htc* might be made using integral solutions for the core velocity, as in Figure 8.2, and use of the Reynolds analogy, equation 2.5.22, with surface shear stress estimated from equation 5.2.14 for free vortex flow and 5.3.12 for Ekman layer flow.

As referenced in Section 5.3.2, Northrop and Owen (1988) and Chew and Rogers (1988) demonstrated good agreement of disc heat transfer measurements and predictions from an integral method. Some sensitivity to disc temperature distribution and the assumptions made to account for this in the integral energy equation was found. Equation 8.1.4 is essentially a simplified version of the fuller energy equation given in equation 5.1.20, and further refinements to better model the effects of disc temperature distribution are given by Chew and Rogers. The degree of complexity justified in engineering applications will depend on the situation considered, accuracy requirements and other modelling uncertainties. As shown in Figure 8.4 by the streaklines in the right centre cavity, practical flows do not exactly conform to the idealised assumptions of the integral models, so CFD is an attractive option for more reliable prediction. However, dependency of heat transfer of surface temperature distribution remains an issue. Lapworth and Chew noted that differences in thermal boundary conditions for their CFD solutions had a greater influence on Nusselt number than differences in disc geometry.

Evaluations of axisymmetric, steady CFD against Northrop and Owen's heat transfer measurements are given by Morse and Ong (1992) and Iacovides and Chew (1993). Morse and Ong used a low Reynolds number $k - \varepsilon$ turbulence model. They showed good agreement with measurements and suggested that improvements could be achieved with further optimisation of the turbulence model. Iacovides and Chew considered a standard $k - \varepsilon$ model with three different near-wall models. These were two mixing length treatments and a one-equation near-wall model. Performance of the best $k - \varepsilon/l_m$ model was similar to that of the $k - \varepsilon/k - l_m$ model. It was stated that Ong and Morse had demonstrated similar or better agreement with data, but some known difficulties with low Reynolds number $k - \varepsilon$ turbulence models were noted. The $k - \varepsilon/k - l_m$ model was recommended for further investigations for consistency with the wider framework of turbulence modelling.

More recent comparisons of CFD with Northrop and Owen's data are shown in Figure 8.5. Here, *Nu* is a local Nusselt number based on radius and disc-to-inlet temperature difference. Results show the level of agreement that is generally achieved between CFD, the integral method and measurements for this test case. The CFD here also confirms that wall functions may be used at high enough Reynolds numbers for this class of flow, as discussed in Section 6.2.3.2. Results are shown from a standard $k - \omega$ SST model (WMURANS) and WMLES with a Smagorinsky sub-grid scale model. Northrop and Owen's data includes direct heat transfer measurements from heat flux gauges and surface heat flux deduced from solutions of the heat conduction equation for the disc with measured surface temperatures as boundary conditions. The two measurement techniques were not entirely independent, owing to the calibration of the gauges for a free disc test case, but the level of agreement is still encouraging. The integral solution shown is from Chew and Rogers. The WMLES was included by Wang *et al.* (2024a) to test this model for heat transfer on a relatively simple test case before considering the more complex buoyancy-driven flows discussed later in this chapter.

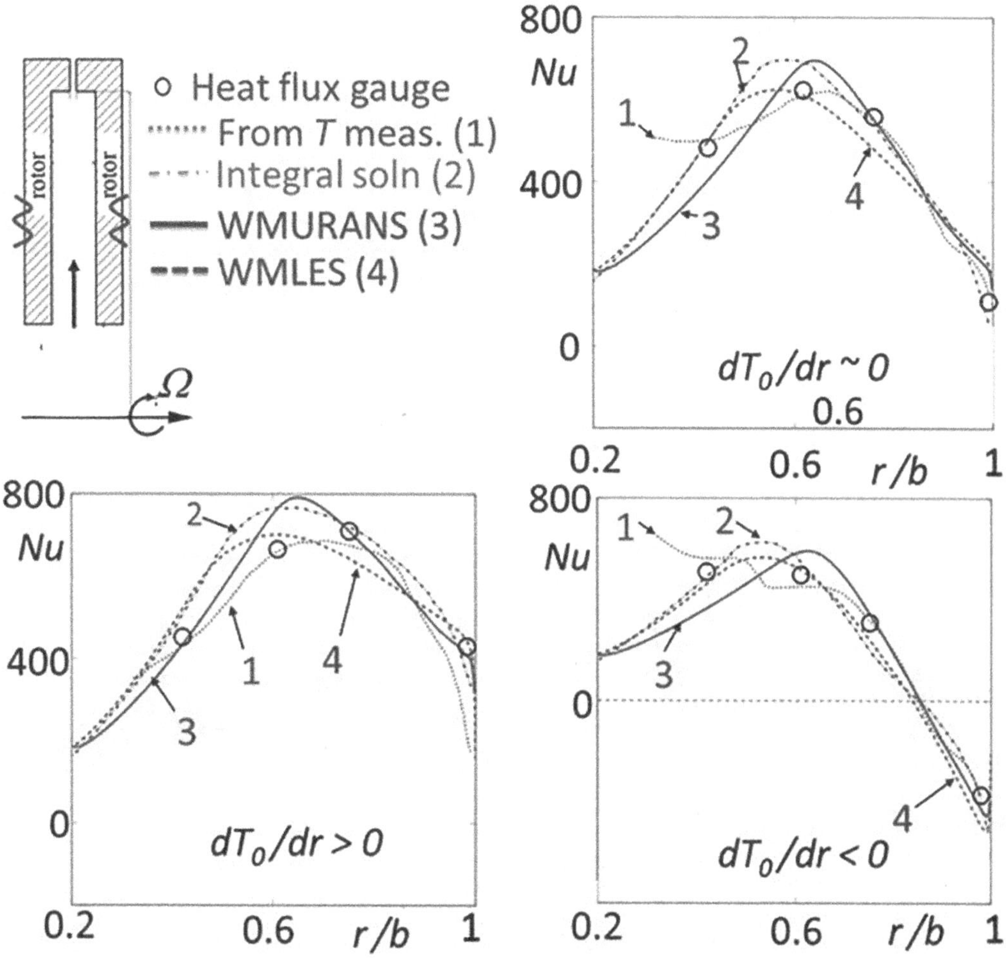

Figure 8.5 Comparison of measured and calculated disc heat transfer for radial outflow in a rotating cavity with three different disc temperature distributions, $Re_\phi = 2 \times 10^6$, $C_W = 7000$. *Source:* Wang *et al.* (2024a) / with permission of American Society of Mechanical Engineers.

Application of CFD in engine thermal modelling is illustrated by Dixon *et al.* (2004) and Sun *et al.* (2010). In both cases CFD heat transfer prediction was coupled with finite element analysis for a low pressure turbine using methods described in Section 6.5. Comparisons with temperature measurements from engine tests was judged to be very encouraging. The example in Figure 8.6 is for a turbine from a more recent three shaft engine than that of Figure 8.4. CFD was applied in the cavities between the intermediate pressure turbine and the first three stages of the low pressure turbine. As shown, there are two cavities between corotating LPT discs and these have radial outflow in the disc boundary layers. The source flow region extends over most of the first LPT cavity but the flow structure in the right-hand cavity suggests the presence of a rotating core with negligible radial velocity. Asymmetry of the cavity flows is also clear with, for example, little ventilation of the region between the third-stage LPT disc and the outer flange on the shroud on the right-hand side of the figure. A forced vortex at LPT speed may form in this region and could be disturbed by buoyancy forces that would not be captured by the steady, axisymmetric CFD model.

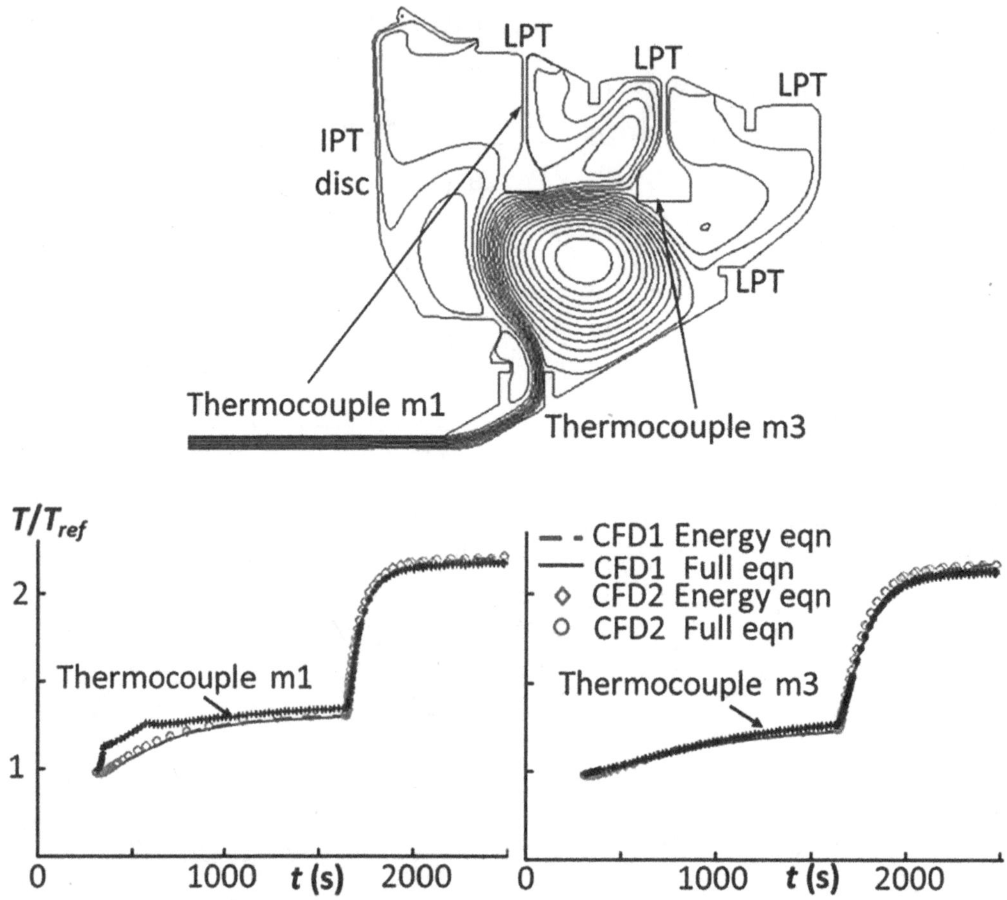

Figure 8.6 Predicted streamlines at maximum power and transient disc temperatures for an engine low pressure turbine. *Source:* Sun *et al.* (2010).

Flow in the cavity between the IPT and first-stage LPT disc is dominated by the pumping of the higher speed IPT disc.

The calculated and measured disc temperatures at two thermocouple locations in Figure 8.6 show mostly good agreement between all the models tested and engine measurements. The test period covers a rapid engine acceleration and stabilisation at high power. Two CFD codes, CFD1 and CFD2, were used with iterative calculation of heat transfer using surface temperatures from FEA for the solid components to ensure consistency. Both CFD codes allowed a frozen flow approximation in which the surface temperature was assumed to only affect the energy equation solution. This approximation worked well. Other CFD boundary conditions and meshes corresponded to either a low or high power condition. The CFD used the $k - \varepsilon$ model with standard wall functions. Apart from the start of the transient for thermocouple m1, calculations and measurements generally agree to within the expected measurement uncertainty. The behaviour of the thermocouple measurements at the start of the cycle is not understood. As problems with instrumentation are not unusual in engine tests, no inference should be drawn from this result. It is included as an example of the unknowns that can occur in practical applications.

8.2 Radial Inflow

Radial inflow in rotating cavities may occur when air for cooling is bled radially inwards from mid stages in axial compressors. Simplified flow structure in a cavity with radial inflow is shown in Figure 5.11 and this class of flow has been discussed in Sections 5.3.3 and 6.2.6. Further details and examples are given here. As for radial outflow above, the focus is on turbulent flow in sufficiently wide, axisymmetric cavities for separate disc boundary layers to form. Dimensional analysis follows that for radial outflow and again the throughflow parameter $\lambda = C_W/Re_\phi^{0.8}$ is useful in correlating results.

Models using the integral method neglect the outlet region and assume radial and axial velocities are small compared to the circumferential velocity component. Examples of assumed flow structures are shown in Figure 8.7. For inlet swirl fractions $v_{\phi,in}/\Omega b < 1$, the outer flow region differs from that for $v_{\phi,in}/\Omega b = 1$ or higher. For $v_{\phi,in}/\Omega b \geq 1$, the flow is relatively simple with predominantly radial inflow throughout. Conservation of angular momentum tends to increase v_ϕ, above disc speed as the fluid moves radially inwards. This leads to radial inflow in the disc boundary layers and, once all the supplied flow has been entrained at radius r_e, the flow comprises constant

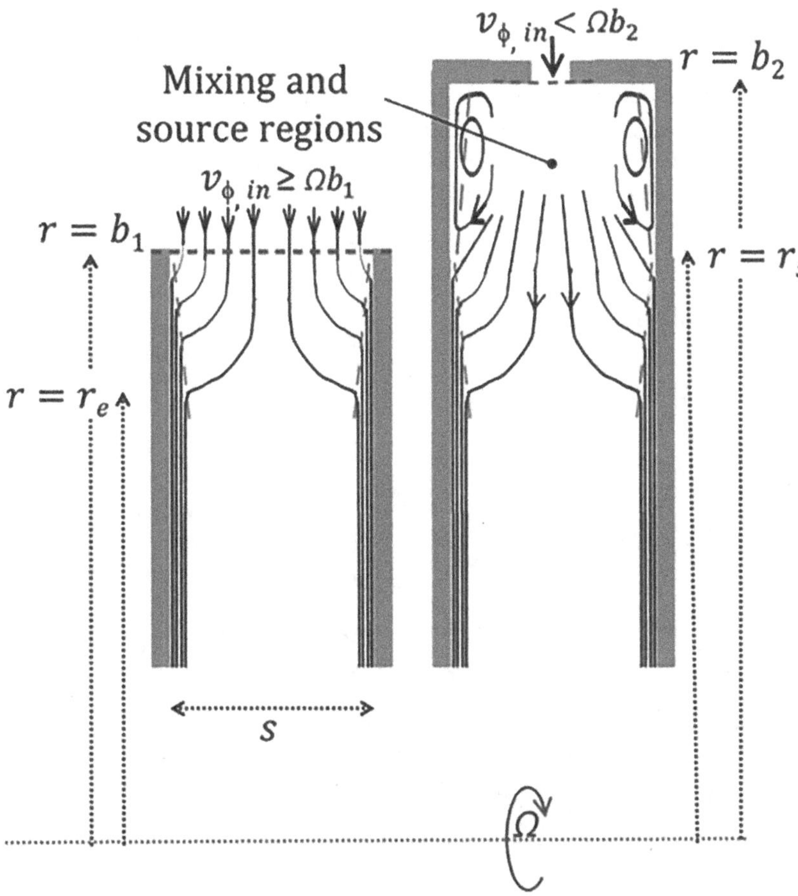

Figure 8.7 Flow structures for a rotating cavity with radial inflow for $v_{\phi,in}/\Omega b \geq 1$ and $v_{\phi,in}/\Omega b < 1$. For $r < r_s$ in the right-hand diagram, the flow structure may be similar to that for the left-hand diagram if $b_1 = r_s$.

mass flow disc boundary layers separated by a rotating core flow. The core rotation continues to increase as the radius reduces, but at a lower rate than for a free vortex. For deswirled inflow, with $v_{\phi,in}/\Omega b < 1$, recirculation is driven by radially outward disc pumping in the outer part of the cavity. At lower radii, the flow reverts to the structure shown for $v_{\phi,in}/\Omega b = 1$. The radius r_s where the fluid velocity approximately matches that on the disc, with $v_{\phi,in}/\Omega r_s = 1$, marks the switch from outflow to inflow in the boundary layers. Flow inboard of this point is similar to that for a hypothetical cavity of reduced radius $b = r_s$ with $v_{\phi,in}/\Omega b = 1$ and the same mass flow and rotational speed.

8.2.1 Integral Solutions for Inlet Swirl Fractions ≥ 1

With $v_{\phi,in}/\Omega b \geq 1$, the integral equation 5.1.14 for the disc boundary layers can be solved numerically starting from specified inlet conditions at $r = b$ and marching radially inwards. Assuming a negligible boundary layer thickness at $r = b$, the principal independent parameters are then inlet swirl fraction and the local mass flow parameter λ_r. With λ_b specified the disc radius ratio, a/b, then gives λ_r at the inner limit of the cavity. The neglect of convection in the linear Ekman layer solution, equation 5.3.13, removes the dependency on inlet conditions but introduces further approximations. As shown below and shown for radial outflow in Figures 5.10 and 8.2, the Ekman layer solution applies in a wider range of conditions than might be expected for a linear approximation.

Figure 8.8 illustrates the integral solutions for radial inflow with $v_{\phi,in}/\Omega b = 1$. Results are given for a range of values of the throughflow parameter λ based on the full cavity mass flow rate that supplies the inflow on two discs. The curves show variation of specific angular momentum in the flow core, normalised by its inlet value, with radius ratio a/b. In the source flow region, the solution assumes a free vortex with $rv_{\phi,in}/\Omega b^2 = 1$, and the integral solution corresponds to that approximated by equation 5.2.16. The radius ratios r_e/b at which the flow is fully entrained into

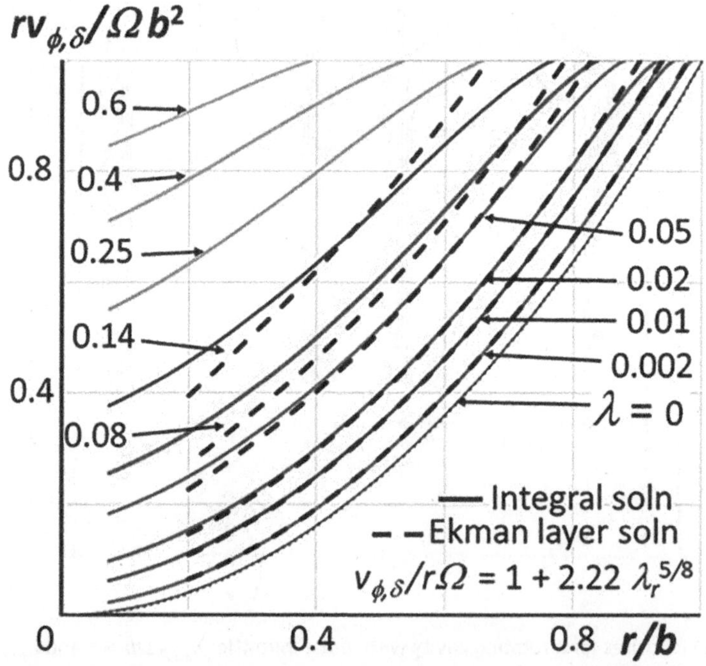

Figure 8.8 Integral solutions for rotating cavities with radial inflow for $v_{\phi,in}/\Omega b = 1$.

the disc boundary layers correspond to solutions of the cubic equation for boundary layer flow rate in 5.2.16. For $\lambda = 0$, the solution corresponds to a forced vortex. The Ekman layer solution is also shown for $\lambda = 0.14$ or lower. At very low values of λ, the flow is a perturbation from the forced vortex, and the Ekman layer solution applies, as expected. Some agreement of the Ekman solution with the full solution in the core flow region remains up to $\lambda = 0.14$ but with increasing differences.

As for radial outflow, disc moments can be estimated from the difference between angular momentum flow at the inlet and outlet. Equation 8.1.1 applies at the outlet $r = a$ if all the flow is entrained into the disc boundary layers. Accounting for inflow, the moment coefficient based on one disc face and half the moment on the shroud is then given by

$$C_M = \lambda Re_\phi^{-0.2}\left[\left(1 - 0.833\left(1 - \frac{v_{\phi,\delta,out}}{\Omega r}\right)\right)\left(\frac{a}{b}\right)^2 - \frac{v_{\phi,in}}{\Omega b}\right]. \tag{8.2.1}$$

If the Ekman layer solution holds at the inner radius, this gives

$$C_M = \lambda Re_\phi^{-0.2}\left[\left(1 + 1.85\lambda^{5/8}\left(\frac{b}{a}\right)^{13/8}\right)\left(\frac{a}{b}\right)^2 - \frac{v_{\phi,in}}{\Omega b}\right]. \tag{8.2.2}$$

Experimental investigations for radial inflow are reported by Firouzian *et al.* (1985, 1986) and Owen *et al.* (1985). These papers include comparisons of velocity measurements with integral solutions. In most of the experiments, flow entered the cavity through holes in a thin rotating cylindrical shroud with inlet swirl below disc speed. Unsteady flow was observed in these conditions. In one configuration, an inlet slot in the shroud was packed with porous foam material to give an inlet swirl fraction of ~1. The core velocity measurements for this configuration are compared with the present integral solutions in Figure 8.9. For these experiments, Re_ϕ is in the range 10^5–3.5×10^5, with C_W in the range 1310–1425. The best agreement between theory and experiment is in the core flow at lower values of λ, where the Eman layer and full solutions are also close. Owen *et al.* noted

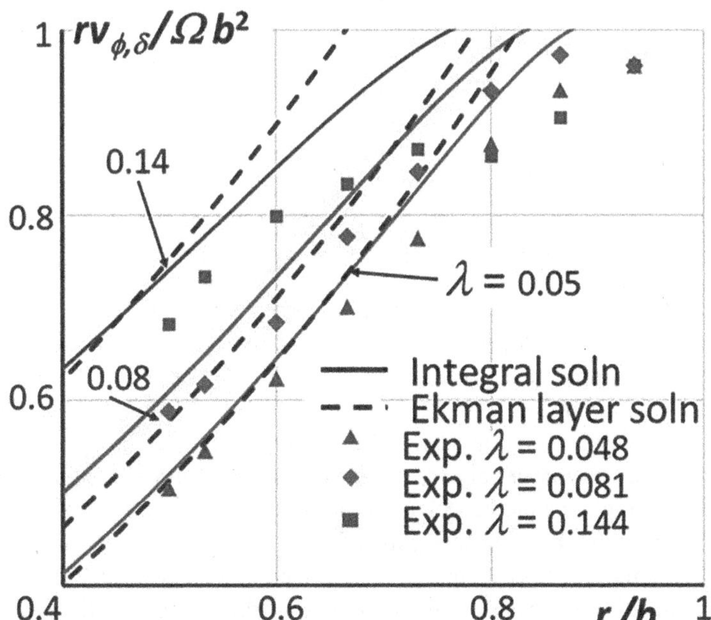

Figure 8.9 Comparison of integral solutions with Owen *et al.*'s (1985) measurements for rotating cavities with radial inflow for $v_{\phi,in}/\Omega b = 1$.

that the agreement of measurement and theory overall was not as good for this experimental configuration as found in some other configurations. It should also be noted that Re_ϕ is relatively low in the experiments, and the best agreement (at low λ) occurs at the highest Re_ϕ, where the turbulent flow assumptions in the integral method are more likely to be valid.

Further comparisons with measurements and consideration of compressible flow effects were given by Chew and Snell (1988), as discussed in Section 5.3.3. Cavity pressure drop predictions corresponding to the above integral solutions are shown in Figure 5.12. This uses a logarithmic scale to accommodate the high pressure gradients that can occur at low radii.

8.2.2 Integral Solutions for Inlet Swirl Fractions < 1

Integral solutions for $v_{\phi,in}/\Omega b < 1$ were described in Section 5.3.3 and compared with Farthing *et al.*'s (1991b) measurements. An effective swirl fraction was estimated from an angular momentum balance in the region $r_s < r < b$ (in Figure 8.7), allowing the use of the integral method in the above subsection for $r < r_s$. Integral solutions were compared to pressure measurements for a rotating cavity with deswirled radial inflow, produced by nozzles in the rotating shroud, angled at 30° to the circumferential direction. Good agreement was obtained at $Re_\phi = 6.1 \times 10^5$, as shown in Figure 5.13.

Assuming the interior flow in the region $r_s < r < b$ (shown in Figure 8.7 for $v_{\phi,in}/\Omega b < 1$) to have a free vortex velocity distribution, an angular momentum balance for this region gives

$$\frac{(rv_\phi)_\delta}{\Omega b^2} = \frac{(rv_\phi)_{in}}{\Omega b^2} + \frac{(C_{M,discs} + C_{M,shroud})Re_\phi^{0.2}}{2\lambda}. \tag{8.2.3}$$

This adapts equation 7.3.7 in Section 7.3.2 to the present case. The angular momentum outside the boundary layers $(rv_\phi)_\delta$ is assumed constant, and the condition of equal disc and fluid velocity at the stagnation radius gives $(rv_\phi)_\delta = \Omega r_s^2$. Using equations 5.2.14 and 5.2.15 to get the moment on the two discs, and estimating shroud moment using the disc friction factor at $r = b$, gives

$$C_{M,discs}Re_\phi^{0.2} = 0.146\left[1 - \left(\frac{r_s}{b}\right)^2\right]\left(1 - e^{-1.15\left(\frac{b}{r_s}-1\right)}\right)$$

$$C_{M,shroud}Re_\phi^{0.2} = 0.336\left(\frac{s}{b}\right)\left(1 - e^{-1.15\left(\frac{b}{r_s}-1\right)}\right)$$

$$\left(\frac{r_s}{b}\right)^2 = \frac{(rv_\phi)_{in}}{\Omega b^2} + \frac{(C_{M,discs} + C_{M,shroud})Re_\phi^{0.2}}{2\lambda} \tag{8.2.4}$$

For given values of the flow parameter λ, shroud width s/b, and inlet swirl $v_{\phi,in}/\Omega b < 1$, these equations may be used to find r_s/b, provided nondimensional angular momentum in equation 8.2.3 remains positive. For example, successive substitution for r_s/b with underrelaxation is an effective solution method and easily implemented on a spreadsheet. The treatment of the shroud moment here differs slightly from that given in Section 5.3.3, using a friction factor based on core velocity rather than a bulk boundary layer velocity. Nevertheless, the solution of equation 8.2.4 together with numerical integration of equation 5.1.14 from r_s (where the swirl fraction is 1) to lower radii gives good agreement with the results in Figure 5.13.

Solutions of equation 8.2.4 are given in Figure 8.10, where $c_{in} = v_{\phi,in}/\Omega b$ is the actual inlet swirl fraction and c_{eff} is an effective inlet swirl fraction defining the strength of the free vortex assumed in the mixing and source flow regions. For the given inlet swirl and λ, these give estimates of r_s/b. The integral solution for disc entrainment for inflow with a free vortex core, represented by

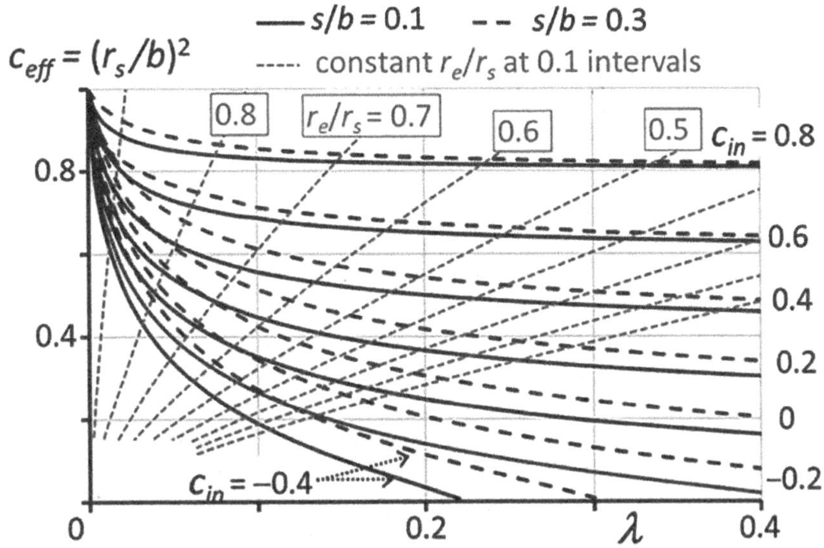

Figure 8.10 Integral solutions for r_s/b and r_e/r_s with deswirled radial inflow.

equation 5.2.16, then gives the equation below, allowing estimates of r_e/r_s

$$\lambda/2(r_s/b)^{2.6} = \left[0.59 - 0.714\frac{r_e}{r_s} - 0.2023\left(\frac{r_e}{r_s}\right)^2 + 0.3242\left(\frac{r_e}{r_s}\right)^3\right]. \tag{8.2.5}$$

Figure 8.10 includes plots of equation 8.2.5 for a range of values of r_e/r_s. The crossing points of these lines with solutions of 8.2.4 give the inner extent of the free vortex, source region. For example, with $s/b = 0.3$, an inlet swirl fraction of 0.4 and $\lambda = 0.2$, the graph gives $r_s/b = \sqrt{0.55} = 0.74$ and $r_e/r_s = 0.5$, and hence $r_e/b = 0.37$. Considering the configuration with $b_1 = r_s$ and $v_{\phi,in} = \Omega b_1$, as in Figure 8.7, selecting $\lambda_1 = \lambda(r_s/b)^{-2.6} = 0.44$ gives a solution corresponding to the region $r < r_s$ in the deswirled configuration. Results in Figure 8.8 for λ_1 confirm the value of r_e/r_s and give solutions for the core velocity at the inner radii.

As for radial outflow and for rotor–stator systems, the core velocity solutions with equation 2.1.15 can be used to estimate radial pressure distributions. Other dependent variables in the integral solution for the mixing and source zones, such as shear stress and boundary layer thickness, are available from the results given in Chapter 5. Equations 8.2.1 and 8.2.2 still apply for moment coefficients if all the flow is entrained into the disc boundary layers. The full boundary layer solutions for $r < r_e$ may be obtained from integration equation 5.1.14 from r_s. As shown in Figure 8.8, the linear Ekman layer solution, given in equation 5.3.13, is a useful approximation over a range of conditions. The validity of the solution is, of course, limited to cases with separate boundary layers and to turbulent flow. The disc moment in equation 8.2.4 assumes the cavity inner radius $a \le r_s$.

8.2.3 CFD and Heat Transfer

Early applications of axisymmetric CFD to radial inflow were reported by Morse (1988) using a tuned $k - \varepsilon$ model and Chew (1987) using a mixing length model, as reported for radial outflow in Section 8.1. These results supported the flow structure assumed above and showed some agreement with measurements and the integral solutions. Morse's results also show some sensitivity to turbulence modelling, particularly at lower radii where core swirl is highest. This may be associated with overprediction of turbulent diffusion in the strongly rotating core flow by some popular transport equation eddy viscosity models, as discussed in Section 6.2.5 and by Kumar *et al.* (2013). However, standard $k-\varepsilon$ model predictions show reasonable agreement with Farthing *et al.*'s measurements for a deswirled configuration, as shown in Figures 6.6 and 6.7 for axisymmetric and 3D CFD. Kumar *et al.* also reported reasonable agreement with heat transfer measurements from Farthing (1988) and Farthing *et al.* (1991a) for a case with an inlet swirl fraction of 1. More recent evaluations of CFD have also considered heat transfer and include LES.

Heat transfer measurements for radial inflow are given by Firouzian *et al.* (1986) for $v_{\phi,in}/\Omega b < 1$ and Farthing *et al.* for $v_{\phi,in}/\Omega b = 1$, with the results collated by Owen and Rogers (1995). Further experimental studies by Gunther *et al.* (2012) include cobbed discs and an axial throughflow between the disc bores and a central shaft. These results illustrate that radial inflow significantly increases heat transfer above that due to axial throughflow alone. The authors emphasise the need for measurements in multiple cavity rigs under conditions similar to engine compressors. Such data was provided by Atkins (2013) who introduced radial inflow to a multiple cavity rig previously used to study axial throughflow (as will be discussed later in this chapter). Focusing on disc temperature measurements, Atkins showed that introducing radial inflow significantly reduced the thermal response time. This indicated a potential method for improving compressor blade tip clearance control.

Evaluation of LES for radial inflow with an inlet swirl fraction of 1 was considered by Onori *et al.* (2016, 2019), comparing simulations with measurements from Firouzian *et al.* and Farthing. The 2016 paper showed that LES successfully predicted Firouzian *et al.*'s measured core velocity and pressures for $s/b = 0.13$, $Re_\phi = 3.45 \times 10^5$ and $C_W = 1300$. Some sensitivity of the disc boundary layer profiles to subgrid scale modelling was shown at this condition. The 2019 paper included heat transfer, comparing with Farthing's measurements for $s/b = 0.27$, $Re_\phi = 1.3 \times 10^6$ and $C_W = 3500$. The computational domain and results for local disc Nusselt number based on radius and disc-to-inlet temperature difference are shown in Figure 8.11. The inner part of the cavity includes vanes, as were fitted in the experiments to reduce vortex strength and limit pressure drop across the system.

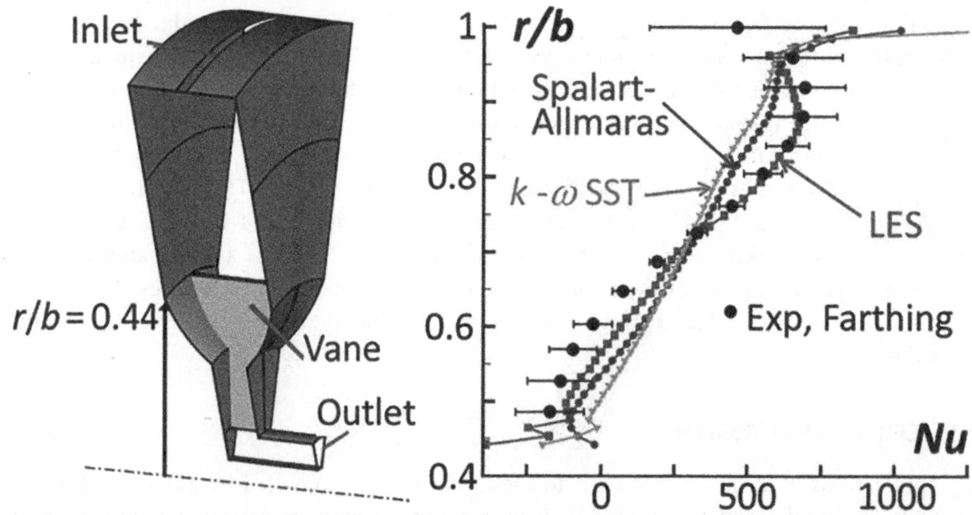

Figure 8.11 Solution domain and comparison of CFD with experimentally derived disc heat flux for radial inflow with $s/b = 0.27$, $Re_\phi = 1.3 \times 10^6$ and $C_W = 3500$. *Source:* Onori *et al.* (2019) / with permission of American Society of Mechanical Engineers (ASME).

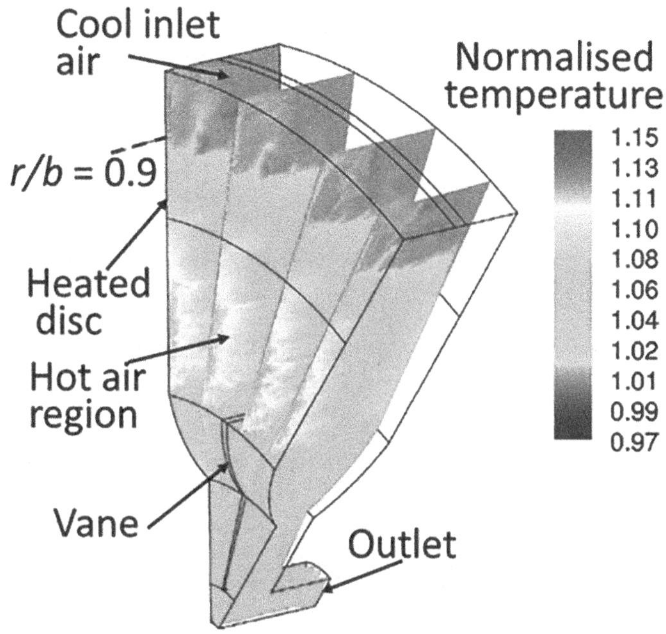

Figure 8.12 Instantaneous temperature contours from LES for radial inflow with $s/b = 0.27$, $Re_\phi = 1.3 \times 10^6$ and $C_W = 3500$. *Source:* Onori *et al.* (2019) / American Society of Mechanical Engineers (ASME).

Just one of the discs was heated, and the LES showed some effects of buoyancy, particularly in the outer part of the cavity. LES shows better agreement with heat transfer measurements than the two popular eddy viscosity models. This was attributed to the effects of turbulence modelling approximations and buoyancy. Some compromises were made on LES mesh quality in the inner part of the cavity, where some LES results fall outside the experimental error bands.

For the conditions in Figure 8.11, $\lambda = 0.045$ and Figure 8.8 indicates $r_e/b{\sim}0.9$. This agrees well with Onori *et al.*'s RANS models and LES. As shown in Figure 8.12, the LES gives a relatively uniform temperature (outside the boundary layers) in the source region $r_e < r < b$. Relatively strong mean radial temperature gradients occur at $r = r_e$. Inboard of this, radial gradients in the rotating core are limited, but an axial temperature gradient develops between the heated and cooler discs. This shows some resemblance to buoyancy-driven flow as illustrated in Figure 3.10 and discussed later in this chapter. As the temperature of the fluid in the heated disc Ekman layer increases, buoyancy changes the balance of forces and drives adjustment of mass flows in the layers on the two discs. This is associated with a weak axial flow across the cavity. A region of heated fluid emanating from the Ekman layer on the heated disc is evident in the figure.

The temperature contours in Figure 8.12 also show that fluid temperature changes significantly within the cavity. Thus, use of the Reynolds analogy with formulae from integral solutions requires estimates of local fluid temperatures. This might be achieved using the integral energy equation, as applied for radial outflow in Figure 8.5. Farthing *et al.* (1991a) found some good agreement of integral solutions with experimentally derived disc heat flux in the constant mass flow layers but not in the source region where they state that there were significant experimental uncertainties. Validation of integral solutions for heat transfer in radial inflow is weaker than that for radial outflow, with the source flow region being subject to uncertainty in both configurations.

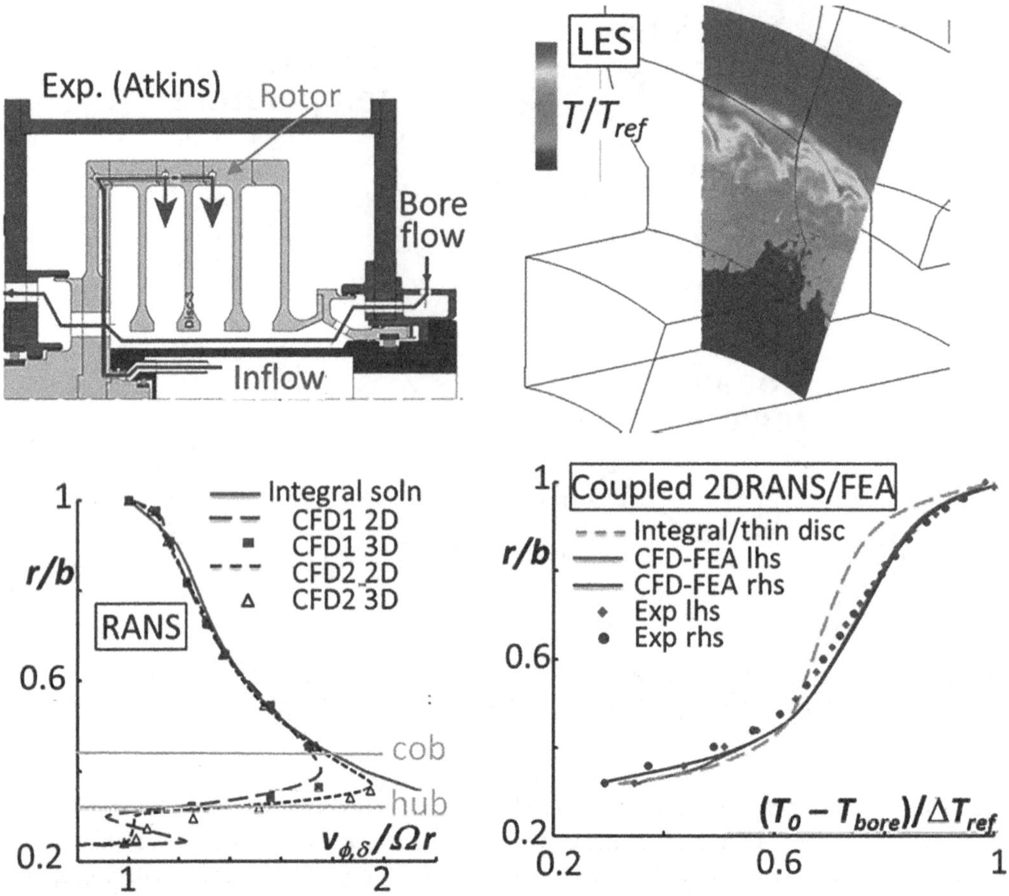

Figure 8.13 Atkins' (2013) experimental configuration, CFD results on a radial inflow cavity mid-axial plane and coupled fluid–solid solutions for disc temperature for radial inflow with $s/b = 0.195$, $Re_\phi = 8.4 \times 10^6$, $C_W = 6200$ and $Ro = 0.44$. *Source:* Sun *et al.* (2015) / with permission of American Society of Mechanical Engineers (ASME).

Sun *et al.* (2015) and Amirante *et al.* (2016) investigated Atkins' mixed axial throughflow/radial inflow configuration using CFD and comparing measured temperatures with predictions from CFD coupled with a finite element model of the discs. Atkins' experimental rig and Sun *et al.*'s results are illustrated in Figure 8.13. This includes temperature contours from LES, assuming circumferentially periodicity on a 22.5° sector, essentially axisymmetric RANS CFD solutions on a 0.2° sector, an integral solution and coupled solid–fluid calculations of disc temperature. Some RANS CFD results were obtained using alternative CFD codes (denoted CFD1 and CFD2). Core swirl velocity predictions are in reasonable agreement from all models, until the outlet is approached at lower radii. The integral solution takes no account of the outlet. The RANS models show sensitivity to turbulence modelling and differ distinctly from the LES in this region (which is not shown here). As reflected in temperature contours, the LES shows a sharp division from rotating cavity core to axial throughflow-dominated regions. Here, the axial throughflow is characterised by the Rossby number $Ro = v_{z,in}/\Omega a = 0.44$. It should though be noted that these simulations may be constrained by the imposition of circumferential periodicity over the sector domain. Amirante *et al.* applied RANS CFD to a 135° sector, showing that at lower inflow rates the cavity flow is destabilised by the thermal stratification.

The disc temperature predictions from coupled CFD-FEA in Figure 8.13 employed axisymmetric RANS CFD solutions over a domain including the two cavities, either side of the central disc, which had thermocouples on both sides. The results agree well with Atkins' measurements. Solutions of a thin disc heat conduction approximation with the integral method showed discrepancies with the measurements. Although less accurate, this very fast model could be used to further examine disc response times. The authors confirmed Atkins' conclusion that using a small radial bleed flow in compressor cavities is effective in reducing thermal response times for the compressor disks and that this could be applied in the management of compressor blade clearance. Amirante *et al.* extended the coupled CFD-FEA solutions to a lower inflow rate, $C_W = 3100$, finding departure from measured disc temperatures as transition from forced convection to buoyancy-driven convection occurred, and cavity flow interaction with the axial throughflow became dominant.

8.2.4 Non-Unique Solutions and Hysteresis

As shown in Figure 5.13, for a rotating cavity with inflow through deswirl nozzles in the rotating shroud, the same pressure difference across the cavity may occur for more than one radial inflow rate. This has important implications for internal air system design, as discussed by Pfitzner and Waschka (2000) and Negulescu and Pfitzner (2001). These papers consider two alternative designs to avoid excessive pressure drop when air is bled radial inwards through a compressor disc cavity. The designs were developed and evaluated for specific engine requirements using elementary modelling, CFD and experiments. Both designs, one using deswirl nozzles, and the other using radial aligned tubes, were reported to operate successfully in service. The deswirl system had advantages in terms of weight and mechanical design, but the flow characteristic exhibited hysteresis in some conditions. Tubed systems have been more widely used.

Transient deswirled system behaviour and hysteresis have been further considered by May *et al.* (2013). A transient, one-dimensional, elementary model of the cavity flow was developed and incorporated within a system model that could include the deswirl nozzles, valves and other features. The model uses a mixed cavity flow assumption for a perfect gas, extending and generalising models previously used for steady and transient flow in rotor–stator cavities. An important feature of the model is the assumption of a vortex in the cavity with $v_\phi \propto r^a$, where a is in the range −1 to 1 and may be specified as a function of other parameters. With the vortex index a specified as a function of λ, the model reproduced Farthing *et al.*'s (1991b) measurements well, with some differences at high values λ, where the cavity flow may rotate in the opposite direction to the discs.

Results from the transient model were compared with measurements from the industrial compressor offtake rig shown in Figure 8.14, alongside example steady operating characteristics and a transient solution. The rig was operated with a nominally constant annulus pressure p_{ann} by opening and closing a downstream valve with fixed backpressure. Some tuning of the model (from values for a simple cavity geometry) to measured steady operating conditions was undertaken. It was suggested that this could be done using CFD for cases where no experiments are available. Tuning at one rotor speed gave a good match to the data at other speeds.

The calculated steady flow characteristics in the bottom graph of Figure 8.14 show the flow rate coefficients for the cavity and the downstream valve as a function of cavity pressure ratio at fixed rotor speed, p_{ann}, and valve backpressure. The valve characteristics vary with opening position, and crossing points with the cavity characteristic give potential solutions for steady flow through the system. These agree well with measurements, considering that differences at low flow rates are likely to be within uncertainty bounds. For pressure ratios between ~1.5 and 2, the cavity characteristic gives three possible flow rates for a given pressure ratio. However, despite careful adjustment of the valve, only the highest and lowest possible steady flow rates were achieved in

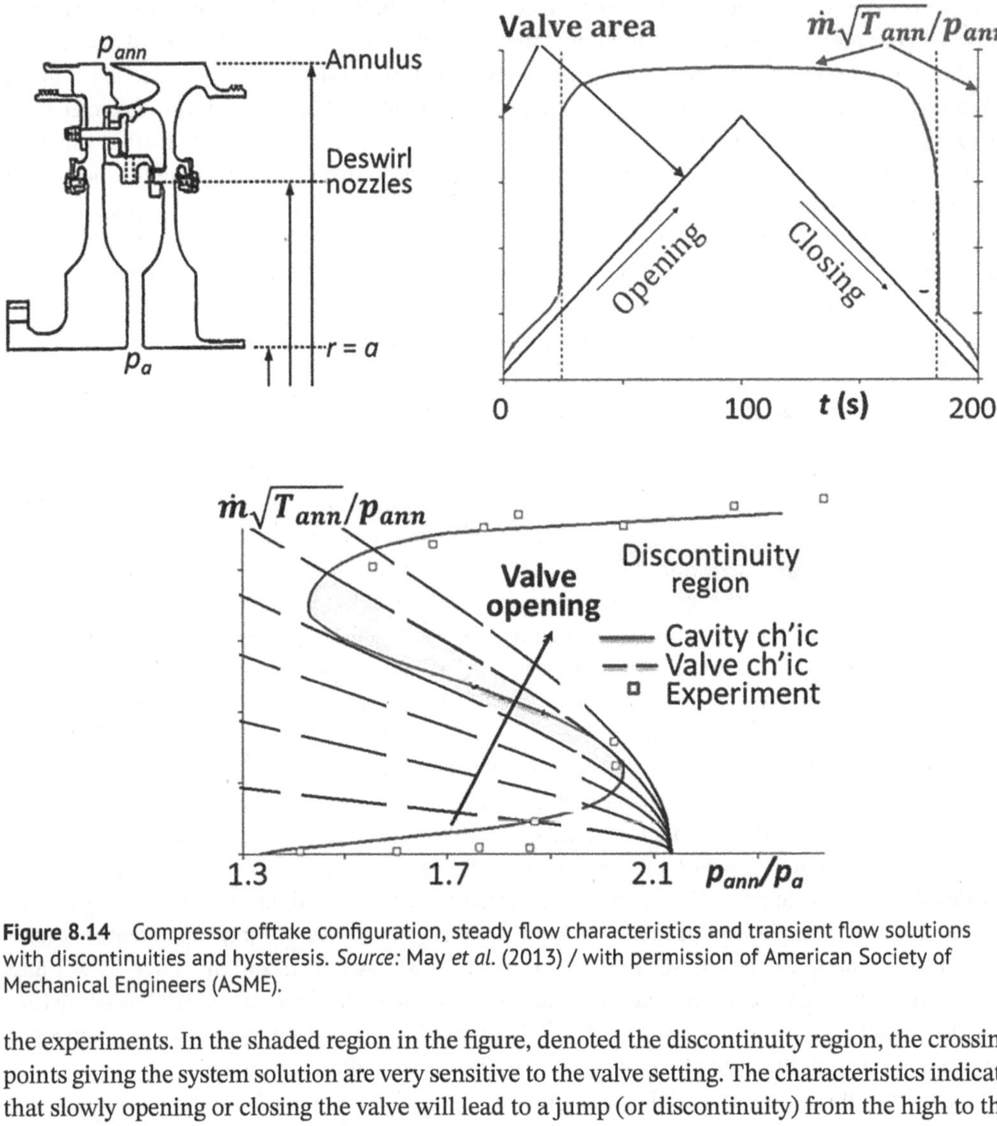

Figure 8.14 Compressor offtake configuration, steady flow characteristics and transient flow solutions with discontinuities and hysteresis. *Source:* May *et al.* (2013) / with permission of American Society of Mechanical Engineers (ASME).

the experiments. In the shaded region in the figure, denoted the discontinuity region, the crossing points giving the system solution are very sensitive to the valve setting. The characteristics indicate that slowly opening or closing the valve will lead to a jump (or discontinuity) from the high to the low flow rate or vice versa.

The transient solution confirms the behaviour suggested by the steady flow characteristics. As the valve opens, the system flow rate follows the low mass flow section of the cavity characteristic until a jump to the high flow section occurs. The jump back to a low flow rate when the valve is closed occurs at a slightly lower valve setting than the jump to the high flow rate (and at a significantly lower pressure ratio). For valve settings between these values, the steady solution depends on how the operating condition is approached. Replacing the downstream valve with a pump that fixes the flow rate would allow the full cavity flow characteristic to be explored. The valve characteristics in Figure 8.14 would be replaced by horizontal lines representing the pump settings and there would be no discontinuity region. Farthing *et al.*'s rig was operated in this way, and the full S-curve cavity characteristic was obtained.

Further studies have considered other methods of allowing radial inflow while avoiding large pressure drops. These include the use of radial fins attached to one of the discs by Chew *et al.* (1989), and a hybrid system using deswirl and short tubes by Wei *et al.* (2020). Shen *et al.* (2022) have applied LES to a deswirl system and further considered elementary modelling.

8.3 Buoyant Flow in Closed Cavities

Closed or very weakly ventilated cavities are found in some engines and are of intrinsic interest for the fundamental study of centrifugally buoyant flows. For a fixed closed cavity geometry, nondimensional flow and heat transfer are expected to depend on the surface temperature distribution and, if the Boussinesq approximation applies, four independent nondimensionals. A suitable choice may be Prandtl number Pr, buoyancy parameter $\beta\Delta T_{ref}$, rotational Reynolds number Re_ϕ and Eckert number Ec. A Rayleigh number $Ra = Pr\beta\Delta T_{ref}Re_\phi^2$ is also used and, in radial convection, is often effective in combining the main effects of changes in $\beta\Delta T_{ref}$ and Re_ϕ. Some important effects of Ec can be captured by the use of the rotary stagnation temperature in place of static temperature, as in equation 2.5.15 and discussed in Section 2.5.1.

8.3.1 Axial Convection

Laminar axial convection in a closed cavity with discs at different temperatures is illustrated in Figure 3.10 and briefly discussed in Section 3.3.3. The analytical, infinite discs solution shown in the figure matches Ekman layers on the two discs to a rotating inviscid core in which equation 3.3.10 applies. Assuming the Ekman layers are thin compared to the cavity width, the first-order solution gives the circumferential velocity at the two Ekman layer boundaries as $u_{\theta,\delta}/\Omega\, r = \pm\, \beta\Delta T/4$, where $\Delta T = T_{z=s} - T_{z=0}$ is the temperature difference across the cavity and β is the thermal expansion coefficient. For $\Delta T > 0$, the positive value of $u_{\theta,\delta}$ corresponds to the hot disc at $z = s$, giving inflow according to equation 3.3.21, and balancing the outflow on the cold disc. The core axial velocity may be estimated from the rate of change with radius of the Ekman layer mass flows. For uniform disc temperatures, this gives

$$u_z/r\Omega = -\beta\Delta T/4\sqrt{r^2\Omega/\nu}. \tag{8.3.1}$$

Substituting into the second of equation 3.3.10 and estimating disc heat transfer on each disc from the near-wall temperature gradient gives

$$\frac{T - T_{z=0}}{\Delta T} = \frac{e^{-\Phi z/s} - 1}{e^{-\Phi} - 1}, \quad \frac{sq_{z=0}}{\kappa\Delta T} = \frac{-\Phi}{1 - e^{-\Phi}}, \quad \frac{sq_{z=s}}{\kappa\Delta T} = \frac{+\Phi}{1 - e^{+\Phi}} \quad \text{with}$$

$$\Phi = \frac{Pr\beta\Delta T}{4}\sqrt{\frac{s^2\Omega}{\nu}}. \tag{8.3.2}$$

The infinite disc solution for heat transfer on the cold disc is compared with Hudson *et al.*'s (1978) measurements for an oil-filled rotating cavity in Figure 8.15. The flow structure in this case is similar to that in Figure 3.10. As shown by the isotherms in Figure 3.10, heat transfer on the cold disc is strongest at lower radii, where the infinite disc solution applies and reduces towards the shroud. In contrast, the overall heat transfer on the hot disc is dominated by the impingement of the recirculating flow at outer radii. At lower radii, the axial temperature gradient near the hot disc (and hence disc heat flux) is very weak, as indicated by the analysis above. The total heat transfer between the two discs (each at a different uniform temperature) with an insulated shroud was measured in the experiment, and this falls a little below that estimated from equation 8.3.2 over the whole of the cold disc.

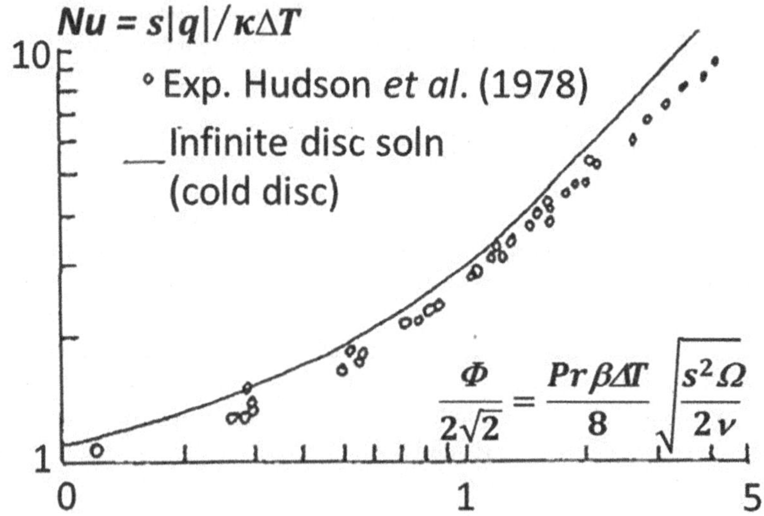

Figure 8.15 Axial convection in an oil-filled rotating cavity, $Pr{\sim}7.2$, $Re_\phi = 1.7 \times 10^6 - 2.7 \times 10^6$, $\beta\Delta T < 0.05$, $a/b = 0$. Measurements are for $s/b = 0.069$ and 0.014.

The infinite disc solution above can be extended to discs with radial temperature gradients by modifying equation 8.3.1 to include the effect of non-zero $d(\Delta T)/dr$ on the mass flow in the Ekman layers. Some CFD and analytical results for such cases, and for combined radial outflow/axial convection, are given by Chew (1985b). However, for positive radial temperature gradients, the stability of the core flow solution, equation 3.3.10, comes into question. A more recent study by Wang *et al.* (2024b) applied LES to an air-filled cavity with higher radius ratio ($a/b = 0.52$) and aspect ratio ($s/b = 0.5$) than the earlier studies. Uniform disc temperatures were assumed with adiabatic inner and outer cylinders. Two conditions were considered: a lower $Re_\phi = 3.1 \times 10^4$ with $\beta\Delta T = 0.17$, and a higher $Re_\phi = 4.1 \times 10^5$ with $\beta\Delta T = 0.1$. A full 360° annulus was modelled. At the lower Re_ϕ, the solution was axisymmetric with no signs of turbulence. The agreement with the infinite disc solution was not as close as for the earlier studies of low a/b configurations with effects of the inner and outer cylinders more prominent.

Results for Wang *et al.*'s higher Re_ϕ condition are shown in Figure 8.16. The instantaneous temperature contours on a plane of constant θ show that an axial temperature gradient occurs across the core as expected, but with evidence of some unsteady flow or weak turbulence. The temperature is higher at low radii because the recirculating flow is heated when it flows inward along the heated disc. There is significant impingement of the recirculating flow at lower radii on the cooled disc as well as at higher radii on the heated disc. The Ekman layers are expected to remain laminar, as the maximum boundary layer Reynolds number estimated using the infinite disc theory ($Re_\delta = \beta\Delta T Re_\phi^{1/2}/4 = 16$) is well below the critical value of 198 for absolute instability given in Table 4.1. The mean Nusselt number distributions on the disc reflect the flow structure with the LES approaching the infinite disc solution away from cylindrical walls. Measurements of mean heat flux for this configuration were reported by Bohn *et al.* (1996). Averaged Nusselt numbers based on total heat transfer across the cavity from the LES are 10.5% lower at the $Re_\phi = 3.1 \times 10^4$ condition and 34.7% higher at the $Re_\phi = 4.1 \times 10^5$ condition than the experimental correlation given by Bohn *et al.* Considering the low level of heat transfer and uncertainties in the radial convection results from the same test rig, as noted in a correction to the correlation by Bohn and Gier (1998). this level of difference was not considered surprising.

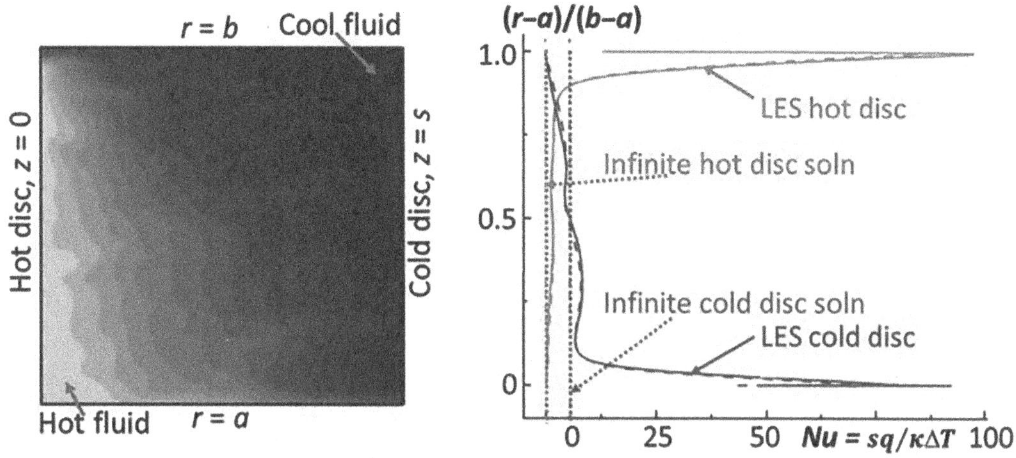

Figure 8.16 Instantaneous temperature contours and mean disc heat transfer from LES for axial convection in an air-filled rotating cavity, $Pr = 0.7$, $Re_\phi = 4.1 \times 10^5$, $\beta\Delta T = 0.1$, $Ra = 1.2 \times 10^{10}$, $a/b = 0.52$ and $s/b = 0.5$. *Source:* Wang *et al.* (2024b) / with permission of American Society of Mechanical Engineers (ASME).

8.3.2 Radial Convection

The development of centrifugal Rayleigh–Bénard convection in the annulus between a cold inner cylinder and a hot outer cylinder is shown in Figure 4.7. Further data from DNS and LES in Figures 6.14 and 6.15 show that while the flow in the main part of the cavity becomes turbulent at moderate Rayleigh numbers, the Ekman layers on the discs remain laminar. These results and much of the previous work in this area are based on a simple annulus geometry with $a/b = 0.52$ and $s/b = 0.5$, as in Bohn *et al.*'s (1995, 1996) experimental studies. The uncertainty regarding experimental accuracy, acknowledged by Bohn and Gier (1998), and constraints on CFD limited progress in earlier research. More recent CFD studies have applied DNS and LES and, with support from new experimental data, a clearer and consistent picture has emerged.

Figure 8.17 shows instantaneous temperature and velocity contours for radial convection in a cavity with the outer cylindrical shroud at a higher, uniform temperature than the inner cylindrical shaft, and adiabatic discs. These are representative of the flow in most of the cavity as effects of the discs are largely confined to the thin Ekman layers. The radial and circumferential velocity components clearly indicate four pairs of counter-rotating, cyclonic/anticyclonic vortices. These and traces of smaller eddies and turbulence are also evident in the temperature contours. Typically, the flow becomes more chaotic as *Ra* increases and the number of vortex pairs may vary and will depend on cavity radius ratio. Time-averaged results show an isothermal core separating boundary layers on the shroud and shaft, as illustrated in Figure 8.18. Note the use of rotary stagnation temperature T^*. At the highest *Ra* considered, this adjusts for Eckert number effects, which give rise to a radial gradient of static temperature in the core. Other results show slightly more variation of core temperature with conditions. This reflects sensitivities to modelling and numerical assumptions. Neglecting the effects of the discs Gao *et al.* (2019) found heat transfer predictions from the Boussinesq and a full compressible model to agree within 10% for $\beta\Delta T \leq 0.2$. The Boussinesq model showed no significant effect of $\beta\Delta T$ for fixed *Ra*. The compressible flow model showed overall Nusselt number dropping by ~10% as $\beta\Delta T$ increased from 0.2 to 0.4. Other studies by Gao and Chew (2021a,b) have examined the effects of the disc

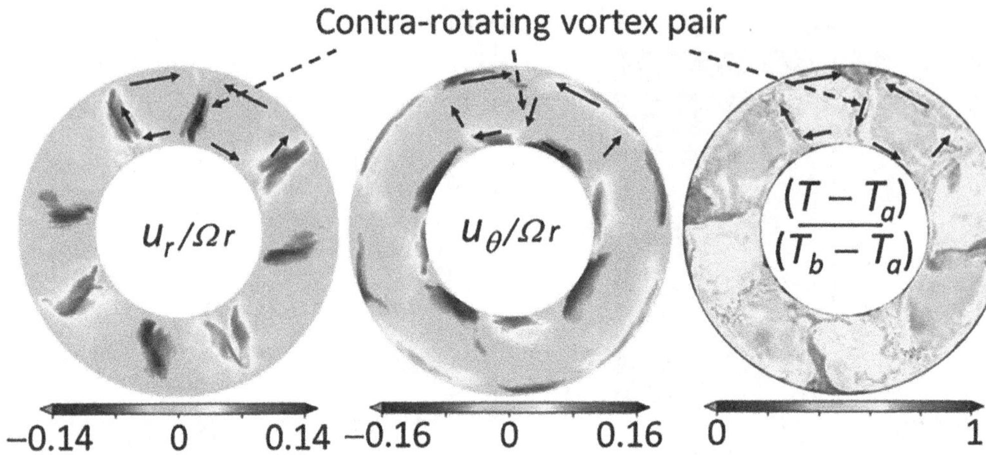

Figure 8.17 Instantaneous velocity and temperature contours on the mid-axial plane from LES for radial convection in an air-filled rotating cavity, $Pr = 0.7$, $Re_\phi = 4.1 \times 10^5$, $\beta \Delta T = 0.1$, $Ra = 1.2 \times 10^{10}$, $a/b = 0.52$, $s/b = 0.5$. *Source:* Gao *et al.* (2020) / with permission of Elsevier. Rotation is anticlockwise.

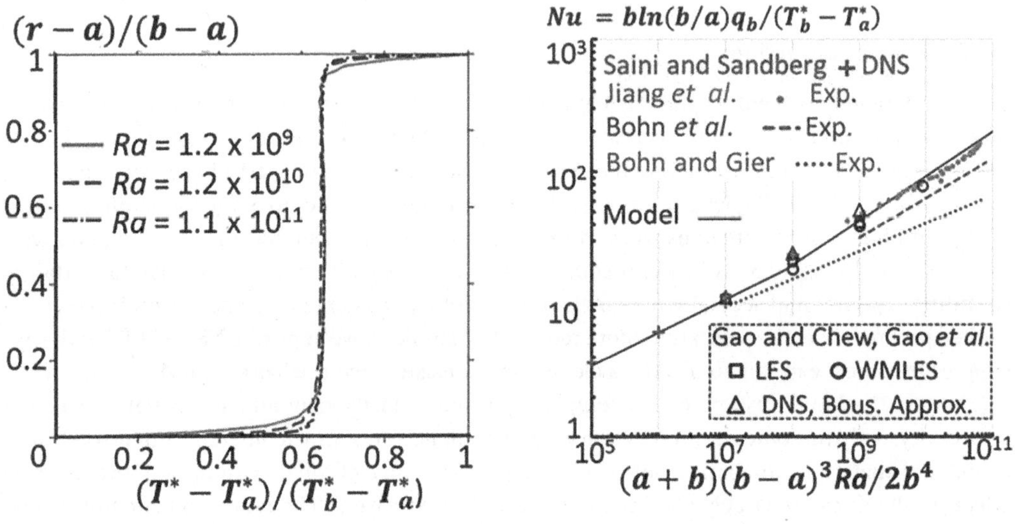

Figure 8.18 Mean core temperature and overall heat transfer from LES for radial convection in an air-filled rotating cavity, $a/b = 0.52, s/b = 0.5$. *Source:* Updated from Gao and Chew (2021a,b).

boundary layers and near-wall modelling in LES. These include DNS, wall resolved LES and wall modelled LES.

The overall heat transfer results in Figure 8.18 show broad consistency in results from a range of sources. These include DNS, wall resolved LES and wall modelled LES from the University of Surrey, DNS from Saini and Sandberg (2020) at the University of Melbourne, experimental correlations from the University of Aachen and measurements from Tsinghua University. Jiang *et al.*'s (2020) experiments are for a water-filled cavity with $a/b = 0.5$, $s/b = 0.5$ and Pr stated to be ~4.3. According to studies of gravitational Rayleigh–Bénard convection, such as that by Hollands *et al.* (1975), this degree of variation in Pr only has a weak effect on Nusselt number at high Ra and so a direct comparison with the other results for air-filled cavities is justified. The axes on the graph correspond to

Nusselt and Rayleigh number definitions given by Bohn *et al.* and adopted in other closed cavity studies. The labelling here follows the present notation. Only the WMLES was applied at the highest Rayleigh number considered with CFD. This reflects its much higher computational efficiency, as will be discussed further for buoyant flows with axial throughflow. At this condition, the Eckert number $Ec = 0.19$ and kinetic energy effects are becoming significant. Both Ra and Nu are based on the rotary stagnation temperature difference in the figure. The difference between Bohn *et al.*'s experimental correlation and the correction to account for heat conduction in the discs by Bohn and Gier indicates the experimental uncertainty. It is encouraging that the correction approaches the other results at high Rayleigh number.

The elementary model for overall heat transfer in Figure 8.18 assumes a uniform core rotary stagnation temperature with heat transfer across the boundary layers on the shroud or shaft estimated from correlations adapted from natural convection under gravity. For adiabatic discs, the energy balance for the closed cavity simply requires that (for statistically steady flow) the mean heat loss by convection from the shroud equals the corresponding heat transfer to the shaft. Adapting Lloyd and Moran's (1974) established correlations obtained from experiments using electrochemical techniques to the present problem

$$\frac{|q_0|s}{2\kappa\left|T^*_{0,r} - T^*_{core}\right|} = 0.54A_1 Ra_{LM}^{1/4} \qquad \text{if } 2.2\times10^4 \leq Ra_{LM} \leq 8\times10^6$$

$$0.15A_2 Ra_{LM}^{1/3} \qquad \text{if } 8\times10^6 \leq Ra_{LM} \leq 1.6\times10^9$$

$$\text{with} \quad Ra_{LM} = Pr\beta\left|T^*_{0,r} - T^*_{core}\right|\left(\frac{\Omega r^2}{\nu}\right)^2\left(\frac{s}{2r}\right)^3 \quad \text{and} \quad r = a \quad \text{or} \quad b. \tag{8.3.3}$$

Here, T^*_{core} is the core temperature and r can be either a or b with the surface temperature $T^*_{0,r}$ chosen appropriately. A_1 and A_2 are scaling factors, introduced to allow matching to data. $A_1 = A_2 = 1$ gives the unscaled correlations. Noting that the energy balance requires $aq_{0.a} = bq_{0.b}$, inspection of the above equations shows that this gives $\left(T^*_{core} - T^*_{0.a}\right) / \left(T^*_{0,b} - T^*_{core}\right) = b/a$. Equivalently, $\left(T^*_{core} - T^*_{0.a}\right) / \left(T^*_{0,b} - T^*_{0,a}\right) = (1 + a/b)^{-1}$ which for $a/b = 0.52$ gives a nondimensional core temperaturer of 0.658. This is reasonably close to the LES values in Figure 8.18. The overall Nusselt number shown by the model in Figure 8.18 was calculated from the above taking the maximum heat transfer from the high and low Ra correlations. This avoids discontinuity in values but shifts the switch between correlations slightly from that given above. The model also extrapolates to higher Ra than Lloyd and Moran's experiments assuming that the scaling $Nu \propto Ra^{1/3}$ is a limiting condition. The scaling of Nu with Ra is consistent between the model and the data. Selecting $A_1 = A_2 = 0.8$ gives a reasonable match to the data values, as shown. As noted by Bohn *et al.* Ra and Re_ϕ were not independently varied in their experiments (or the CFD modelling of these experiments). However, agreement with the model and Jiang *et al.*'s data indicates that Nu is principally dependent on Ra, which combines the effects of Re_ϕ and $\beta\Delta T$.

Bohn *et al.*'s experimental conditions are considered further in Figure 8.19, adapted to the present notation from Gao and Chew (2021a). The operating line shown assumes the cavity to be at atmospheric pressure (at lower Ra) or operating at the maximum rig speed of 3500 rpm (at higher Ra). Assuming the rig pressure never fell below 1 atm, this gives the upper limit of the Eckert number for the rig at each value of $\beta\Delta T$. The figure indicates that exchange of kinetic and internal energy is significant for $Ec > 0.2$, at which point the rotary stagnation temperature difference across the cavity is 10% less than the static temperature difference. The highest Ra CFD condition ($Ra = 1.1 \times 10^{11}$) in Figure 8.18 falls just below this line, and Gao and Chew's (2021b) LES shows a distinct positive radial gradient in mean static temperature for this case.

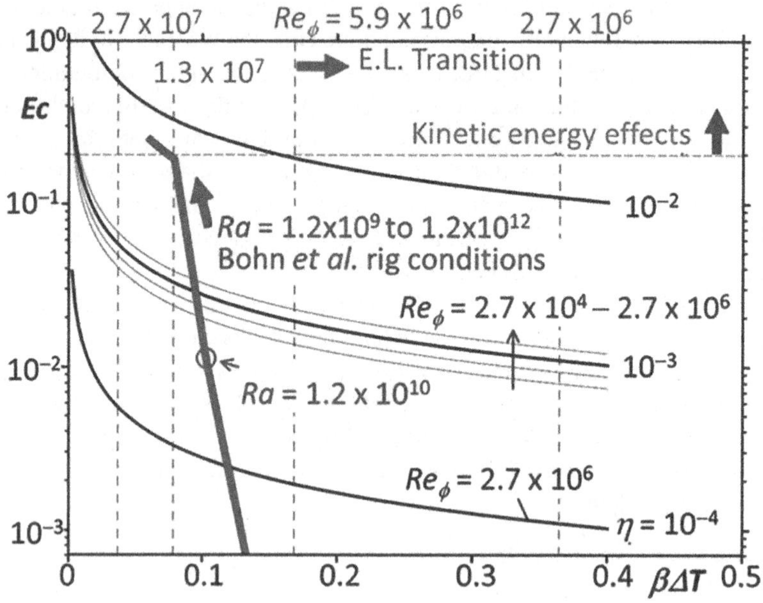

Figure 8.19 Bohn *et al.*'s operating conditions for radial convection in an air-filled rotating cavity, $a/b = 0.52$, $s/b = 0.5$, showing criteria for effects of kinetic energy exchange, Ekman layer transition and viscous heating. *Source:* Gao and Chew (2021a) / American Society of Mechanical Engineers (ASME).

The vertical dashed lines (and associated labels) in Figure 8.19 give combinations of Re_ϕ and $\beta\Delta T$ at which the transition from laminar to turbulent Ekman layers is estimated to occur. These assume representative core velocity fluctuations of $0.2\sqrt{\beta\Delta T}\Omega b$ as deduced from simulations, and a critical boundary layer Reynolds number $Re_\delta \sim 200$ as in Table 4.1. At conditions to the right of these lines, the Ekman layers are expected to be intermittently turbulent for the Re_ϕ shown. The layers will undergo local transition to turbulence and re-laminarisation as the relative core velocity changes. The maximum value of Re_ϕ in the experiments is about 5.5×10^6, so the Ekman layers are expected to remain laminar at these conditions. However, higher values of $\beta\Delta T$ and Re_ϕ occur in engines. This is discussed further for rotating cavities with axial throughflow for which there is considerable interest in Ekman layer transition.

A further effect identified by Gao *et al.* is that frictional dissipation in the Ekman layers led to noticeable temperature rises in the core near the discs. This effect is shown in Figure 8.19 by lines of constant η for $Re_\phi = 10^6$ and was called Ekman layer scrubbing. Here, η is the ratio of internal energy generation by friction from both discs to the shroud total heat transfer. For $\eta = 0.001$, a range of Re_ϕ is considered. These results were estimated from the laminar Ekman layer solution given in Section 3.3.2 with core velocity fluctuations of $0.2\sqrt{\beta\Delta T}\Omega b$, as used above. Although values of η are quite small, the mean relative velocity is zero and so the heat generated can accumulate until temperature gradients are sufficient for diffusion to balance dissipation.

As shown in Bohn *et al.*'s experiment, it is difficult to avoid some disc heat transfer in practice. This is further illustrated in experiments by Jackson *et al.* (2022a) and is important in determining disc temperature distributions. Figure 8.20 shows Jackson *et al.*'s cavity geometry, typical disc temperature measurements and temperature distributions from a thermal model given by Liu (2024). From fast response pressure measurements, Jackson *et al.* deduced that three or four vortex pairs occurred in the cavity, depending on Ra, and these structures rotated within 1% of cavity speed.

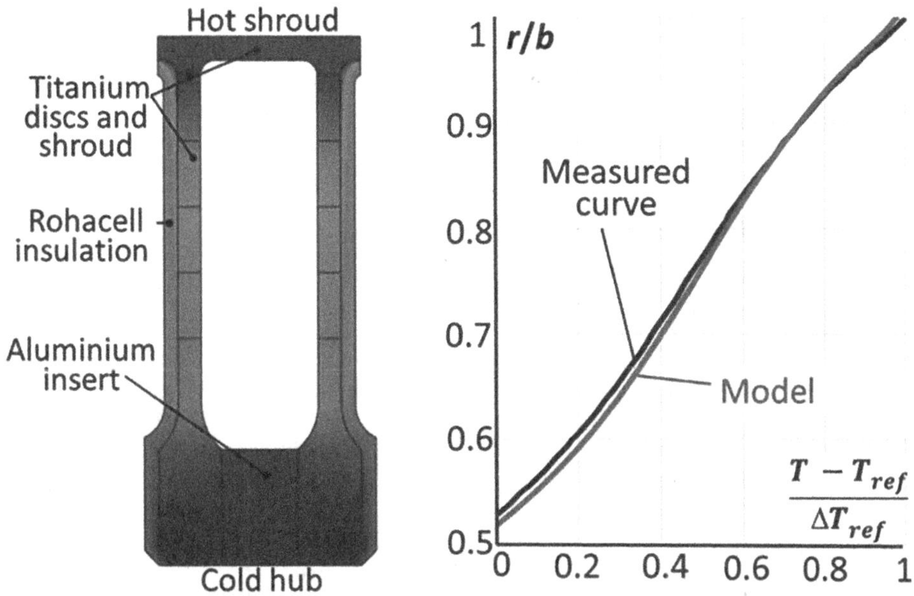

Figure 8.20 Jackson *et al.*'s closed, air-filled rotating cavity, $a/b = 0.45$, $s/b = 0.17$, with typical temperature contours from Liu's thermal model and comparison of disc temperatures for $Re_\phi = 7 \times 10^5$, $\beta \Delta T = 0.125$ and $Ra = 4.3 \times 10^{10}$.

They also found that nondimensional radial disc temperature distributions over the experimental range all fell close to a single curve. The thermal model applies a commercial finite element solver to the axisymmetric geometry shown with specified temperature on the outer shroud boundary, and convective boundary conditions elsewhere. At the hub, a standard duct flow *Nu* correlation is used with a measured throughflow air temperature to model the experimental condition. An isothermal core flow exchanging heat with interior surfaces is assumed in the cavity, such that the net interior heat exchange is zero. Convection on the cylindrical boundaries to the cavity is modelled using correlations similar to equation 8.3.3. The disc heat transfer assumes conduction from the core temperature to the disc across a laminar Ekman layer giving $q_0 = \kappa(T_0 - T_{core})\sqrt{\Omega/\nu}/\pi$. This assumption is discussed and justified further later in this chapter. A similar convective condition was applied to the Rohacell insulation attached to the outer disc faces. These form boundaries to two other (cooler) closed cavities in the experimental rig for which plausible core flow temperatures were selected. Internal radiative heat transfer between cavity surfaces was also included in the model.

The comparison in Figure 8.20 shows that the thermal model reproduces the measured disc temperature distributions well. Sensitivity studies for the convective boundary conditions found that inclusion of heat transfer through the 5-mm layer of Rohacell insulation enabled a good match to the disc temperature distribution. As shown by the temperature contours, the Rohacell is effective in restricting heat transfer, but the heat loss was still found significant compared to heat transfer on the internal disc face owing to the larger temperature difference. This is consistent with more basic analysis. The thermal resistance of a 5-mm Rohacell layer is around $0.17\ \text{K. m}^2\text{. W}^{-1}$, which is about $180Re_\phi^{-1/2}$ times the value for conduction through air across an Ekman layer for the rig with $b = 0.24$ m. Considering possible air-to-temperature differences in the rig supports the conclusion from Liu's model, perhaps differing with Nicholas *et al.*'s (2025) report that their calculated disc temperature for the same rig was 'relatively insensitive' to the upstream and downstream cavity temperatures. Some other assumptions in Nicolas *et al.*'s model also differ from Liu's. The comparison illustrates the inherent uncertainty in deducing surface heat fluxes from temperature measurements, and the difficulty in specifying convective boundary conditions in engine or test rig thermal modelling. With low levels of disc heat transfer, the air-filled closed configuration is particularly challenging.

8.3.3 Mixed Axial/Radial Convection

The combined effect of temperature difference between the two discs and between the shroud and shaft has been considered in LES by Wang *et al.* (2024b). The results in Figure 8.21 may be compared to those for axial convection in Figure 8.16 and radial convection in Figure 8.17. The solution, with the shroud and left-hand disc at a higher uniform temperature than the opposite walls, displays features seen separately for axial and radial convection. An axial temperature gradient occurs and is associated with a weak mean axial flow from the hot to the cold disc. Temperature contours on the mid-axial plane show plumes similar to those for radial convection. These account for reduced

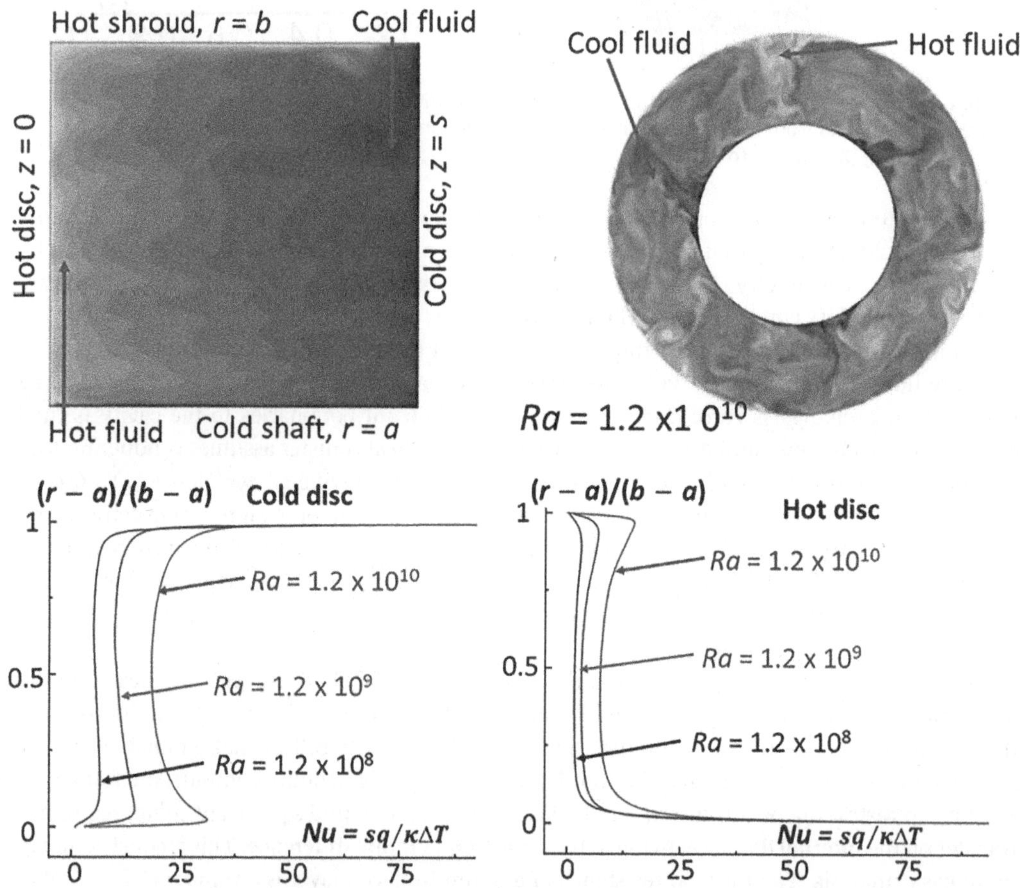

Figure 8.21 Instantaneous temperature contours on azimuthal and mid-axial planes, and mean disc heat transfer from LES for mixed axial/radial convection in an air-filled rotating cavity, $Pr = 0.7, a/b = 0.52, s/b = 0.5$. *Source:* Wang et al. (2024b) / with permission of American Society of Mechanical Engineers (ASME).

radial temperature gradients compared to the axial convection case. Overall, shroud heat transfer was found to increase by around 20% compared to the radial convection cases. Shaft overall heat transfer was less affected. As shown in the figure, disc heat transfer was increased considerably above the low levels found for axial convection. However, the axial flow transferring fluid from the hot disc to the cold disc continues to enhance heat transfer on the cold disc compared to the hot disc.

A further simulation reported by Wang (2024) uses a different surface temperature distribution, with T_0 varying linearly from its lowest value at a corner on the shaft to a mid-value at the adjacent corners and to a high value at the opposite corner. There are significant changes to the heat transfer with changes in direction of the heat flux on both discs. This sensitivity to surface temperature distribution makes estimation of disc heat transfer particularly difficult without recourse to LES. As for radial convection, and axial throughflow discussed below, WMLES could be useful in allowing computations at higher Ra.

8.4 Rotating Cavities with Axial Throughflow

Multistage axial compressors, such as the IPC and HPC in Figure 1.1, often have rotating cavities with axial throughflow of air between the disc bores and an inner shaft. The interaction of the throughflow with the buoyancy-driven flow in the disc cavities can be complex and has been subject to considerable research. In their review of previous work, Owen and Long (2015) describe the progress made and note the potential of LES for future work. More recent studies have exploited this potential, combining with experimental research to give better understanding and predictive capability for the flow and heat transfer. The emphasis here is on the more recent studies.

For a fixed geometry and assuming the Boussinesq approximation applies, nondimensional flow and heat transfer are expected to depend on the surface temperature distribution and seven independent nondimensionals. A possible choice of independent parameters is Prandtl number Pr, buoyancy parameter $\beta \Delta T_{ref}$, rotational Reynold number Re_ϕ, Eckert number Ec, Rossby number $Ro = v_{z,in}/\Omega a$, inlet swirl fraction $v_{\phi,in}/\Omega a$ and shaft to disc speed ratio Ω_{shaft}/Ω. The last three parameters extend those used for the closed cavity in Section 4.3 and neglect the possible effect of nonuniform inlet profiles for the axial throughflow. Here the fluid inlet temperature replaces the shaft temperature for the closed cavity. As for the closed cavity, Rayleigh number $Ra = Pr\beta \Delta T_{ref} Re_\phi^2$ is also used, and some important effects of Ec can be captured by the use of the rotary stagnation temperature in place of static temperature. In addition, other parameters have been used, attempting to combine effects from Ro and $\beta \Delta T$. A buoyancy number, defined here as $Bo = Ro/\sqrt{\beta \Delta T}$, and representing the ratio of throughflow to buoyancy velocity scales, is used later in this chapter.

8.4.1 Flow Structure

The flow structure summarised in Figure 8.22 was deduced from LES by Gao and Chew (2022) of the research rig described by Jackson *et al.* (2021) for $3.2 \times 10^5 \leq Re_\phi \leq 2.2 \times 10^6$, $0.4 \leq Ro \leq 0.8$, $0.11 \leq \beta \Delta T \leq 0.25$ and $0.12 \leq Ec \leq 0.312$. It is consistent with Jackson *et al.*'s (2021, 2022b) measurements and earlier experimental evidence. A dominant large-scale flow structure rotates at a different speed than the rotor, as identified through flow visualisation in Farthing *et al.*'s (1992) experiments and confirmed from fast response pressure measurements by Jackson *et al.* (2022b). The LES showed the main flow to be turbulent with highly unsteady laminar Ekman layers on the discs responding to the passing flow structure, and Rayleigh–Bénard convection cells forming

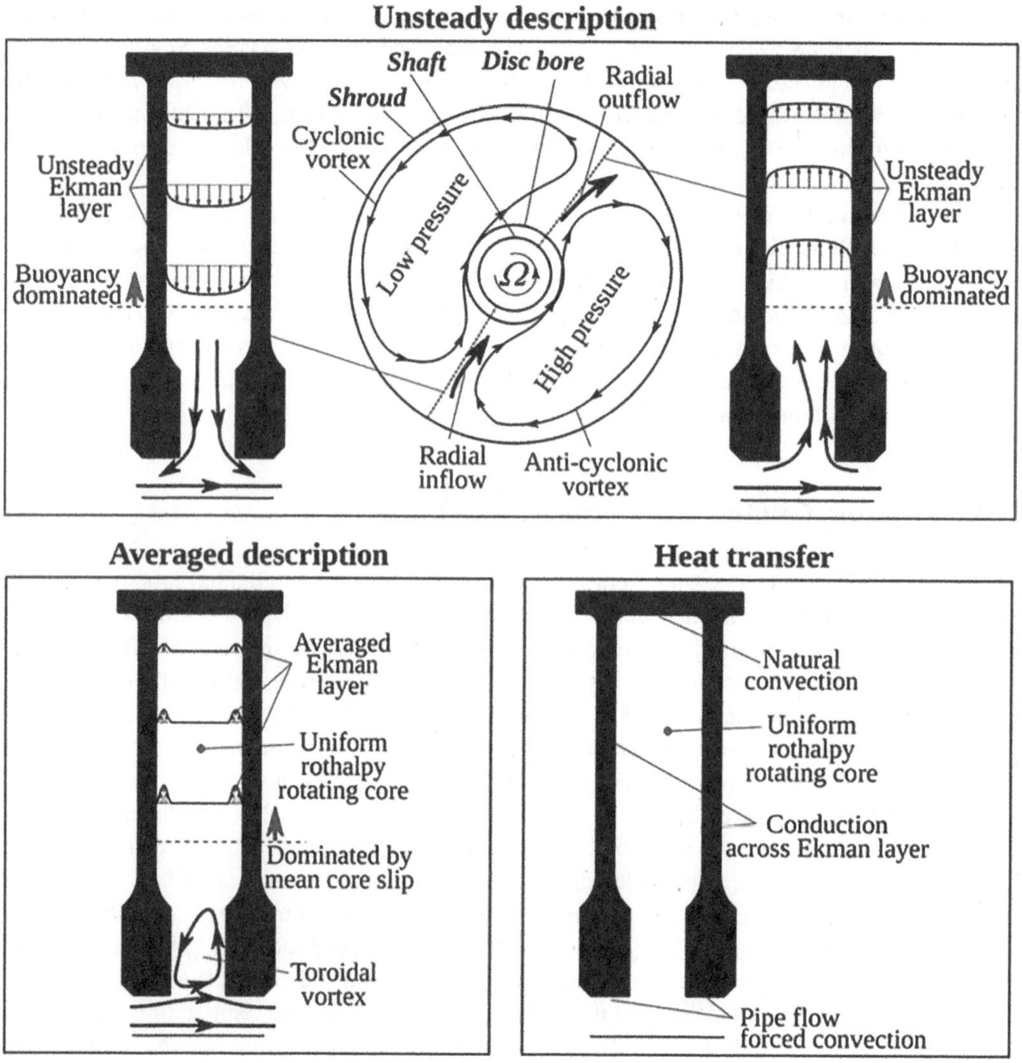

Figure 8.22 Summary of flow and heat transfer mechanisms in a rotating cavity with axial throughflow. Gao and Chew (2022) / American Society of Mechanical Engineers (ASME).

near the shroud. Mass and energy exchange between the cavity and throughflow is dominated by the radial outflow and inflow plumes. A second radial outflow plume emerged intermittently in the simulation. The mean circumferential velocity in the cavity core flow is lower than disc speed, as also shown in earlier measurements and LES in Figure 6.13. In this case, the slip is greater than in the previous study due to lower inlet swirl. The negative slip drives a very small mean radial outflow in the Ekman layers. The time-averaged flow in the cavity shows a uniform mean rothalpy (or rotary stagnation temperature) core with zero mean radial velocity in the outer part of the cavity. Near the disc bores, the mean flow shows a toroidal vortex driven by the throughflow. This is similar to the flow expected for high *Ro* or low and negative $\beta \Delta T$ when the axial throughflow will dominate, generating a main toroidal vortex as in the figure and weaker counter-rotating vortices at high radii in the cavity.

As indicated in the bottom right section of Figure 8.22, scaling of Nu with Ra for shroud heat transfer is consistent with centrifugal natural convection driven by the core-to-shroud temperature difference. This is also consistent with Long and Childs' (2007) measurements. With large-scale unsteadiness and turbulence in the core, the laminar Ekman layers provide the main resistance to disc heat transfer. Mean convection is very low in the Ekman layers and so heat conduction from the mean core temperature to the disc across the laminar Ekman layers is the dominant mechanism. At lower disc radii, effects of the throughflow may become significant, particularly on the downstream disc where flow impingement occurs. Heat transfer on the disc bores was found to be reasonable represented by correlations for pipe flow.

Details of the flow are dependent on cavity geometry and operating conditions. Examples are given by Owen and Long and in more recent studies. For example, Sun *et al.* (2022, 2024) show LES with two to four vortex pairs in a higher radius ratio (a/b) cavity and coupling of the unsteady flow between two cavities in a dual cavity configuration. They also confirmed similar pressure fluctuation levels to those measured by Jackson *et al.* Pernak *et al.* (2023, 2025) experimentally demonstrated the strong effect of surface temperature distribution and changes in strength of the large-scale unsteadiness with Rossby number. Wang *et al.* (2025b) extended studies with LES to $Re_\phi = 10^7$, showing effects of Ekman layer transition to turbulence on disc heat transfer. Parametric dependencies are discussed further in section 8.4.4.

8.4.2 Elementary Modelling

An elementary treatment of the cavity heat transfer, relevant at stabilised high power conditions, is indicated by the averaged flow and heat transfer descriptions shown in Figure 8.22. A well-mixed, uniform rothalpy core with centrifugal Rayleigh–Bénard convection to the shroud and heat conduction to the discs across laminar Ekman layers is assumed. In addition, energy transfer between the core and the axial throughflow is associated with mass flow exchange. In most studies to date, the heat transfer from the disc cob faces is relatively low and the exact placement of the inner boundary of the core makes little difference. The core is assumed to extend to the disc bore radius here. For conditions with higher cob heat transfer, it may be necessary to move this boundary radially outwards and treat the cob heat transfer separately.

A fully mixed core, with rotary stagnation temperature T^*_{core}, is assumed to exchange fluid with the axial throughflow at rate $\dot{m}_{\text{ex}}$. With inlet temperature T^*_{in} and cavity outflow temperature T^*_{core}, an energy balance then gives the following equation:

$$\dot{m}_{\text{ex}} C_p \left(T^*_{core} - T^*_{in}\right) = Q = Q_{\text{sh}} + Q_{disc} \tag{8.4.1}$$

where Q is the total heat transfer from the shroud and two discs. At high Ra conditions, equation 8.3.3 for shroud heat transfer and the relation giving $q_0 = \kappa(T_0 - T_{core})\sqrt{\Omega/\nu}$ for Ekman layer conduction give

$$Q_{\text{sh}} = 0.15 A_2 \beta^{1/3} \text{Re}_\phi^{2/3} Pr^{1/3} \kappa \left(T^*_{sh} - T^*_{core}\right)^{4/3} 2\pi s \tag{8.4.2}$$

$$Q_{disc} = B\, \text{Re}_\phi^{1/2}\, \kappa \left(T^*_{disc,av} - T^*_{core}\right) 2(b^2 - a^2)/b. \tag{8.4.3}$$

Here, area-averaged temperatures are used for the discs and shroud, assuming relatively weak variation of shroud temperature over the surface and neglecting the effect of temperature and pressure on fluid properties. A_2 and B are adjustment factors, as might be used in matching thermal models to test data.

Defining nondimensional core and area-averaged disc temperatures, respectively, as $X = \left(T^*_{core} - T^*_{in}\right)/\Delta T^*$ and $Y = \left(T^*_{disc} - T^*_{in}\right)/\Delta T^*$ with $\Delta T^* = T^*_s - T^*_{in}$ equations 8.4.1–8.4.3 give

$$m^* = \frac{\dot{m}_{ex}}{2\pi s_a a \rho_{ref}\sqrt{\beta\Delta T^*}\Omega b} = \frac{0.15A_2 sb(1-X)^{4/3}}{as_a X(\beta\Delta T^*)^{1/6} Re_\phi^{1/3} Pr^{2/3}} + \frac{B(b^2-a^2)(Y-X)}{\pi a s_a X\sqrt{\beta\Delta T^*} Re_\phi^{1/2} Pr} \tag{8.4.4}$$

where s_a is the cavity width at the inner radius. The nondimensional exchange flow rate m^* uses the velocity scale $\sqrt{\beta\Delta T^*}\Omega b$ associated with buoyancy and the area available for flow exchange at the disc bore radius. It is related to the exchange flow rate as a fraction of the axial throughflow rate through the equation

$$m^* = \frac{\left(a^2 - r_{sh}^2\right)Ro}{2s_a b\sqrt{\beta\Delta T^*}}\left(\frac{\dot{m}_{ex}}{\dot{m}_{in}}\right) = \frac{\left(a^2 - r_{sh}^2\right)Bo}{2s_a b}\left(\frac{\dot{m}_{ex}}{\dot{m}_{in}}\right) \tag{8.4.5}$$

where r_{sh} is the shaft radius and $Bo = Ro/\sqrt{\beta\Delta T^*}$ is a buoyancy number.

For a given geometry, surface temperature distribution and operating conditions, equation 8.4.4 gives the exchange flow rate m^* as a function of the core temperature X, or *vice versa*. Thus, specifying either m^* or X at a given operating point would give estimates of the disc and shroud heat transfer. Measurements of core temperature were reported for a cavity between plane discs with $a/b = 0.093$ and $s/b = 0.314$ by Long and Tucker (1994). These indicate core temperatures X in the range 0.2–0.3 for cases with a heated shroud and unheated discs and in the range 0.3–0.4 for cases with heated shrouds and discs. Based on a more engine-representative multi-cavity geometry with shroud heating, $a/b = 0.319$ and $s_a/b = 0.117$, Long and Childs (2007) deduced values of X in the range 0.2–0.4 by matching shroud heat transfer measurements to a natural convection correlation over a relatively broad range of operating conditions with Re_ϕ in the range 1.1×10^6–7.7×10^6 and Ro in the range 0.27–6.7.

Broad agreement of the elementary model of with Jackson *et al.*'s (2021) measurements with $a/b = 0.291$ and $s_a/b = 0.108$ is illustrated in Figure 8.23. Each point on the graph corresponds to an experimental condition for which shroud heat transfer measurements and disc heat transfer deduced from temperature measurements were available. For each experimental condition, core

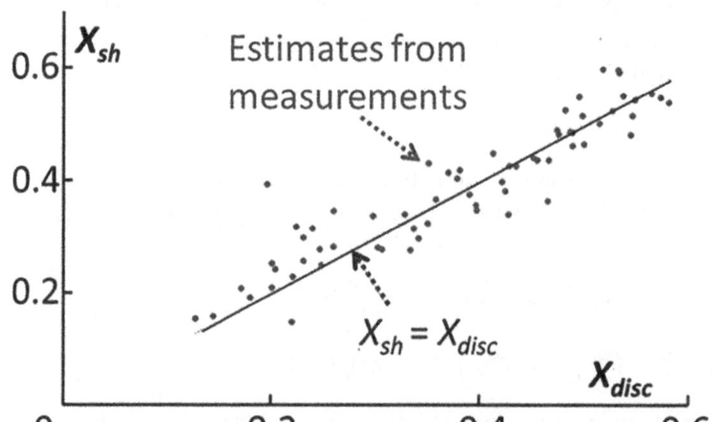

Figure 8.23 Estimated nondimensional core temperature from Jackson *et al.*'s shroud and disc heat transfer measurements. *Source:* Following Wang et al. (2025b).

temperature was estimated independently from either the shroud or the disc heat transfer using equations 8.4.2 or 8.4.3. Taking $B = 1$ in 8.4.3 and adjusting A_2 to 0.61 in 8.4.2 gave the results shown with a linear best fit giving $X_{sh} = X_{disc}$. Wang *et al.* (2025b) also showed that $A_2 = 0.61$ gave a reasonable fit to shroud heat transfer estimated from LES for this configuration.

Further discussion of the elementary model above and how such methods can be used in engine thermal modelling is given by Sun *et al.* (2024) and later in this chapter. This includes estimation of mass flow exchange m^* and tailoring of the model to different engine configurations and conditions using CFD. Other workers have given alternative models. Notably, several models have been considered at the University of Bath, with calibration against experimental data. Examples are given by Owen and Tang (2015) and Tang and Owen (2017, 2023). Assumptions in earlier models differ with the flow structure identified in Figure 8.22, but good agreement with experimental work was demonstrated. Later models have much in common with the simpler model presented above and also show agreement with measurements. Nicholas *et al.* (2025) use the Ekman layer conduction assumption for disc heat transfer and include a mass exchange between the cavity and throughflow. Their calculated mean core static temperatures are uniform at low *Ec* conditions. At higher *Ec*, radial variation of core static temperature is attributed to allowance for compressibility in a forced vortex at rotor speed. This approximates constant rotary stagnation temperature.

8.4.3 Computational Fluid Dynamics

As discussed in Section 6.3.2, RANS CFD models show variable performance for this class of flow, and LES is judged to be substantially more reliable. Comparisons of RANS and LES are further considered by Wang *et al.* (2024a, 2025b) with examples suggesting that differences may become more apparent at high Reynolds numbers. Evaluations of WMLES against Jackson *et al.*'s (2021, 2022b) measurements have shown encouraging results. This is illustrated in Figure 8.24, which includes the rig working section, the CFD domain, shroud Nusselt numbers from available WMLES studies, and comparison of disc heat transfer and air temperatures with measurements at a maximum rig speed condition of 8000 rpm. The Nusselt number plot includes equation 8.3.3 with $A_1 = A_2 = 1$ and the higher *Ra* correlation with $A_2 = 0.61$. This latter curve gives a reasonable fit to the LES data as mentioned above. The level of scatter is comparable to that implied for experimental results in Figure 8.23. Here, Nu_{sh} and Ra_{LM} use the core temperature which was not measured in the experiments. However, estimating core temperature from experimentally derived disc heat transfer (assuming Ekman conduction), Nicholas *et al.* (2025) give experimentally derived values of Nu_{sh}. These are close to the WMLES at higher Ra_{LM} but show higher values at low Ra_{LM}. Direct comparisons of WMLES with measurements at specific operating conditions by Wang *et al.* (2024a,b) show differences in shroud heat transfer of up to 22% at a condition thought to be particularly sensitive to operating conditions. At the high speed test condition shown in the figure, WMLES was within 2% of the measurement. Reasonable agreement with disc heat transfer and air temperature measurements is shown in the figure at this condition. It should be noted that the experimental disc heat transfer is deduced from temperature measurements using thin disc theory and is subject to significant uncertainty at high and low radii. RANS model results for this case differ considerably more from the measurements than the WMLES.

The simulation with highest Rayleigh number for which the shroud *Nu* is given in Figure 8.24 was designed to investigate more extreme operating conditions. This case corresponds to a hypothetical condition obtained by increasing cavity pressure for one of Jackson *et al.*'s tests at $Re_\phi = 3 \times 10^6$ to give $Re_\phi = 10^7$. The CFD code used was highly parallelised and employed finite

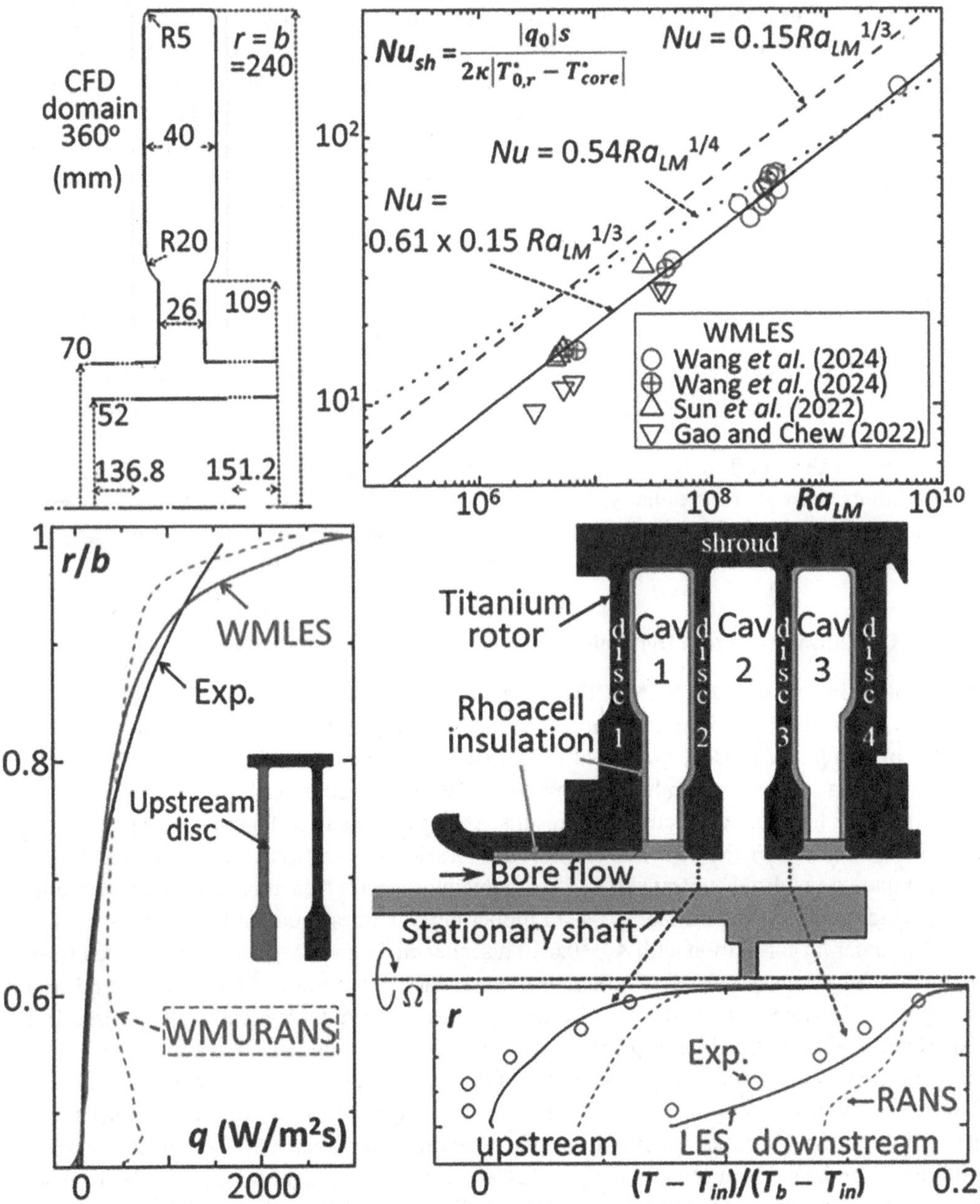

Figure 8.24 Heat transfer for Jackson *et al.*'s test rig with $Re_\phi = 3 \times 10^6$, $\beta\Delta T^* = 0.16$ and $Ro = 0.2$ at the experimental condition shown, $a/b = 0.29$, $s/b = 0.17$, $s_a/b = 0.11$. *Source:* Wang *et al.* (2025b) / with permission of American Society of Mechanical Engineers (ASME).

volume discretisation and a pressure-based solver for the incompressible flow equations with the Boussinesq approximation for buoyancy. Eckert number effects were modelled through use of the rotary stagnation temperature in place of the static temperature. This code had been shown to be considerably more efficient for this problem than a density-based solver applied to the full compressible gas flow equations. An advantage of the incompressible flow solver was that it allowed use of a larger time step than the compressible flow solver, with this being related to the relative velocity rather than the speed of sound.

A particular issue with CFD for this class of flow is the modelling of the Ekman layers. The simulations in Figure 8.24 use standard wall functions with the meshing requirement that the near-wall spacing is no more than the Ekman depth $\sqrt{\nu/\Omega}$. This results in the wall function having little effect in the Ekman layers for all but the highest Re_ϕ conditions considered. The rather course resolution of the Ekman layers appears sufficient to capture the most important effects. Gao and Chew (2022) show that at a test condition with $Re_\phi = 2.2 \times 10^6$, the velocities within the Ekman layer are close to the laminar solution and that the main temperature gradient is across the Ekman layer thickness. Further simulations and measurements confirm the dominance of conduction across the laminar Ekman layer on disc heat transfer. Figure 8.19 for a closed cavity test rig can also be used to estimate when Ekman layer transition is likely to be important in cavities with axial throughflow. This indicates that the Ekman layers will remain laminar for Jackson *et al.*'s test conditions but transition is likely for the simulated condition with $Re_\phi = 10^7$. Consideration of the wall function suggests that WMLES will also indicate transition at this condition. Calculating wall shear stress from the laminar Ekman layer solution, equation 3.3.20, the nondimensional wall distance at the Ekman depth $n^+ = (2Re_\phi)^{1/4}(u_{x,\delta}/\Omega b)^{1/2}$. Considering a critical boundary layer Reynolds number of 200, as for the closed cavity in Section 8.3, this gives $n^+ \sim 17$ at the critical condition. For a classical boundary layer profile, this is in the buffer region between the laminar sublayer and the turbulent logarithmic layer. Hence the wall function will increase the wall shear stress above the value for laminar flow.

Figure 8.25 shows mean disc heat transfer and instantaneous contours of boundary layer Reynolds number from WMLES for high Re_ϕ conditions. The heat transfer results are compared with values corresponding to conduction between core and disc temperatures across the Ekman layers. This agrees reasonably well with the simulations at $Re_\phi = 2.3 \times 10^6$ and at higher radii for $Re_\phi = 3 \times 10^6$ but severely underpredicts the simulation results at $Re_\phi = 10^7$. The modified Ekman

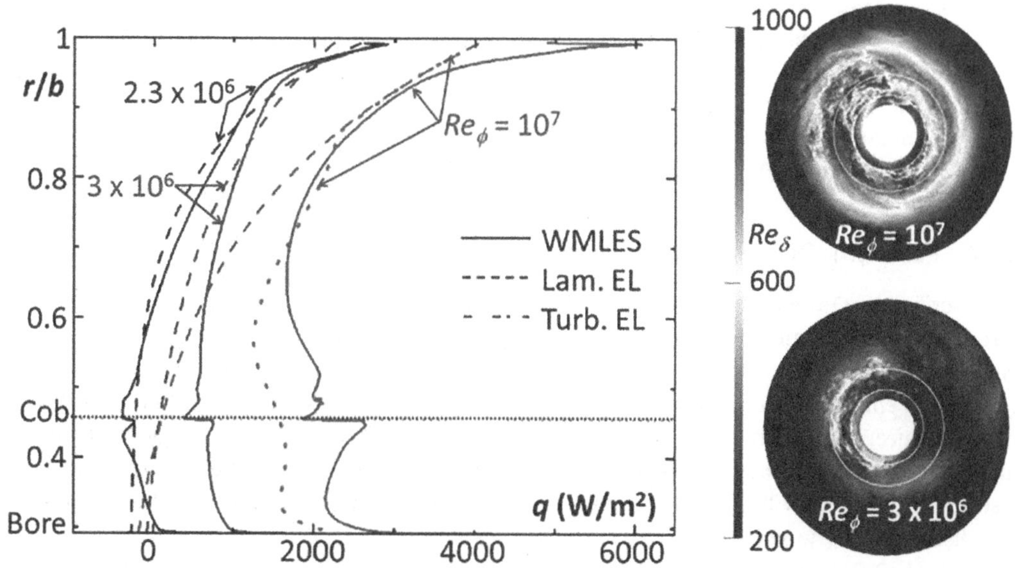

Figure 8.25 Disc heat transfer and boundary layer Reynolds number for axial throughflow with $\beta\Delta T = 0.24$, $Ro = 0.2$. *Source:* Wang *et al.* (2025b) / with permission of American Society of Mechanical Engineers (ASME).

layer conduction curve uses a local core temperature and applies a correction from turbulent Ekman layer theory. This takes better account of the isothermal core being restricted to higher radii and transition to turbulence. The correction for turbulence multiplies the laminar heat Ekman layer conduction value by $(Re_\delta/180)^{0.6}$ when this is greater than 1. This factor is the ratio of circumferential wall shear stress for a turbulent Ekman layer according to equation 5.3.12 and that in laminar Ekman layers. The modification brings the theoretical prediction much closer to the WMLES at lower radii. Underestimation of the correction may be associated with averaging of nonlinear effects and neglect of the radial velocity component in calculating Re_δ.

The contours of instantaneous boundary layer Reynolds numbers (Re_δ) in Figure 8.25 further indicate that turbulent Ekman layers are expected at lower radii. These are based on just the relative circumferential velocity component in the core. The higher boundary layer Reynolds numbers at low radii regions are to a large extent associated with slip of the core flow relative to the rotating disc. Values of Re_δ for $Re_\phi = 10^7$ are clearly well above the critical value for Ekman layers of $\sim$200. For $Re_\phi = 3 \times 10^6$, there are also some high values of Re_δ, and these may account for the increase in disc heat above the laminar Ekman layer conduction values for this case. In the simulations shown, no inlet swirl is imposed in the throughflow, which is consistent with the experimental condition but will differ from many engine conditions. Further simulations with inlet swirl by Wang *et al.* (2025a) suggest the transition to turbulence at lower radii may be delayed but that transition enhances disc heat transfer at higher radii.

Studies using wall resolved LES have been restricted to lower Reynolds numbers. Pitz *et al.* (2019) extended their closed cavity configuration to include axial throughflow, as in Figure 6.14, with $Re_\phi = 1.1 \times 10^5$, $\beta\Delta T = 0.134$, $Ro = 0.2$ or 0.4, $Ra = 1.2 \times 10^9$ and $Ec = 0$. For this relatively wide cavity with $s/b = 0.5$, the impingement of axial throughflow on the downstream disc strongly affects heat transfer. Saini and Sandberg (2021) used LES to investigate axial throughflow in a rotating cavity for the scaled engine cavity geometry studied experimentally by Atkins and Kanjirakkad (2014). In their simulations, $Re_\phi = 7.8 \times 10^4$, $\beta\Delta T = 0.085$ or 0.17, $Ro = 4.5$, $Ra = 5.7 \times 10^8$ or 7.3×10^6 and $Ec = 0.21$ or 0.42. The results show that the near-shroud region is dominated by the centrifugal buoyancy-induced flow similar to that in a closed cavity, but with the low Re_ϕ case giving reduced turbulence and possibly partially laminar flow at outer radii. Due to the high value of Ro, the axial throughflow is more dominant at lower radii. The reported core temperatures and shroud heat flux indicate relatively weak flow exchange between the throughflow and the outer part of the cavity for these conditions. This is consistent with the weak dependency of measured disc temperatures on buoyancy for $Re_\phi < 10^6$ identified by Atkins and and Kanjirakkad.

8.4.4 Parametric Dependencies

As described by Owen and Long (2015), it is well established from experimental studies that the disc temperature distribution, Ro, $\beta\Delta T_{ref}$ and Re_ϕ have significant effects on heat transfer. Some correlation of shroud heat transfer measurements has been achieved using the Rayleigh number to combine effects of $\beta\Delta T_{ref}$ and Re_ϕ. It is also clear that the buoyancy effects diminish as $\beta\Delta T_{ref}$ or Re_ϕ reduce, and that mass exchange between the throughflow and the cavity depends on Ro and other parameters. For example, Owen and Long report that changing the shaft diameter had a strong effect on velocity measurements in the inner part of the cavity and shroud heat transfer. The discussion below focuses on more recent studies.

Sun *et al.* (2024) considered parametric dependencies for an engine geometry cavity using fully compressible WMLES, as previously applied to the test configuration in Figure 8.24, and the elementary model of Section 8.4.2. The engine geometry has higher radius and width ratios than the test rig with $a/b = 0.38$, $s/b = 0.22$ and $s_a/b = 0.15$, and smaller cobs. For most conditions, Ro was 0.7 or 1, $\beta\Delta T^*_{ref}$ was 0.44 and Re_ϕ was 5.4×10^5. These Ro values avoided problems with reversed flow at the inlet and exit and the relatively low value of Re_ϕ limited the computational requirements. The estimated inlet swirl and surface temperature distributions also differed from the rig (which had thicker discs insulated on one side). The simulations showed two or more outflow plumes in the main flow structures rather than the single-lobed structures shown in Figure 8.22. Scaling of Nu_{sh} with Ra_{LM} was similar to that in Figure 8.24 but with higher values. It was concluded that the model gives a useful basis for engineering calculations involving correlation and extrapolation of WMLES and experimental results. The model could be tuned using WMLES at steady state conditions for new configurations and to account for variations from the heat transfer correlations used in Section 8.4.2. This would allow engine temperature prediction through a transient operating cycle using the tuned elementary model.

Figure 8.26 shows the exchange flow rate m^* obtained by using the surface heat flux from WMLES or experiment in equation 8.4.1. The WMLES results are as given by Sun *et al.* (2024), and the experimental results were deduced from data described by Jackson *et al.* (2022b). Plotting against Bo has collapsed the test rig data to some extent although there is a spread of results around $Bo = 0.75$. There is reasonable agreement between experiment and WMLES. From equation 8.4.5, assuming a fixed geometry and constant ratio of exchange to throughflow rate $\dot{m}_{ex}/\dot{m}_{in}$ gives a linear dependency of m^* on Bo. This is illustrated for the test rig with $\dot{m}_{ex}/\dot{m}_{in} = 0.2$ and 0.4 in the figure. For all conditions considered, incomplete mixing of the throughflow into the cavity limits heat transfer.

The highest value of m^* for the engine cavity, just above 0.1 in Figure 8.26, is for a modified geometry with reduced bore diameter for the downstream disc. Impingement increases heat transfer on disc cob and enhances mass exchange in this case. Otherwise, for $1 \le Bo \le 1.6$, there is some agreement in the level of m^* in the two configurations. At higher values of Bo, buoyancy effects weaken. This, and differences in geometry and operating conditions, may contribute to the larger differences in m^* for the two configurations for $Bo \approx 2.1$ and 3.1. Wang *et al.* (2025a) have conducted further WMLES for the test geometry using an incompressible flow solver with the Boussinesq approximation. The simulations are for $Ro = 0.4$ with $8 \times 10^5 \le Re_\phi \le 10^7$ and $0.79 \le Bo \le 1$.

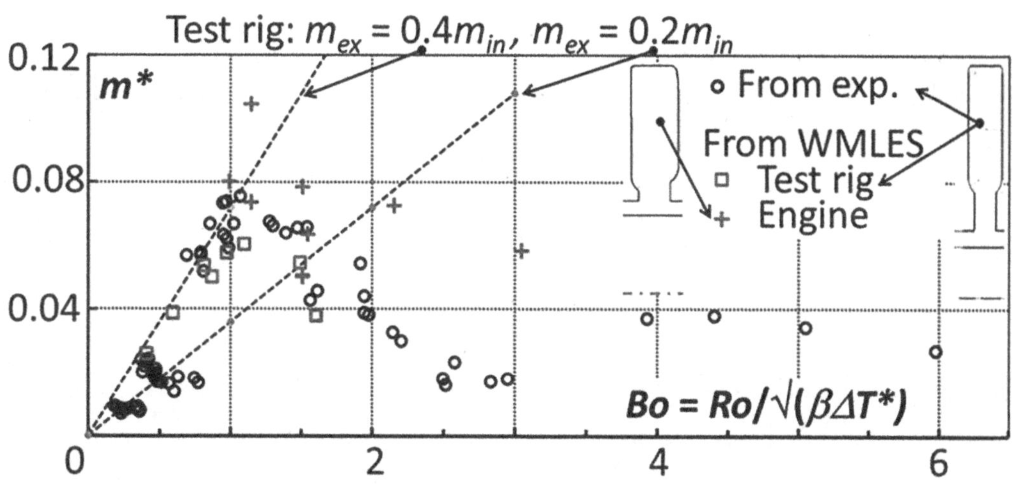

Figure 8.26 Estimated mass flow exchange rate for axial throughflow from WMLES and Jackson *et al.*'s experiments.

The results confirm the sensitivity of m^* in this range of Bo, and highlight dependence on inlet swirl in the throughflow.

Large changes in $\beta\Delta T_{ref}$ and the form of the surface temperature distribution can occur following engine accelerations or decelerations, and industrial thermal models must capture the effects on heat transfer. Sun *et al.* (2024) found that the elementary model, assuming a constant nondimensional exchange flow rate, reproduced the effects of a change in surface temperature, with the radial gradient remaining positive, in their WMLES results reasonably well. Pernak *et al.* (2025) show experimental studies for uniform and radially decreasing surface temperature distributions with $\beta\Delta T_{ref} \approx 0$ or -0.12. The isothermal case shows little heat transfer and no clear flow structure in the outer part of the cavity at $Ro = 0.2$, but flow structures are evident from fast response pressure measurements at higher Ro. For $\beta\Delta T_{ref} = -0.12$, disc heat transfer was reported at lower levels than for positive $\beta\Delta T_{ref}$, with rotating flow structures at all values of Ro studied.

Nicholas *et al.* (2025) and Wang (2024) have further investigated the effects of changing cavity geometry. The configurations considered are shown in Figure 8.27. Nicholas *et al.* fitted aluminium inserts to the disc cobs in Jackson *et al.*'s configuration to halve or fill the gap. The latter case produced the closed cavity geometry of Figure 8.20. The measured disc temperatures for the case with reduced gap are closer to those of the closed cavity than those for the original gap. This is attributed to much reduced mass exchange between the throughflow and cavity. Pressure measurements also indicate differences with modulation between two and three vortex pairs in the cavity for the reduced gap. Wang applied WMLES to the test rig and engine geometries of Figure 8.26 and variations of the test rig to engine gap and radius ratios. His results, for $Ro = 0.4$, $\beta\Delta T^* = 0.2$, $Re_\phi = 2.33 \times 10^6$, $\upsilon_{\phi,in} = 0$ and a temperature profile matching a test rig condition, show limited sensitivity of flow structure and heat transfer to the geometry changes in WMLES studies. Nu_{sh} remained close to the correlating line in Figure 8.24. The disc Nusselt number for the increased a/b and engine geometries was noticeably higher than for the baseline case. This was attributed to a reduced extent of the isothermal core.

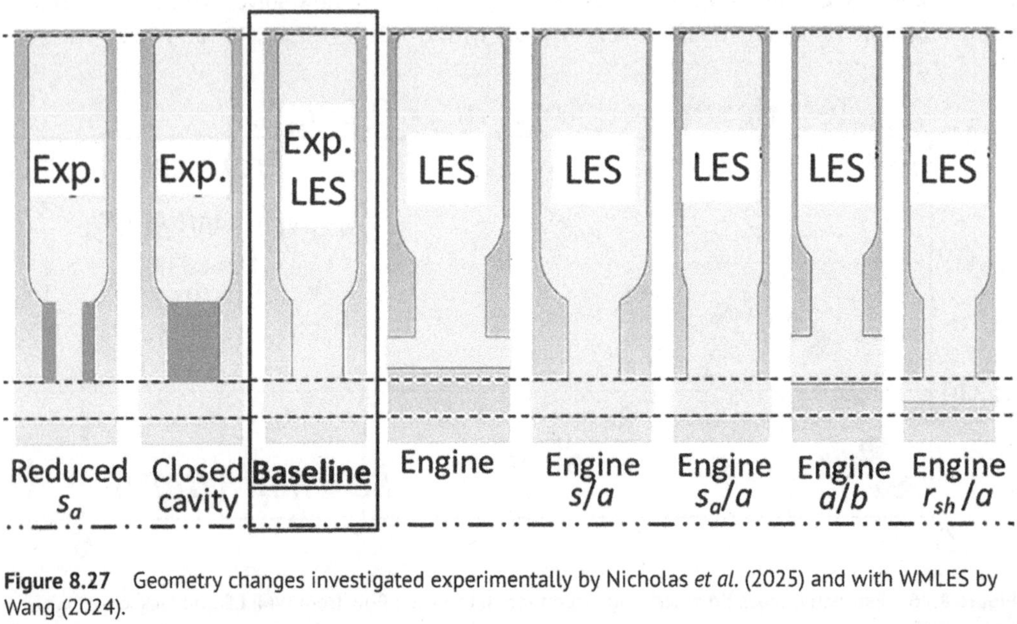

Figure 8.27 Geometry changes investigated experimentally by Nicholas *et al.* (2025) and with WMLES by Wang (2024).

Wang found shroud Nusselt number to be more sensitive to surface temperature distribution and inlet swirl than to the geometric changes studied. Using the surface temperature distribution from Sun *et al.*'s study with the test rig geometry increased Nu_{sh} to about 20% above the correlation in Figure 8.24. As already mentioned, inlet swirl also had noticeable effects. Positive inlet swirl increased shroud heat transfer. At high Re_ϕ, inlet swirl affected Ekman layer transition in the simulations.

8.5 Concluding Remark

As for rotor–stator cavities, research over several decades with experimentation, analysis and CFD has greatly advanced understanding of corotating disc cavity flows. Integral methods and RANS CFD models are useful for cavities with radial throughflow, and first estimates of flow characteristics can be made using the charts given in Sections 8.1 and 8.2. For radial inflow, strong pressure gradients can develop, buoyancy effects have been found significant in some studies, and LES improves on RANS CFD in some cases. For forced convection, many of the closing remarks for Chapter 7 apply here. A further observation that applies to both categories of flow is that careful application of thermal modelling to research facilities can help considerably with interpretation of results and exposing uncertainties.

Understanding of buoyancy-driven flows has benefited from advanced CFD using DNS and LES. WMLES has proved effective in extending predictive capability to high Reynolds number engine conditions. These have allowed details of the turbulence and Ekman layers to be examined. In many circumstances, the main cavity flow is turbulent, while the Ekman layers remain laminar and disc heat transfer is largely controlled by the thermal resistance of these layers. Further progress can be expected as research continues in this area. Although analysis strongly indicates that transition to turbulence in the highly unsteady Ekman layers occurs at high Reynolds number, this has not yet been thoroughly examined. There are also outstanding questions regarding the heat transfer in the disc cob region and the interaction between cavity flows in multicavity systems.

References

Amirante, D., Sun, Z., Chew, J. W., Hills, N. J., and Atkins, N. R., 2016. Modeling of compressor drum cavities with radial inflow. Proceedings of the ASME Turbo Expo, Paper GT2016-56505. V05AT15A010. ASME. https://doi.org/10.1115/GT2016-56505.

Atkins, N. R., 2013. Investigation of a radial-inflow bleed as a potential for compressor clearance control. ASME Paper GT2013-95768. https://doi.org/10.1115/GT2013-95768.

Atkins, N. R., and Kanjirakkad, V., 2014. Flow in a rotating cavity with axial throughflow at engine representative conditions. ASME Turbo Expo, Paper GT2014-27174. https://doi.org/10.1115/GT2014-27174.

Bohn, D. and Gier, J., 1998. The effect of turbulence on the heat transfer in closed gas-filled rotating annuli for different Rayleigh numbers, ASME Paper No. 98-GT-542. https://doi.org/10.1115/98-GT-542.

Bohn, D., Deuker, E., Emunds, R., and Gorzelitz, V., 1995. Experimental and theoretical investigations of heat transfer in closed gas-filled rotating annuli. J. Turbomach., 117, 175–183. https://doi.org/10.1115/1.2835635.

Bohn, D., Emunds, R., Gorzelitz, V., and Krüger. U., 1996. Experimental and theoretical investigations of heat transfer in closed gas-filled rotating annuli II. J. Turbomach., 118(11–19), 1996 https://doi.org/10.1115/1.2836599.

Chew, J. W., 1985a. Prediction of flow in a rotating cavity with radial outflow using a mixing length turbulence model. Proceedings International Conference on Laminar and Turbulent Flow, Swansea, Pineridge Press, 318–329.

Chew, J. W., 1985b. Computation of convective laminar flow in rotating cavities. J. Fluid Mech., 153, 339–360. https://doi.org/10.1017/S0022112085001288.

Chew, J. W., 1987. Computation of flow and heat transfer in rotating disc systems. 2nd ASME-JSME Thermal Engineering Conference, Vol. 3, 361–367.

Chew, J. W., 1990. Prediction of rotating disc flow and heat transfer in gas turbine engines. Proceedings 3rd International Symposium on Transport Phenomena and Dynamics of Rotating Machinery, eds J. H. Kim and W. J. Yang, Hemisphere, 145–160.

Chew, J. W. and Rogers, R. H., 1988. An integral method for the calculation of turbulent forced convection in a rotating cavity with radial outflow. Int. J. Heat Fluid Flow, 9, 37–48. https://doi.org/10.1016/0142-727X(88)90028-8.

Chew, J. W. and Snell, R. J., 1988. Prediction of pressure distribution for radial inflow between co-rotating discs. ASME paper 88-GT-127. https://doi.org/10.1115/88-GT-61.

Chew, J. W., Farthing, P. R., Owen, J. M., and Stratford, B., 1989. The use of fins to reduce the pressure drop in a rotating cavity with a radial inflow. ASME J. Turbomach., 111, 349–356. https://doi.org/10.1115/1.3262279.

Dixon, J. A., Verdicchio, J. A., Benito, D., Karl, A., and Tham, K. M., 2004. Recent developments in gas turbine component temperature prediction methods, using computational fluid dynamics and optimization tools, in conjunction with more conventional finite element analysis techniques. Proc. Inst. Mech. Eng. Part A: J. Power Energy, 218, 241–155. https://doi.org/10.1243/0957650041200641.

Farthing, P. R., 1988. The effect of geometry on flow and heat transfer in a rotating cavity. D. Phil. thesis, University of Sussex, UK.

Farthing, P. R., Long, C. A., and Rogers, R. H., 1991a. Measurement and prediction of heat transfer from compressor discs with a radial inflow of cooling air. ASME Paper 91-GT-053. https://doi.org/10.1115/91-GT-053.

Farthing, P. R., Chew, J. W., and Owen, J. M., 1991b. The use of deswirl nozzles to reduce the pressure drop in a rotating cavity with a radial Inflow. ASME J. Turbomach., 113, 106–114. https://doi.org/10.1115/1.2927727.

Farthing, P. R., Long, C. A., Owen, J. M., and Pincombe, J. R., 1992. Rotating cavity with axial throughflow of cooling air: Flow structure. ASME J. Turbomach., 114, 237–246. https://doi.org/10.1115/1.2927991.

Firouzian, M., Owen, J. M., Pincombe, J. R., and Rogers, R. H., 1985. Flow and heat transfer in a rotating cavity with a radial inflow of fluid—Part 1: The flow structure. Int. J. Heat Fluid Flow, 6, 228–234. https://doi.org/10.1016/0142-727X(85)90054-2.

Firouzian, M., Owen, J. M., Pincombe, J. R., and Rogers, R. H., 1986. Flow and heat transfer in a rotating cavity with a radial inflow of fluid —Part 2: Velocity, pressure and heat transfer measurements. Int. J. Heat Fluid Flow, 7, 21–27. https://doi.org/10.1016/0142-727X(86)90037-8.

Gao, F. and Chew, J. W., 2021a. Ekman layer scrubbing and shroud heat transfer in centrifugal buoyancy-driven convection. ASME J. Eng. Gas Turbines Power, 143, 071010. https://doi.org/10.1115/1.4050366.

Gao, F. and Chew, J. W., 2021b, 2021. Evaluation and application of advanced CFD models for rotating disc flows. Proc. Inst. Mech. Eng., Part C: J. Mech. Eng. Sci., 235(23), 6847–6864. https://doi.org/10.1177/09544062211013850.

Gao, F. and Chew, J. W., 2022. Flow and heat transfer mechanisms in a rotating compressor cavity under centrifugal buoyancy-driven convection. ASME J. Eng. Gas Turbines Power, 144, 051010. https://doi.org/10.1115/1.4052649.

Gao, F., Chew, J. W., and Pitz, D. B., 2019. Numerical study of buoyancy-driven flow in a closed rotating annulus. Proceedings of Global Power and Propulsion Society, ISSN-Nr: 2504-4400, GPPS-BJ-2019-0034. https://gpps.global/wp-content/uploads/2021/02/GPPS-BJ-2019_paper_34.pdf.

Gao, F., Pitz, D. B., and Chew, J. W., 2020. Numerical investigation of buoyancy-induced flow in a sealed rapidly rotating disc cavity. Int. J. Heat Mass Transfer, 147, 118860. https://doi.org/10.1016/j.ijheatmasstransfer.2019.118860.

Günther, A., Uffrecht, W., and Odenbach, S., 2012. Local measurements of disk heat transfer in heated rotating cavities for several flow regimes. ASME J. Turbomach., 134, 51016 https://doi.org/10.1115/1.4003965.

Hollands, K. G. T., Raithby, G. D., and Konicek, L., 1975. Correlation equations for free convection heat transfer in horizontal layers of air and water. Int. J. Heat Mass Transfer, 18, 879–884. https://doi.org/10.1016/0017-9310(75)90179-9.

Hudson, J. L., Tang, O., and Abell, S., 1978. Experiments on centrifugally driven thermal convection in a rotating cylinder. J. Fluid Mech., 86, 147 https://doi.org/10.1017/S0022112078001044.

Iacovides, H. and Chew, J. W., 1993. The computation of convective heat transfer in rotating cavities. Int. J. Heat Fluid Flow, 14, 146–154. https://doi.org/10.1016/0142-727X(93)90022-F.

Jackson, R. W., Luberti, D., Tang, H., Pountney, O. J., Scobie, J. A., Sangan, C. M., Owen, J. M., and Lock, G. D., 2021. Measurement and analysis of buoyancy-induced heat transfer in aero-engine compressor rotors. ASME J. Eng. Gas Turbines Power, 143, 061004. https://doi.org/10.1115/1.4049100.

Jackson, R. W., Tang, H., Scobie, J. A., Owen, J. M., and Lock, G. D., 2022a. Measurement of heat transfer and flow structures in a closed rotating cavity. ASME J. Eng. Gas Turbines Power, 144, 051005. https://doi.org/10.1115/1.4053392.

Jackson, R. W., Tang, H., Scobie, J. A., Pountney, O. J., Sangan, C. M., Owen, J. M., and Lock, G. D., 2022b. Unsteady pressure measurements in a heated rotating cavity. ASME J. Eng. Gas Turbines Power, 144, 041017. https://doi.org/10.1115/1.4053390.

Jiang, H., Zhu, X., Wang, D., Huisman, S. G., and Sun, C., 2020. Supergravitational turbulent thermal convection. Sci. Adv., 6 https://doi.org/10.1126/sciadv.abb8676.

Kumar, B. G. V., Chew, J. W., and Hills, N. J., 2013. Rotating flow and heat transfer in cylindrical cavities with radial inflow. J. Eng. Gas Turbines Power, 135, 032502. https://doi.org/10.1115/1.4007826.

Lapworth, B. L. and Chew, J. W., 1992. A numerical study of the influence of disc geometry on the flow and heat transfer in a rotating cavity. ASME J. Turbomach., 114, 257 https://doi.org/10.1115/1.2927993.

Liu, V., 2024. Thermal modelling for a compressor drum test rig. Student project report, Mechanical Engineering, University of Surrey.

Lloyd, J. R. and Moran, W. R., 1974. Natural convection adjacent to horizontal surface of various planforms. ASME J. Heat Transfer, 96, 443–447. https://doi.org/10.1115/1.3450224.

Long, C. A. and Childs, P. R. N., 2007. Shroud heat transfer measurements inside a heated multiple rotating cavity with axial throughflow. Int. J. Heat Fluid Flow, 28, 1405–1417. https://doi.org/10.1016/j.ijheatfluidflow.2007.04.009.

Long, C. A. and Tucker, P. G., 1994. Shroud heat transfer measurements from a rotating cavity with an axial throughflow of air. ASME J. Turbomach., 116, 525–534. https://doi.org/10.1115/1.2929441.

May, D., Chew, J. W., and Scanlon, T. J., 2013. Prediction of deswirled radial inflow in rotating cavities with hysteresis. ASME J. Turbomach., 135, 041025. https://doi.org/10.1115/1.4007741.

Morse, A. P., 1988. Numerical prediction of turbulent flow in rotating cavities. ASME J. Turbomach., 110, 202–212. https://doi.org/10.1115/1.3262181.

Morse, A. P., 1991. Application of a low Reynolds number $k - \varepsilon$ turbulence model to high-speed rotating cavity flows. ASME J. Turbomach., 113, 98–105. https://doi.org/10.1115/1.2927743.

Morse, A. P. and Ong, C. L., 1992. Computation of heat transfer in rotating cavities using a two-equation model of turbulence. ASME. J. Turbomach., 114, 247–255. https://doi.org/10.1115/1.2927992.

Negulescu, D. and Pfitzner, M., 2001. Secondary air systems in aeroengines employing vortex reducers. Proceedings of the ASME Turbo Expo. https://doi.org/10.1115/2001-GT-0198.

Nicholas, T. E. W., Pernak, M. J., Lock, G. D., Scobie, J. A., and Tang, H., 2025. Mass and heat exchange in rotating compressor cavities with variable cob separation. ASME J. Eng. Gas Turbines Power, 147, 011018. https://doi.org/10.1115/1.4066247.

Northrop, A. and Owen, J. M., 1988. Heat transfer measurements in rotating-disc systems part 2: The rotating cavity with a radial outflow of cooling air. Int. J. Heat Fluid Flow, 9, 27–36. https://doi.org/10.1016/0142-727X(88)90027-6.

Onori, M., Amirante, D., Hills, N. J., and Chew, J. W., 2016. LES validation for a rotating cylindrical cavity with radial inflow. Proceedings of the ASME Turbo Expo, https://doi.org/10.1115/GT2016-56393.

Onori, M., Amirante, D., Hills, N. J., and Chew, J. W., 2019. Heat transfer prediction from large eddy simulation of a rotating cavity with radial inflow. ASME J. Eng. Gas Turbines Power, 141, 121002. https://doi.org/10.1115/1.4045150.

Owen, J. M. and Long, C. A., 2015. Review of buoyancy-induced flow in rotating cavities. ASME J. Turbomach., 137, 111001. https://doi.org/10.1115/1.4031039.

Owen, J. M. and Rogers, R. H., 1995. Flow and heat transfer in rotating disc systems, Vol. 2: Rotating cavities, Taunton, UK: Research Studies Press.

Owen, J. M. and Tang, H., 2015. Theoretical model of buoyancy induced flow in rotating cavities. ASME J. Turbomach., 137, 111005. https://doi.org/10.1115/1.4031353.

Owen, J. M., Pincombe, J. R., and Rogers, R. H., 1985. Source–sink flow inside a rotating cylindrical cavity. J. Fluid Mech., 155, 233–265. https://doi.org/10.1017/S0022112085001793.

Pernak, M., Nicholas, T., Carnevale, M., Lock, G. D., Tang, H., and Scobie, J., 2025. Flow and heat transfer in rotating compressor cavities with inverted shroud-throughflow temperature differences. ASME J. Turbomach., 147, 071005. https://doi.org/10.1115/1.4067075.

Pernak, M. J., Nicholas, T. E. W., Williams, J. T., Jackson, R. W., Tang, H., Lock, G. D., and Scobie, J. A., 2023. Experimental investigation of transient flow phenomena in rotating compressor cavities. ASME J. Turbomach., 145, 121005. https://doi.org/10.1115/1.4063507.

Pfitzner, M. and Waschka, W., 2000. Development of an aero engine secondary air system employing vortex reducers. 22nd ICAS Congress, Harogate, UK. https://www.icas.org/icas_archive/ICAS2000/PAPERS/ICA0511.PDF.

Pitz, D. B., Chew, J. W., and Marxen, O., 2019. Effect of an axial throughflow on buoyancy-induced flow in a rotating cavity. Int. J. Heat Fluid Flow, 80, 108468. https://doi.org/10.1016/j.ijheatfluidflow.2019.108468.

Saini, D. and Sandberg, R. D., 2020. Simulations of compressibility effects in centrifugal buoyancy-induced flow in a closed rotating cavity. Int. J. Heat Fluid Flow, 85, 108656. https://doi.org/10.1016/j.ijheatfluidflow.2020.108656.

Saini, D. and Sandberg, R. D., 2021. Large-eddy simulations of high Rossby number flow in the high-pressure compressor inter-disk cavity. ASME J. Turbomach., 143, 111002. https://doi.org/10.1115/1.4050951.

Shen, W., Wang, S., Zhang, X., and Liang, X., 2022. Large-eddy simulation and mathematical model of vortex breakdown and pressure drop in a cavity with tubeless vortex reducer. Eng. Appl. Comput. Fluid Mech., 16, 1344–1363. https://doi.org/10.1080/19942060.2022.2091662.

Sun, Z., Amirante, D., Chew, J. W., and Hills, N. J., 2015. Coupled aerothermal modeling of a rotating cavity with radial inflow. ASME J. Eng. Gas Turbines Power, 138, 032505. https://doi.org/10.1115/1.4031387.

Sun, Z., Chew, J. W., Hills, N. J., Volkov, K. N., and Barnes, C. J., 2010. Efficient finite element analysis/computational fluid dynamics thermal coupling for engineering applications. ASME. J. Turbomach., 132, 031016. https://doi.org/10.1115/1.3147105.

Sun, Z., Chew, J. W., Gao, F., Bristot, A., and Williams, J., 2024. Parametric dependence and elementary modelling for compressor disc cavity heat transfer. Proc. Inst. Mech. Eng., Part C: J. Mech. Eng. Sci., 238, 2474–2488. https://doi.org/10.1177/09544062231190282.

Sun, Z., Gao, F., Chew, J. W., and Amirante, D., 2022. Large eddy simulation Investigation of low Rossby number buoyant flow in rotating cavities. ASME J. Eng. Gas Turbines Power, 144, 121023. https://doi.org/10.1115/1.4055686.

Tang, H. and Owen, J. M., 2017. Effect of buoyancy-induced rotating flow on temperature of compressor discs. ASME J. Eng. Gas Turbines Power, 139, 062506. https://doi.org/10.1115/1.4035400.

Tang, H. and Owen, J. M., 2023. Plume model for buoyancy-induced flow and heat transfer in closed rotating cavities. ASME J. Turbomach., 145, 011005. https://doi.org/10.1115/1.4055449.

Virr, G. P., Chew, J. W., and Coupland, J., 1994. Application of computational fluid dynamics to turbine disk cavities. ASME. J. Turbomach., 116, 701–708. https://doi.org/10.1115/1.2929463.

Wang, R., 2024. Advanced modelling of flow and heat transfer in rotating disc cavities. PhD thesis, University of Surrey. https://doi.org/10.15126/thesis.901371.

Wang, R., Gao, F., Chew, J. W., Marxen, O., and Sun, Z., 2024a. Advanced modeling of flow and heat transfer in rotating disk cavities using open-source computational fluid dynamics. ASME J. Eng. Gas Turbines Power, 146, 061022. https://doi.org/10.1115/1.4063989.

Wang, R., Gao, F., Chew, J. W., and Marxen, O., 2024b. Large-eddy simulation of axial, radial and mixed centrifugal convection in a closed rotating cavity. Int. J. Heat Mass Transfer, 227, 125559. https://doi.org/10.1016/j.ijheatmasstransfer.2024.125559.

Wang, R., Chew, J. W., Gao, F., and Marxen, O., 2025a. Effect of axial throughflow swirl on flow and heat transfer in a compressor disc cavity. ASME. J. Turbomach., 147, 121017. https://doi.org/10.1115/1.4069800.

Wang, R., Gao, F., Chew, J. W., and Marxen, O., 2025b. Flow and heat transfer in a rotating disc cavity with axial throughflow at high speed conditions. ASME J. Turbomach., 147, 091005. https://doi.org/10.1115/1.4067437.

Wei, S., Mao, J., Yan, J., Han, X., Tu, Z., and Tian, R., 2020. Experimental study on a hybrid vortex reducer system in reducing the pressure drop in a rotating cavity with radial inflow. Exp. Therm. Fluid Sci., 110, 109942. ISSN:0894-1777 https://doi.org/10.1016/j.expthermflusci.2019.109942.

9

Rim Sealing

The importance of outlet boundary conditions for rotor–stator systems with radial outflow was noted in Chapter 7. This has received considerable research attention for axial turbines in which ingestion of hot mainstream gas into disc cavities can cause overheating. In engines, the clearance between the rotor and stator must be sufficient to accommodate relative movements, and the flow required to prevent or control ingestion through the clearance may be more than that needed for cooling. This incurs penalties to engine efficiency associated with the removal of sealing air from the compressor and reintroduction into the main turbine gas path. Examples of aeroengine turbine rim seals are indicated in Figures 1.1 and 1.2. A wide variety of seal and cavity geometries occur in practice. This chapter summarises and explains the current understanding of the flow physics and predictive capability and illustrates design approaches.

A widely adopted measure of rim seal performance is sealing effectiveness as detected from gas concentration measurements. Typically, the cavity purge or seal flow is seeded with carbon dioxide, and effectiveness ε is defined in terms of the gas concentration c as

$$\varepsilon = (c_{ext} - c)/(c_{ext} - c_{in}) \tag{9.1}$$

with subscripts *in* and *ext* denoting conditions at the cavity inlet and external to the cavity. This effectiveness is a function of position in the cavity and is often measured from gas samples extracted through tappings in the stator. To further characterise the seal (for a particular test configuration), the minimum nondimensional cavity flow rate $C_{W,min}$ or $U_{C,min}/\Omega b$ required to prevent ingestion (at a measurement point) is frequently reported. Here U_C is the mean velocity through the seal calculated from continuity at the minimum seal clearance s_C using the purge flow inlet flow rate and density. (Some published papers use the minimum axial clearance, which may be greater than the minimum clearance used here, or incorporate the annulus flow density.) In practice, ε becomes weakly dependent on the purge flow rate as it approaches unity and $C_{W,min}$ can depend significantly on the curve fitting method and measurement position used. It is generally assumed that reducing seal flow requirements will be beneficial to engine performance. Further consideration of the effects of seal design and flow rate on turbine performance may include turbine stage efficiency or loss coefficients.

Reviews of previous studies of turbine rim sealing are given by Johnson *et al.* (1994), Scobie *et al.* (2016a) and Chew *et al.* (2019). Johnson *et al.* describe the complexity of the problem, recognising disc pumping, annulus flow asymmetries, seal geometry, turbulent transport in overlapping platform regions, eccentricity and manufacturing tolerances as significant factors. Classifying ingress as externally-induced, rotationally-induced or combined, Scobie *et al.* fitted data for sealing effectiveness from various research groups using methods based on early studies and applied at the University of Bath over several years. The methods are described as very useful in ranking the effectiveness of different seals. Chew *et al.* emphasised the intrinsically unsteady nature of many ingesting rim seal flows. While sealing effectiveness for particular configurations is often reasonably correlated using a simple velocity ratio and measured minimum sealing flow, the correlations do not capture the known flow mechanisms and so have limited generality. The reviews also note that comparisons of computational fluid dynamics (CFD) predictions with measurements have often shown disparities. Since these reviews were published, the importance of intrinsic unsteadiness in the flow has been more widely recognised and a fuller understanding of the flow mechanisms is emerging.

9.1 Flow Mechanisms

Figure 9.1 shows a turbine stage including the rim seal and disc cavity upstream of the rotor. This is based on a research facility at the University of Oxford and is a simplified representation of an aeroengine single-stage high-pressure (HP) turbine. The cavity purge flow joins the mainstream turbine flow through the rim seal, and some hot mainstream gas may flow radially inwards through the seal. The main annulus flow near the seal has a mean swirl velocity above rotor speed, has circumferential pressure and velocity asymmetries associated with the stationary vanes and rotating blades, contains cooling flow ejected upstream and is subject to 3D secondary flows. Some further discussion of flow in the cavity shown in the figure was given in Section 6.2.5, including comparison of pressure measurements with CFD in Figure 6.4. At low purge flow rates, the rotor–stator disc cavity is dominated by recirculation driven by disc pumping and may be susceptible to rotating flow modes, as discussed in Section 7.1.4. Disc pumping and annulus pressure asymmetries are highlighted in the figure and have long been recognised as potential drivers of mainstream gas ingestion. When ingress occurs interaction of the purge and mainstream flow

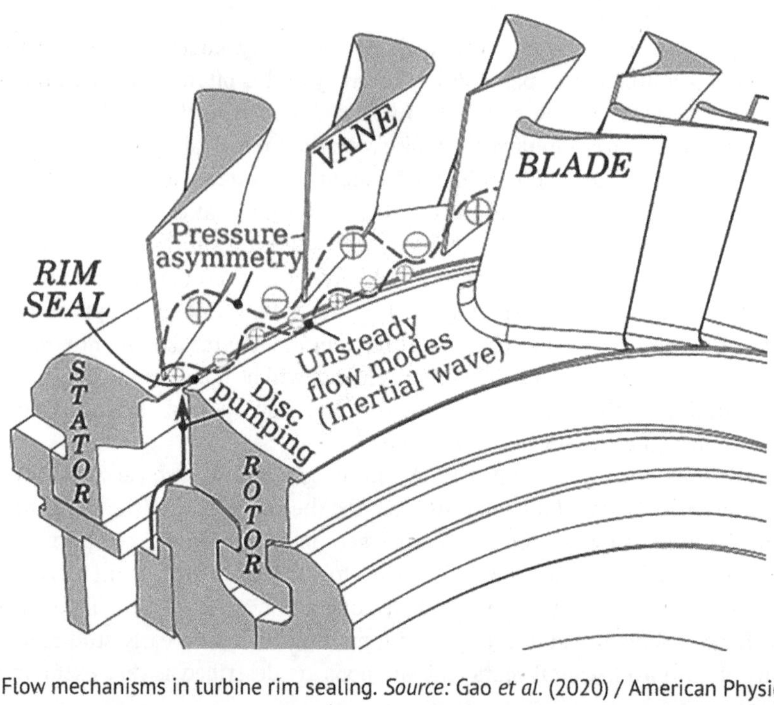

Figure 9.1 Flow mechanisms in turbine rim sealing. *Source:* Gao *et al.* (2020) / American Physical Society.

generates rotating, unsteady flow modes within the seal. Further interactions, such as coupling with rotating flow modes associated with vanes or blades, may also occur. Disc pumping and pressure-driven effects are considered in the next two subsections. A third subsection considers broader effects of flow interactions, including the unsteady flow modes indicated in the figure.

9.1.1 Disc Pumping

As indicated in Figure 3.5, an open rotor–stator cavity in a quiescent environment can be expected to draw in flow to satisfy disc pumping. In this case, Stewartson-type flow occurs, with the disc boundary layer approximating to that of a free disc. For a closed or partially closed cavity, disc pumping induces recirculation of the fluid. A rotating core then develops to balance the radial flows in the rotor and stator disc boundary layers with the supplied purge flow, and any flow ingested from the environment. This is Batchelor type flow, as in Figure 3.5.

A further illustration of disc cavity flow is given in Figure 9.2. In this case, the purge flow is assumed sufficient to prevent ingestion through a simple axial seal adjacent to the rotor. For this condition, the purge flow accelerates to pass through the narrow rim seal gap, requiring a pressure drop across the seal from the outer cavity to the environment. Bayley and Owen (1970) defined, the minimum purge flow rate to that at which the pressure drop is zero and obtained an empirical correlation for $C_{W,min}$. This is written in the original and two other forms as follows:

$$C_{W,min} = 0.61(s_c/b)Re_\phi \tag{9.1.1a}$$

$$U_{c,min}/\Omega b = 0.0971 \tag{9.1.1b}$$

$$\lambda_{min} = 0.61(s_c/b)Re_\phi^{0.2} \tag{9.1.1c}$$

The second equation above shows that the minimum sealing flow is independent of viscosity. Equation 9.1.1c shows consistency with scaling of the cavity flow deduced from integral methods, as discussed in Section 7.1.2, provided that seal clearance is scaled with the disc boundary layer thickness $\delta \propto Re_\phi^{-0.2}$.

Neglecting the effect of circumferential velocity, it may be hypothesised that for $C_W > C_{W,min}$, the pressure drop across the seal may be proportional to the difference between the dynamic head of the seal flow and a fraction of the dynamic head of the radial flow in the disc boundary layer. In the limit

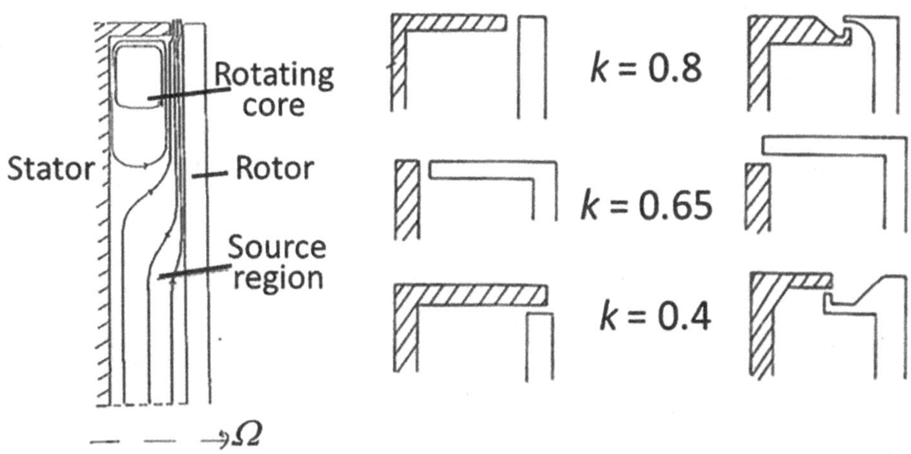

Figure 9.2 Pumping flow for a simple axial seal with $C_W > C_{W,min}$ and model coefficients for different seals.

$C_W = C_{W,min}$, the seal flow is then given by $U_{C,min} = kU_{d,m}$, where k is an empirical constant and $U_{d,m}$ is representative of the boundary layer radial velocity approaching the seal. Chew (1991) estimated $U_{d,m}$ from integral momentum solutions for rotor–stator disc cavity flows. The values of λ_{min} deduced varied linearly with seal clearance (as in Equation 9.1.1c) for values of λ less than ~0.1. For a simple rotor–stator cavity, the integral solution gives $U_{d,m}/\Omega b \approx 0.121$ for these conditions, giving

$$U_{c,min}/\Omega b \approx 0.121k \tag{9.1.2}$$

Values of k deduced by Chew *et al.* (1992) from published experimental data for different seal types are given in Figure 9.2. Considering the flow paths, plausible explanations for the variation of k can be given. The top two seals are fed directly by the pumped flow and so have the highest values of k. The rotating shroud of the middle two seals turns the flow reducing k, but the pumped fluid has high circumferential velocity and centrifugal force, so the subsequent turn to the radial direction

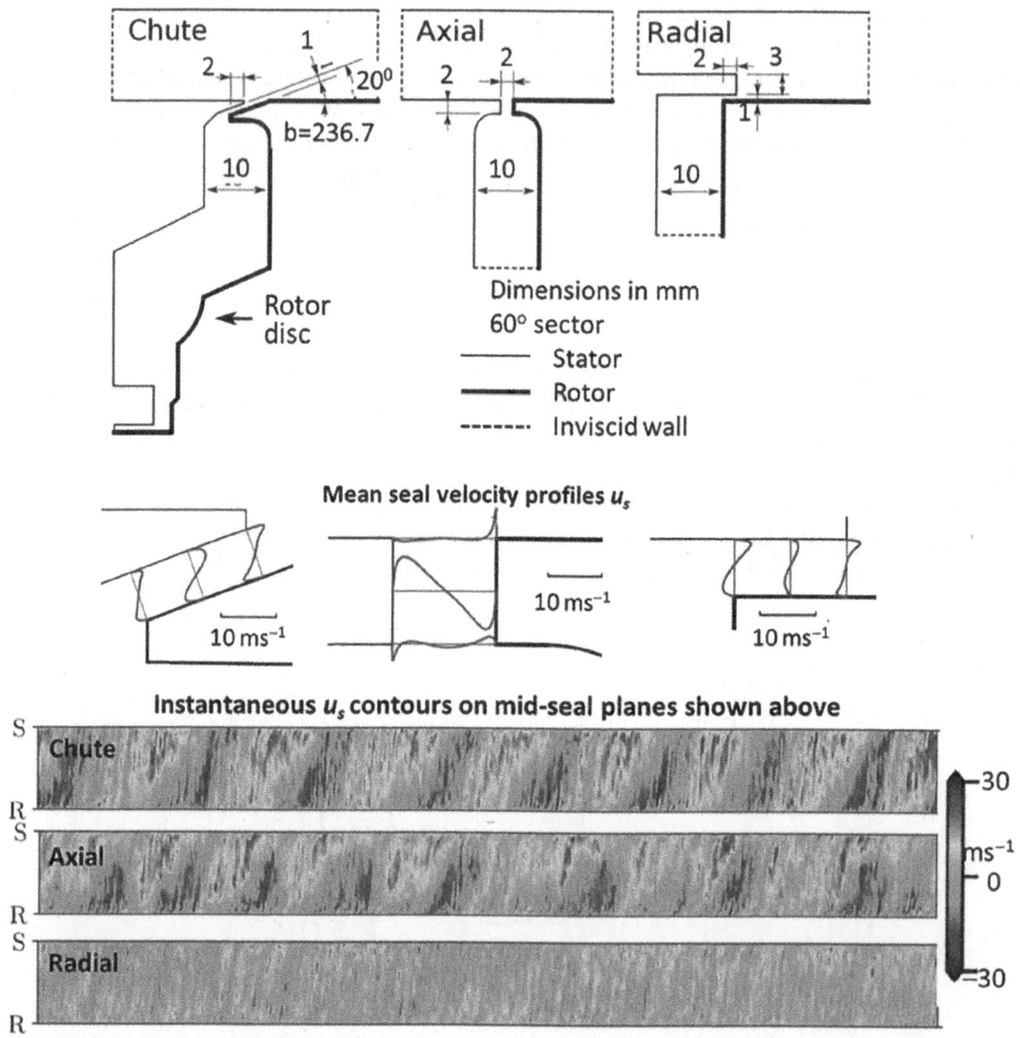

Figure 9.3 LES for closed rim seal configurations, $Re_\phi = 2.6 \times 10^6$. *Source:* Gao et al. (2020) / American Physical Society.

in the left-hand seal offers little further resistance. In the two bottom seals, the pumped flow will impinge on the stator and must turn sharply into the seal.

As purge flow is reduced below the minimum sealing rate, ingress (or ingestion) occurs and the radial inflow and outflow through the seal interact. This produces 3D, unsteady flows, as illustrated in Figure 9.3. In these simulations, there is no flow through the boundaries of the CFD domains shown for chute, axial and radial seals. Hence, flow through the seals is due solely to the disc rotation. Results are given for mean and instantaneous values of the velocity component u_s aligned with the seal boundaries in an $r - z$ plane. The instantaneous contour plots for the chute and axial seals show alternating inflow and outflow regions distributed circumferentially within the seals. Note that the seal clearance is stretched in these plots. The large-scale structures have a circumferential length scale of order $10s_c$ and fluctuate somewhat. For this configuration, the structures rotate at between 40% and 50% of rotor speed. The instantaneous velocities in the structures are considerably higher than the mean velocities, for which profiles are shown at the seal inlet, mid-seal and seal exit position. Flow in the radial seal is more uniform circumferentially. Turbulence at smaller length scales is visible for all three seals. The radial seal shows no sign of Taylor vortices as can occur in the space between a rotating inner cylinder and stationary outer cylinder. This is consistent with the high Reynolds number, featureless turbulence regime described by Grossmann *et al.* (2016).

As discussed in Section 3.3.4, rotating flows are susceptible to inertial waves associated with the restoring nature of the Coriolis force. This is clearly the case for the chute and axial seals. The radial seal, which limits the radial fluid motion with the seal geometry, restricts the Coriolis force and suppresses the inertial wave. The flows in Figure 9.3 would not occur without disc rotation, but similar effects may arise in stationary cavity flows interacting with a swirling external flow. This was demonstrated using RANS CFD by Tang *et al.* (2017) for a stationary shroud cavity supplying cooling air to the main annulus upstream of the turbine blades through a chute-shaped channel.

9.1.2 Pressure-driven Ingestion

In the absence of rotation and purge flow, circumferential variation of pressure in the external flow will produce inflow through the seal where there is high external pressure and outflow where there is low external pressure. Campbell (1978) and Abe *et al.* (1979) were amongst the first to highlight this ingress mechanism in the open literature. Phadke and Owen (1988) investigated the effect by creating asymmetries in a nominally axial external flow using sections of honeycomb in the annulus. $C_{W,min}$ was correlated with the magnitude of the pressure asymmetry. The mechanism is illustrated in Figure 9.4, which assumes a sinusoidal circumferential distribution of annulus pressure and includes purge flow. Increasing purge flow pressurises the cavity with minimum sealing flow expected to correspond to the point at which the cavity pressure reaches the maximum pressure in the annulus. The simple "orifice model" assumes that the seal flow at each circumferential position is characterised by a discharge coefficient C_d. Assuming a constant C_d with the sinusoidal pressure distribution then gives the following estimate for the minimum sealing flow:

$$\frac{U_{c,min}}{\Omega b} = C_d \frac{2}{\pi} \sqrt{\frac{2 \Delta p_\phi}{\rho (\Omega b)^2}} \tag{9.1.3}$$

Here Δp_ϕ is the minimum-to-maximum annulus circumferential pressure difference. This equation agrees with that given by Hamabe and Ishida (1992), who also considered triangular (or sawtooth) and square wave pressure distributions.

Comparing equations 9.1.2 and 9.1.3 may suggest that the pressure asymmetries typically generated by turbine vanes and blades will be a dominant driving mechanism for ingestion in many

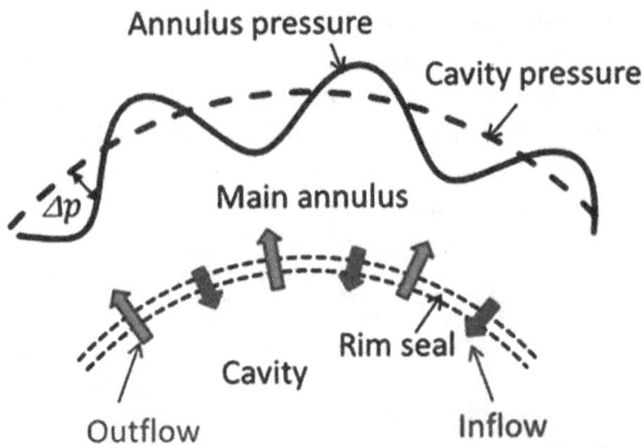

Orifice model: $U_{seal} = \text{sgn}\,(\Delta p)C_d\sqrt{2|\Delta p|/\rho}$

Figure 9.4 Illustration of pressure-driven ingestion.

cases. However, neither of the equations takes account of the high circumferential velocity of flow upstream of a turbine blade. In highly swirling flow, an element of fluid may be subject to alternating positive and negative radial pressure gradients as it moves circumferentially within the seal. This was discussed by Chew *et al.* (1994) and Gentilhomme *et al.* (2003) with modelling of the inertial effects indicating their tendency to reduce ingestion, depending on the seal geometry and asymmetry wavelength. Dawson *et al.* (2022) discuss the effect in further detail showing important effects of small amplitude, long wavelength circumferential pressure asymmetries. As discussed by Phadke and Owen, these are difficult to avoid in experiments. Dawson *et al.* demonstrate significant effects of such asymmetries on ingestion in full-stage turbine experiments and CFD.

9.1.3 Mainstream, Seal and Cavity Flow Interactions

Figure 9.5 shows measurements and CFD calculations for a rim seal (or gap) in the second stage of a two-stage turbine. The circumferential velocity in the annulus leaving the vanes is about $1.8\Omega b$. Although the seal is close to both the rotor blades and stationary vanes and subject to the associated pressure asymmetries, Cao *et al.* (2004) found that other effects dominated. RANS CFD studies showed alternating inflow and outflow regions rotating at slightly below disc speed, as in the bottom right of the figure. Pressure measurements confirmed the CFD. This is illustrated by the measurements from a fast response transducer in the gap near the root of the vanes. The spectrogram at the top right of the figure shows results for a range of rotor speeds at constant pressure level and ratio. The main frequencies detected increase slightly with speed and can be compared with the engine order traces that are just visible as diagonal straight lines. The results indicate that the frequency is a stronger function of annulus swirl velocity than rotor speed. The phase angle between measurements at different annular positions confirms that the structures rotate just below disc speed and indicate 11–14 inflow/outflow pairs or nodal diameters (ND). The unsteadiness was much stronger at the blade root than at the rotor hub position shown in the figure. The spectrogram at the bottom left of the figure shows some random flow behaviour. The number of flow structures around the annulus may vary with time.

Many other studies with fast response instrumentation, unsteady RANS CFD and LES have found similar rotating flow structures in the seal and disc cavity to those in Figures 9.3 and 9.5.

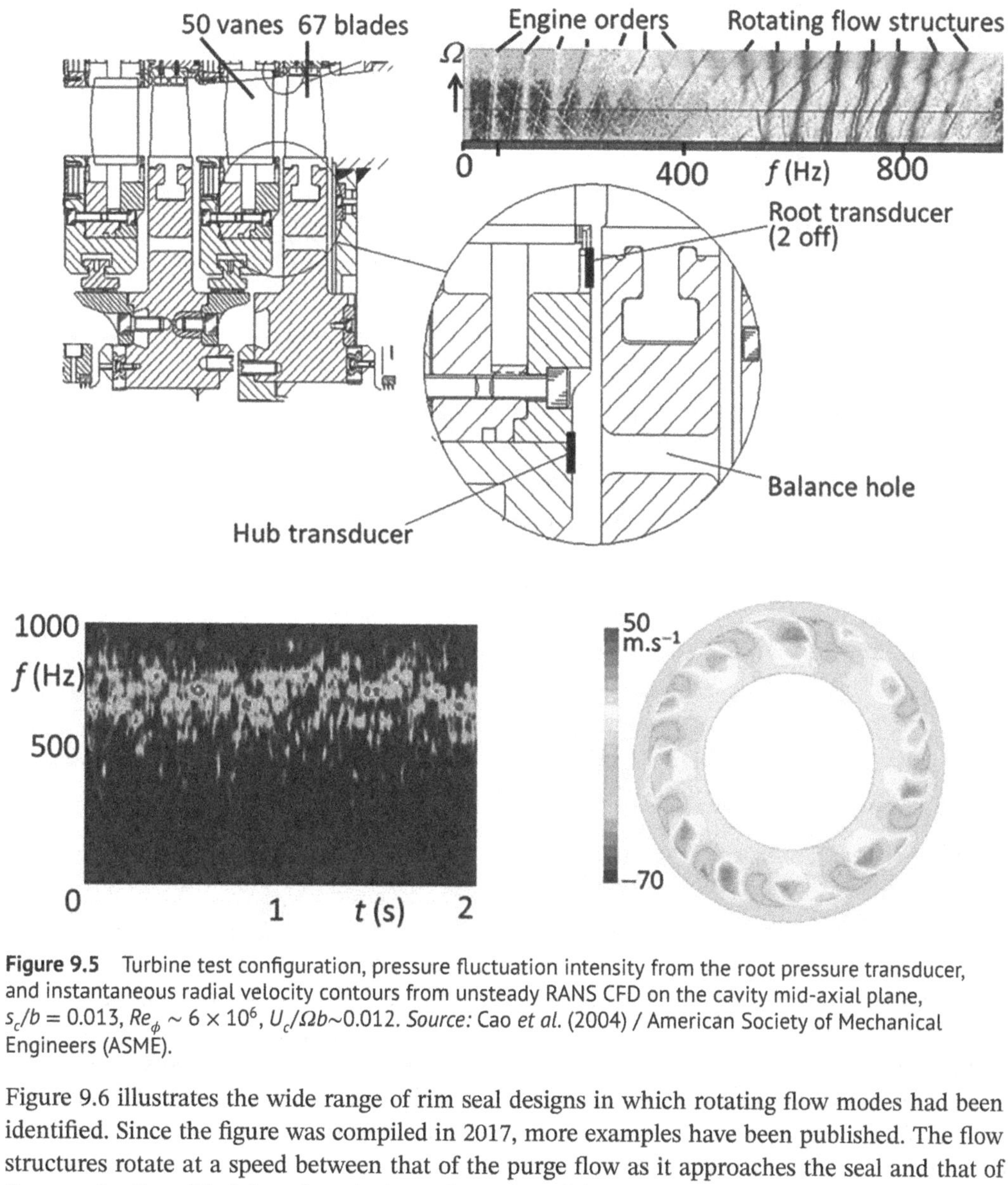

Figure 9.5 Turbine test configuration, pressure fluctuation intensity from the root pressure transducer, and instantaneous radial velocity contours from unsteady RANS CFD on the cavity mid-axial plane, $s_c/b = 0.013$, $Re_\phi \sim 6 \times 10^6$, $U_c/\Omega b \sim 0.012$. *Source:* Cao *et al.* (2004) / American Society of Mechanical Engineers (ASME).

Figure 9.6 illustrates the wide range of rim seal designs in which rotating flow modes had been identified. Since the figure was compiled in 2017, more examples have been published. The flow structures rotate at a speed between that of the purge flow as it approaches the seal and that of the annulus flow. Their length scale depends on that of the seal or surrounding geometry. Seals in which the minimum clearance is radial often also have wider axial gaps, and these may account for the intrinsic unsteadiness in these cases. In many configurations, the unsteady flow structures clearly dominate the flow exchange between the cavity and the outer annulus.

Calculated and measured ingestion levels also reflect the importance of the rotating flow modes generated by interaction of the annulus and purge flows. Further to Cao *et al.*'s axial gap computations, Boudet *et al.* (2005, 2006) reported large sector CFD solutions with axisymmetric geometry and boundary conditions for the chute seal, denoted "b" in Figure 9.6. This allows isolation of the unsteady effects from the influence of vanes and blades. These studies found higher levels of ingestion than predicted by smaller sector models that included vanes and blades but suppressed the intrinsic unsteadiness. This is consistent with experiments in which the distance between upstream

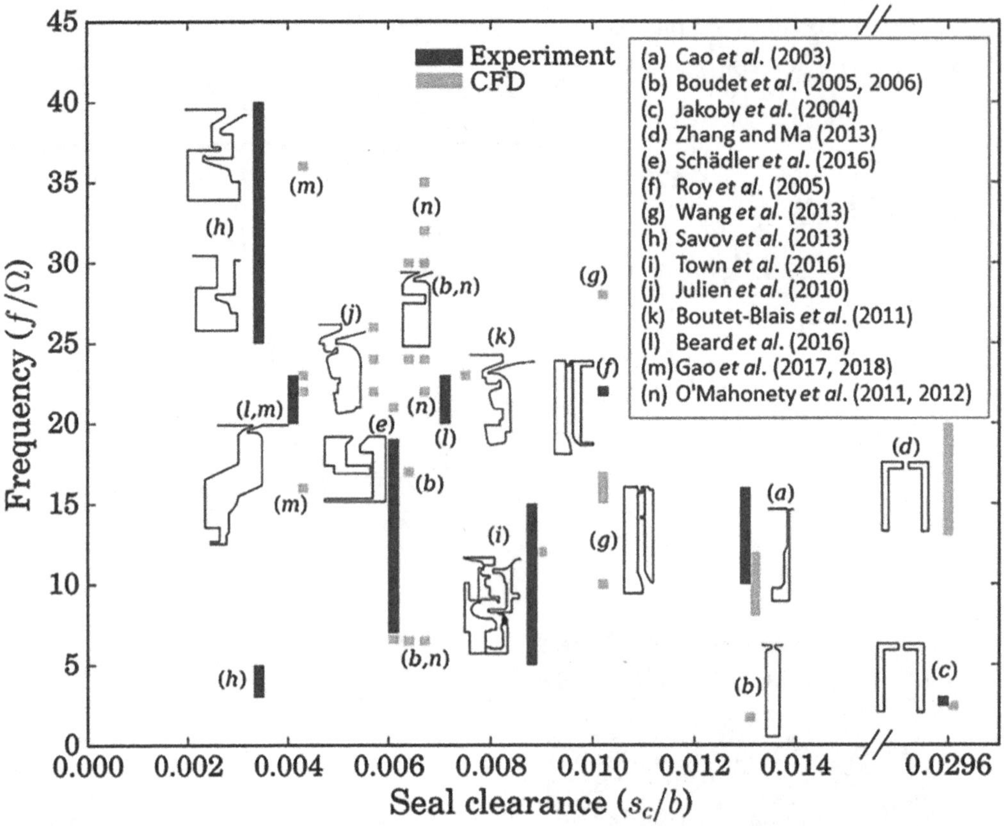

Figure 9.6 Examples of experimental and CFD evidence for non-blade passing-related rotating flow modes. Chew *et al.* (2019) / with permission of SAGE Publications.

vanes and the rim seal was varied. The static pressure asymmetry due to the vanes decays quite rapidly with downstream distance, but measurements show a relatively weak corresponding reduction in ingestion. In their turbine stage experiments, Hualca *et al.* (2020) found that varying the vane axial position only affected the ingestion in conditions when interaction with the rotor blades increased ingestion. In experiments for a cavity downstream of the rotor, Graikos *et al.* (2022) adjusted the annulus flow while keeping the rotor speed and purge flow rate fixed. Examining sealing effectiveness for two blade designs, including data from some severe off-design conditions, it was concluded that the difference in circumferential velocity between the annulus and seal flow was the main driver for ingestion.

The rotating modes in the seal may extend into the cavity or interact with other modes. Wang *et al.*'s (2013) RANS CFD solutions for a full turbine stage indicated different rotating flow modes in the seal and rim cavity. Their cavity geometry is denoted "g" in Figure 9.6. The seal mode had 15 to 17 ND, compared to 12ND in the outer cavity. The inner cavity showed a weak 4ND mode. There are similarities with the 3 lobed cavity mode in Figure 7.14 that interacts with flow around 18 boltheads on the rotor. Pogorelov *et al.*'s (2019) LES also identified different modes in their axial seal and cavity. In their case, a 3ND acoustic wave was identified. Bru Revert *et al.*'s (2023) measurements in the disc cavity of Figure 9.1 give experimental confirmation that inertial and acoustic waves can interact. The effect mode interactions have on ingestion is not entirely clear, although any resonance reinforcing the seal modes could be expected to encourage ingestion. Zhou *et al.*'s (2011) ingestion measurements show some improvements in sealing for smaller rim cavities (with higher a/b) that may be less susceptible to low ND modes. However, geometry and inlet conditions could also clearly affect the flow in the cavity and approaching the seal so there may be other factors involved. The decay of higher ND modes and prevalence of low ND modes for cavities with low a/b is consistent with the analysis for inertial waves in Section 3.3.4.

Interaction of the purge and main annulus flows affects both sealing performance and turbine efficiency. Schädler *et al.* (2016) reported unsteady flow measurements in a stage-and-a-half turbine with seal geometry "e" in Figure 9.6. Their aerodynamic measurements in the annulus show interaction of the purge flow with secondary flow generated by the upstream vane. The unsteady hub cavity modes were linked to aerodynamic losses and were also considered to have a significant effect on noise generation for the turbine. Other studies, such as Boudet *et al.'s* (2006) RANS CFD for seal geometry "b" in Figure 9.6, show the purge flow also interacting with the rotor blade passing. Seal egress was shown to align with radially outwards secondary flows generated on the vane and blade suction surfaces. Ingress occurred between these zones. Pressure signals show rotating modes in the rim cavity at 12 %, 44 %, 56% and 100% of blade passing frequency 54 Ω. These reflect interaction of a dominant seal mode at 44% with the blade passing. The lower frequency, corresponding to a mode with lower ND, was strongest in the inner cavity and could be detected at the vane leading edge in the annulus. Hualca *et al.*'s (2020) pressure measurements for a radial clearance seal also show effects of blade passing on the cavity modes. Introducing blades led to frequency scattering in the cavity mode and, at certain purge flow rates, a shift in the dominant frequency to half the blade passing frequency. This was associated with an increase in ingestion.

The examples given above illustrate the complex flow interactions that can occur in turbine rim seals. These depend on seal, turbine and cavity design. Most examples are for rim seals upstream of a high-pressure turbine disc. Further configurations in which rim sealing is important occur in other turbomachines. Cavities downstream of the rotor will be subject to annulus flow with lower swirl and cavity recirculation driven by disc pumping is in the opposite direction (relative to the annulus flow). Hösgen *et al.* (2023, 2024) have found significant interaction of low ND modes for upstream and downstream cavities in their 1.5 stage turbine simulations. Merli *et al.*'s (2023) pressure measurements in a downstream cavity showed a dominant signal at blade passing frequency. Although no unrelated rotating flow modes were identified, a weaker signal at ~63% associated with the interaction of high- and low-pressure turbine blade passing was found. Da Valle *et al.* (2025) found little evidence of cavity modes in the downstream rim seal of a low-pressure turbine.

9.2 Dimensional Analysis and Elementary Modelling

Considering the observed flow mechanisms described above the simplest plausible representation of an axial turbine rim seal might assume circumferentially and radially uniform main annulus flow approaching the seal and neglect fluid property variations. For a fixed geometry, extending the dimensional analysis for rotor–stator cavities in Sections 7.1.2 and 2.3.2 indicates that sealing effectiveness will depend on five independent parameters. As $U_C/\Omega b$ has been shown to be useful in correlating measurements, this parameter is selected to characterise the purge flow rate rather than C_W or λ. Re_ϕ and inlet swirl fraction $v_{\phi,in}/\Omega r$ are retained from the simple cavity analysis. Reynolds number effects have been shown to be weak in many studies, with examples given by Scobie *et al.* (2016a) and later in this chapter. In many circumstances, the inlet swirl fraction might remain fixed or have little influence on the cavity flow. Thus, the mean seal velocity may be the dominant parameter in correlating results without annulus flow, as in equations 9.1.1b and 9.1.2.

A suitable choice of the two parameters representing the annulus flow is the flow coefficient $v_{z,ann}/\Omega b$ and the annulus flow swirl ratio $v_{\phi,ann}/\Omega b$, where $v_{z,ann}$ and $v_{\phi,ann}$ are annulus inlet values of the axial and circumferential velocity components. The ratio $v_{\phi,ann}/v_{z,ann}$ is the tangent of the annulus flow angle to the axial direction. If the swirl is generated by fixed upstream vanes, the flow angle might be considered constant, reducing the number of independent parameters by one. A typical value of the exit flow angle for a high-pressure turbine vane would be ~70°. As shown by Scobie *et al.*'s survey of test facilities, most research on full-stage turbines has a flow coefficient around 0.5. This is representative of high-pressure turbines. Low-pressure turbines may have higher flow coefficients of ~1. In full-stage turbines, the annulus flow drives the rotor. If velocity triangles for the blades remain unchanged over operating conditions, $v_{z,ann}/\Omega b$, and $v_{\phi,ann}/\Omega b$ will also be unchanged. Then, as for cases without annulus flow, the value of $U_{C,min}/\Omega b$ may characterise a particular configuration.

Equation 9.1.3 has been used to correlate results in various studies including annulus flow with fixed vanes upstream of the seal. Noting that Δp_ϕ will generally be proportional to the annulus flow dynamic head, this can be rewritten as

$$\frac{U_{c,min}}{\Omega b} = A\frac{v_{\phi,ann}}{\Omega b} \quad \text{or} \quad \frac{U_{c,min}}{\Omega b} = B\frac{v_{z,ann}}{\Omega b} \tag{9.2.1}$$

where $A = \frac{2C_d\sqrt{K}}{\pi} sin\alpha, B = Atan\alpha, \alpha = arctan(v_{\phi,ann}/v_{z,ann})$ and $K = 2\Delta p_\phi/\rho(v_{\phi,ann}^2 + v_{z,ann}^2)$ are assumed constant for a given configuration. Ωb is a common term that can be cancelled in these equations. This is to be expected as equation 9.1.3 assumes that ingestion is driven solely by the external flow. In these circumstances, Ωb can be replaced by $v_{z,ann}$ as the characteristic velocity in the dimensional analysis. The analysis then implies that a single value of $U_{C,min}/v_{z,ann}$ may characterise a particular configuration, as also implied by equation 9.2.1. Many studies have suggested using Δp_ϕ as a basis for extrapolating laboratory results to engine conditions, as in equation 9.1.3. Equation 9.2.1 offers alternative extrapolations based on the mean annulus flow rather than the pressure asymmetry. The physical interpretation of the factors A and B in these equations changes, but their values can be estimated from the minimum sealing flow, flow coefficient and flow angle for the laboratory experiment.

When ingestion occurs, complex, unsteady flows often develop within the seal. An exception is purely pressure-driven flow, as described in the previous section, and Figure 9.4. With assumed values for the discharge coefficient and known annulus pressure distribution, the equation given in the figure can be used to estimate the seal flow at any circumferential position and hence the egress and ingress mass flows. Assuming ingress and egress to be at the annulus inlet and fully mixed cavity conditions then gives $\varepsilon = \dot{m}/(\dot{m}_{ing} + \dot{m})$, where $\dot{m}$ and $\dot{m}_{ing}$ are the purge flow supply and ingress mass flow rates, respectively. This is known as the orifice model and has been considered in many studies. Attempts to use the model to predict sealing effectiveness using correlations or measurements of discharge coefficients, such as that by Gentilhomme *et al.* (2003), showed discrepancies with experiments. Adjusting discharge coefficients to match test data for specific configurations has had more success.

At the University of Bath, substantial use has been made of an orifice model assuming a triangular wave pressure asymmetry with different discharge coefficients for seal egress and ingress. With this approach, sealing effectiveness depends on $U_C/U_{C,min}$ and the ratio of C_d for ingress to that for egress, denoted Γ, as follows:

$$U_c/U_{c,min} = \frac{\varepsilon}{[1 + \Gamma^{-2/3}(1-\varepsilon)^{2/3}]^{3/2}} \tag{9.2.2}$$

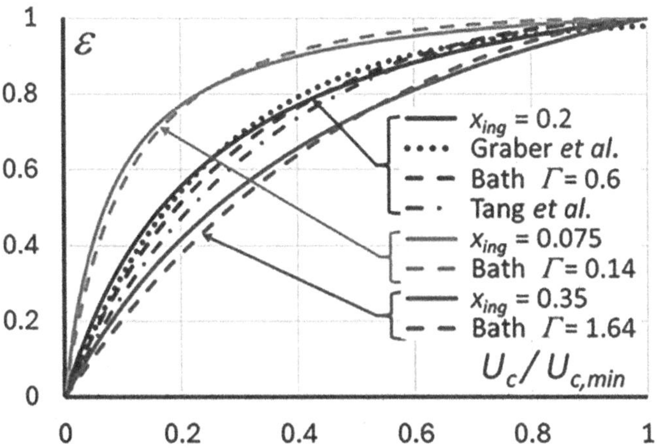

Figure 9.7 Examples of fitting curves for sealing effectiveness.

Figure 9.7 shows examples of effectiveness profiles from this equation for representative values of Γ reported by Scobie *et al.* For comparison, curves assuming that the ingested flow rate is a fraction x_{ing} of the shortfall between the supplied flow and the sealing minimum are also shown. These curves are given by equation (9.2.3), which is derived using the fully mixed cavity assumption:

$$\varepsilon = U_c/[(1 - x_{ing})U_c + x_{ing}U_{c,min}] \tag{9.2.3}$$

Considering disc pumping flow, as in Section 9.1.1 and Figure 9.2, $x_{ing} = 0.2$ was found to give a reasonable representation of the data then available (Chew *et al.*, 1992). The comparisons show that similar curves can be generated with either method. The spread of the curves obtained by varying Γ or x_{ing} is representative of that from different experimental configurations. Other relations have also been used in empirical correlations of sealing effectiveness. These include turbulent diffusion models developed by Graber *et al.* (1987) and Savov and Atkins (2017), and a wave model introduced by Tang *et al.* (2024). The figure includes Graber *et al.*'s curve $\varepsilon = 1 - e^{-\alpha\lambda}$, with α constant and $U_{C,min}$ corresponding to the point, where $\varepsilon = 0.98$, and Tang *et al.*'s more recent result. The proximity of Tang *et al.*'s relation in Figure 9.7 to correlations proposed more than 30 years earlier reflects similar behaviour in the wide range of rim sealing tests conducted over the intervening period.

The above analysis neglects many complicating factors but provides a basis for engineering evaluations. Estimates of sealing effectiveness may be used directly to gauge cavity temperatures or to obtain ingress and egress flow rates. These flow rates can be incorporated into fully mixed models of rim rotor–stator cavities, as illustrated for preswirl chambers in Section 7.3.2. Care is needed in extrapolating from laboratory to engine conditions. Uncertainties may arise due to differences in turbine design, seal and cavity geometry, Mach number, purge to annulus flow density ratio, large-scale circumferential asymmetries, interactions with other components, geometric tolerances, flow inlet variations and additional leakage paths around the blade roots. Some of these effects are illustrated in the following section.

9.3 Sealing Effectiveness

9.3.1 Measurements with No or Weak Annulus Flow

Many turbines use double seals in which the outer seal is exposed to the main annulus flow, and the inner seal will operate in a more benign environment. For inner seals, disc pumping may be

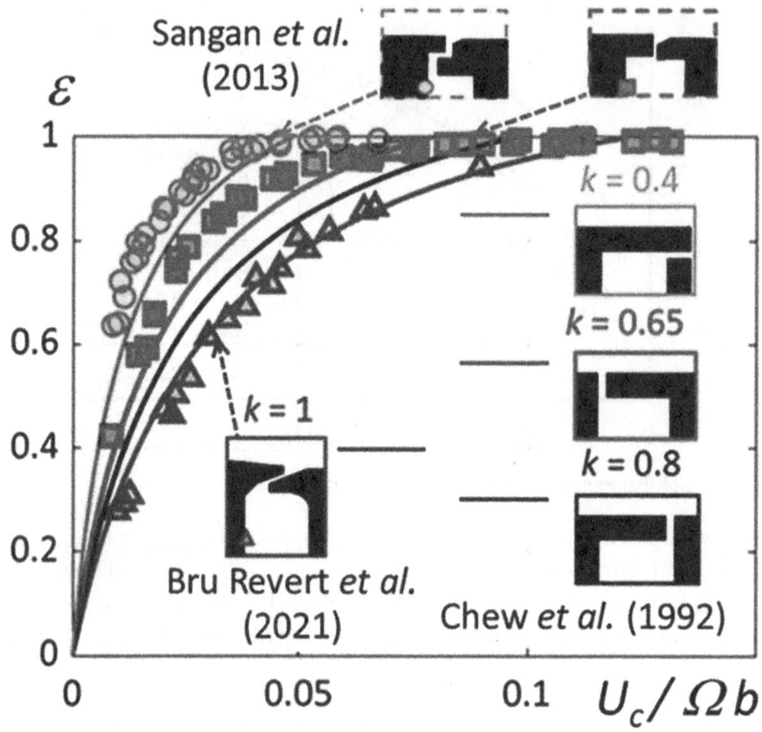

Figure 9.8 Examples of sealing effectiveness at high radii on the stator as a function of non-dimensional flow rate. *Source:* Adapted from Bru Revert *et al.* (2021a) / American Society of Mechanical Engineers (ASME).

expected to be the dominant mechanism driving ingestion. Experiments focussing on this effect have considered seals exposed either to quiescent, open laboratory conditions or to weak flow in an enclosing annulus. Various studies have examined the effect of operating conditions and seal geometry on sealing effectiveness. As illustrated and discussed in Sections 9.1.1 and 9.2, the performance of a given seal geometry has been found to depend principally on the purge flow rate $U_C/\Omega b$. Selecting a concentration measurement position on the stator near the seal, various studies have reported the variation of effectiveness with $U_C/\Omega b$, and a comparison of some results is given in Figure 9.8.

The solid lines in Figure 9.8 are correlations based on early studies using equations 9.1.2 and 9.2.3 with $x_{ing} = 0.2$. Curves for $k = 0.4$, 0.65 and 0.8 represent results for seals shown in Figure 9.2. Sangan *et al.*'s (2013b) measurements for their radial and axial clearance seals are for $Re_\phi = 5.3 \times 10^5$, 8.2×10^5 and 9.7×10^5. These are correlated with $k \approx 0.3$ and 0.6, although the curves are not shown. Considering the qualitative explanations for ranking of seals given in Section 9.1.1 and differences in seal geometry the lower values compared to earlier results for radial and axial seals (with $k = 0.4$ and 0.65) appear reasonable. Sangan *et al.*'s radial seal has a more tortuous flow path than the other seals, reducing k. Turning of the pumping flow into Sangan *et al.*'s axial seal will not be assisted by impingement on the stator as in the earlier axial seal.

Bru Revert *et al.*'s (2021a) measurements for a chute seal for $Re_\phi = 3.8 \times 10^5$ and 1.5×10^6 are correlated with $k = 1$. The radiused corner on the rotor might reduce pressure losses in the pumped flow as it approaches the seal compared to most other seals shown, but this does not account for the comparison with the axial seal with $k = 0.8$. In further tests, Bru Revert *et al.* (2021b) modified the chute geometry axially displacing the rotor disc rearwards. As a result, the overlap in the chute seal was lost and the minimum clearance increased by a factor 1.85. Plotted on the same basis as Figure 9.8 effectiveness for the modified configuration was closer to the curve for $k = 0.8$. This may reflect influence of the rotating flow structures that form in the chute and are shown in Figure 9.3. It is important to note that the comparison in Figure 9.8 is based on seals having the same minimum clearance. Based on the same axial clearance, the effectiveness for the 20° chute seal is higher than all the axial seals in the figure.

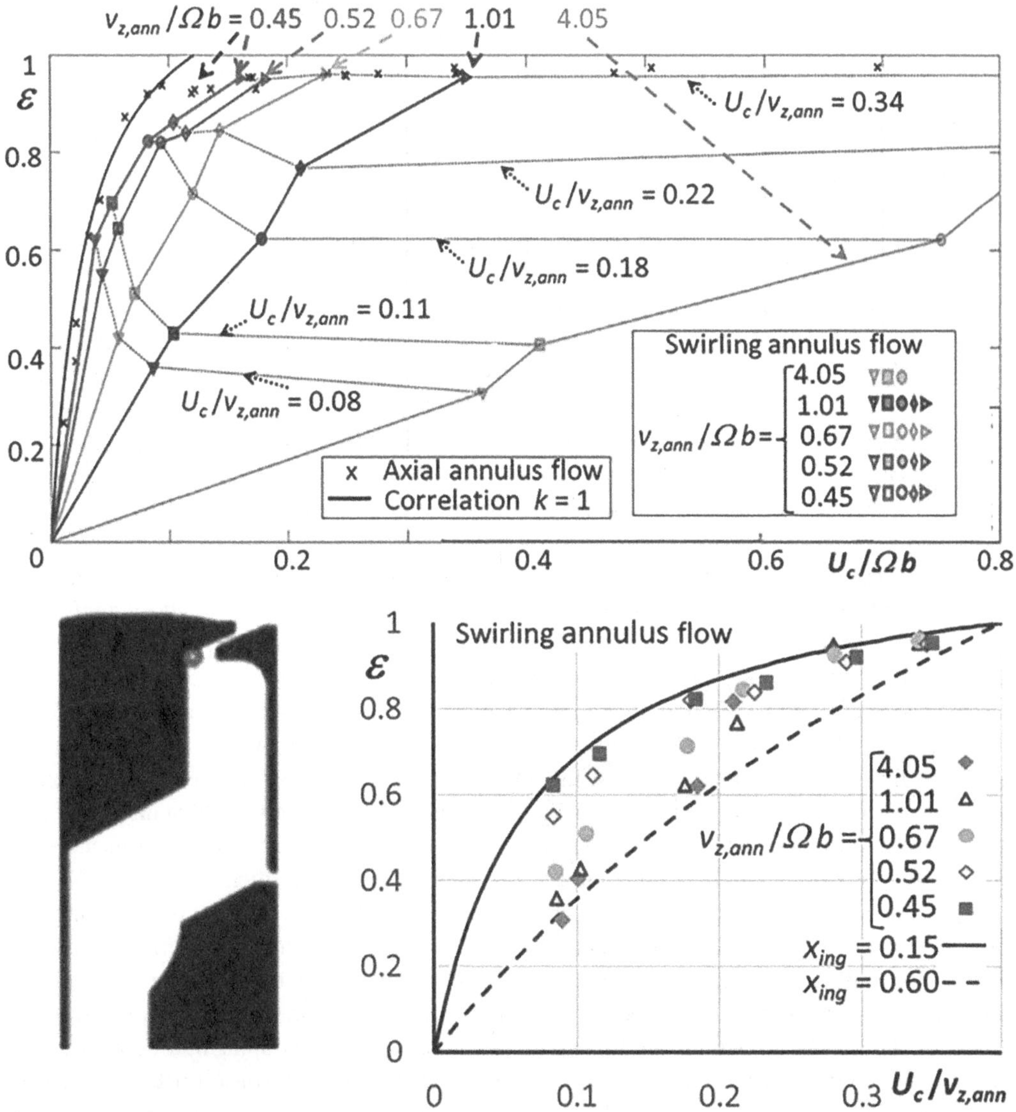

Figure 9.9 Sealing effectiveness at $r/b = 0.99$ against purge flow normalised with rotor speed (top) or annulus flow velocity (bottom) for different seal-to-axial velocity ratios and flow coefficients, $s_c/b = 0.0042$. *Source:* Bru Revert *et al.* (2021a) / American Society of Mechanical Engineers.

9.3.2 Measurements with Annulus Flow

Results from Bru Revert *et al.*'s chute seal with an imposed annulus flow are given in Figure 9.9. The top graph includes the disc pumping correlation with $k = 1$ to represent conditions without external flow, and measurements with an axial annulus flow without vanes or blades fitted. These show the axial external flow having little effect at low purge flow rates but producing a persistent low level of ingestion at higher flows. All other results in both graphs have a swirling annulus flow generated by vanes fitted upstream of the seal, as in Figure 9.1. No rotating blades were fitted for these experiments, which were all conducted at the same nominal annulus flow rate and mean axial velocity component. The test conditions with vanes used a series of values of purge flow rates at five different rotor speeds. In the top graph, test points at the same rotor speed are connected by solid lines and points at the same purge flow rate are connected by dotted lines. As the annulus flow is the same in all cases, these approximate to lines of constant flow coefficient $v_{z,ann}/\Omega b$ and constant purge flow rate normalised by annulus velocity $U_C/v_{z,ann}$. For this fixed vane configuration, sealing effectiveness might be expected to depend principally on these two nondimensionals. Their product gives the abscissa $U_C/\Omega b$ of the top graph. For ingestion driven entirely by the external flow ε is expected to be principally a function of just $U_C/v_{z,ann}$.

For ε lower than ~0.7 the top graph in Figure 9.9 shows clear trends. Consider the variation in ε as rotor speed increases and flow coefficient decreases (going from right to left along a dotted line in the top graph). There is little change in effectiveness as $v_{z,ann}/\Omega b$ reduces from 4.05 to 1.1, but further reduction improves sealing with ε approaching values obtained without external flow. For $\varepsilon > 0.8$, trends are less clear. The bottom graph shows some correlation of ε with $U_C/v_{z,ann}$ for these high ε conditions indicating dominance of external flow. The graph also confirms the strong effect of flow coefficient at lower values of ε. The two curves in the bottom graph are from equation 9.2.3 with $U_{C,min}/v_{z,ann} = 0.4$ and two different values of x_{ing} giving the ingestion flow rate as a fraction of the shortfall between the supplied flow and the sealing minimum. Noting that $U_{C,min}/v_{z,ann}$ can also be adjusted to fit data, the spread of results is comparable to that of the fitting curves in Figure 9.7.

Bru Revert *et al.* found that measurements in Figure 9.9 for $v_{z,ann}/\Omega b = 1.01$ and 4.05 could be correlated with Hamabe and Ishida's (1992) pressure-driven ingestion model with a sinusoidal annulus pressure distribution and $C_d = 0.6$. However, whether the ingress is principally driven by the pressure asymmetry or other interactions of the purge and annulus flows is unclear. In their studies of seal geometries "h" in Figure 9.6, Savov *et al.* (2017) included measurements for a configuration without blades but with vanes fitted. Varying rotor speed with the other operating conditions fixed, they also found that sealing effectiveness generally reduced as flow coefficient increased. For their single lipped seal, a point was reached at high flow coefficients where the trend reversed and effectiveness (particular at a lower radius in the cavity) started to increase with flow coefficient. Savov *et al.* attributed the influence of flow coefficient to the difference in swirl velocities between the purge and annulus flows affecting the interaction. The implied increase in x_{ing} with flow coefficient in Figure 9.9 might well be associated with higher annulus velocities strengthening rotating flow structures and turbulent mixing. Angular momentum will be conserved apart from the effect of viscous drag and higher shear will generate stronger turbulence. For conditions with low differences in swirl velocity disc pumping may dominate.

Both Bru Revert *et al.* and Savov *et al.* associate ingestion with intrinsic unsteadiness. Bru Revert *et al.* (2023) identify the interaction of inertial and acoustic waves as a possible source of some variations in results. Savov *et al.*'s measurements show secondary variations with Mach and Reynolds number and the addition of simulated rotor blades. The qualitative agreement in trends between the two independent experimental studies despite quite different configurations

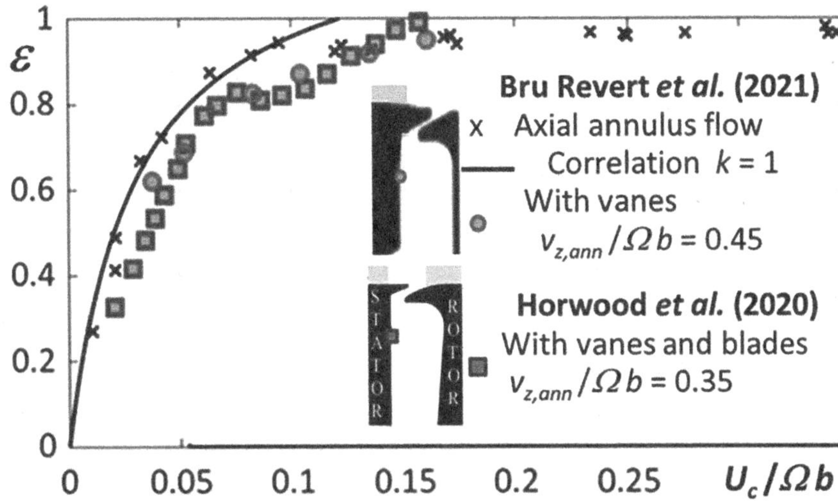

Figure 9.10 Sealing effectiveness at $r/b = 0.96$ of two chute seals under different external flow conditions. *Source:* Bru Revert *et al.* (2021a) / American Society of Mechanical Engineers.

suggests that similar observations will also apply to other seal geometries. A further observation by Savov and Atkins (2017) is that their static rim seal ring was made of 16 individual rim seal pieces introducing some steps in the stator surface between pieces. Later measurements with a single machined ring are stated to give higher sealing performance. This highlights manufacturing tolerance as another source of uncertainty and variation in seal performance.

Savov *et* al. simulated rotor blade effects by fitting rotating teardrop-shaped blockages in place of blades. These had quite a small effect on ingestion for their double lip seal, but increased ingestion for the single lip seal, particularly for $0.8 < \varepsilon < 0.95$. Hualca *et al.*'s (2020) effectiveness measurements in the outer cavity for a double radial clearance seal also show increased ingestion due to rotor blades for higher values of ε. Associated pressure measurements confirm the interaction of the unsteady seal flow with the rotating blade pressure field at these conditions. Horwood *et al.*'s (2020) measurements for a chute seal show similar effects. Their results are compared with those of Bru Revert *et al.* with vanes only in Figure 9.10. The detailed seal design, vane design and flow coefficient differ in the two studies, but the effectiveness levels are quite close. Horwood *et al.*'s data shows an inflection at $\varepsilon \approx 0.8$. Effectiveness measurements on other full-stage turbines by Gentilhomme *et al.* (2003), Clark *et al.* (2018) and Roy *et al.* (2021) show similar effects. Other studies have shown inclusion of rotor blades can produce different effects. For example, Bohn *et al.* (2000) reported that the presence of blades increased ingestion for one configuration but decreased it for another. They also found that annulus pressure asymmetry increased with introduction of the blades, suggesting that the strong rotor–stator interaction combined with unsteady seal flows could be a particular factor in this configuration.

The above observations relate to turbine rim seals upstream of the rotor. Annulus flows downstream of the rotor generally have low or negative swirl. Graikos *et al.*'s (2022) effectiveness measurements show strong sensitivity to levels of (low positive and negative) swirl near the design points for their two blade geometries and in off-design conditions for (positive) swirl above $\sim 0.5\,\Omega b$. For positive annulus swirl between about 30% and 50% of disc speed, effectiveness reaches a maximum and sensitivity is reduced. This corresponds to annulus swirl levels close to expected values for the purge flow. In attempting to correlate effectiveness measurements using an orifice model,

Scanlon *et al.* (2004) found that data for simple axial seals from the University of Aachen correlated reasonably well for the upstream cavity but not for the downstream cavity. They suggested this may reflect differences in the way the disc pumping flow interacted with the seal flow. Balasubramanian *et al.* (2011) presented measurements for ingestion for a downstream radial seal. For the conditions considered, high levels of ingestion were found in the rim cavity and it was suggested that the minimum sealing flow rate would be higher than that given by Bohn and Wolff's (2003) correlation based on the annulus pressure asymmetry. This may be consistent with low effectiveness at the design point for a blade with negative outlet swirl shown by Graikos *et al.*

Graikos *et al.*'s study includes measurements for a downstream cavity with no rotor blades fitted. In this case, the annulus swirl and flow coefficient reached similar levels to those for the upstream seal considered by Hualca *et al.* in the same test rig. This allows upstream and downstream double radial clearance seals to be compared at similar bulk annulus flow conditions. At the two purge flow rates reported the downstream seal has lower effectiveness. This may be due the different direction of the cavity recirculation, but it should be noted that the annulus hub boundary layers will also differ for the two cases. Notwithstanding these effects, the blade outlet flow angle or swirl is a key parameter to be considered.

While similar effects of annulus flow on effectiveness are found in different configurations, exact performance depends on numerous factors and may have to be evaluated for individual designs. The results presented above are based on single sample points on the stator. Significant variations are found within the cavity and are discussed in the next subsection. For high radius purge flow injection, sealing can be dependent on the inlet arrangement and inlet swirl. Leakage flows through gaps in the stator and rotor assemblies also modify the sealing effectiveness. Examples and discussion of these points are given by Clark *et al.* (2017, 2018), Patinios *et al.* (2018) and Darby *et al.* (2021).

9.3.3 Effectiveness Distribution in the Cavity

For low purge flow rates, the outer part of the cavity is expected to be dominated by recirculation driven by disc pumping, as in Figure 9.2. If ingested flow is assumed to fully mix with the recirculating flow as it enters the cavity the measured effectiveness will be constant along the stator in the outer region of the cavity. A higher effectiveness may be expected on the rotor as this is shielded by the pumped flow originating from the purge inlet. Early experimental and CFD studies, for example by Dadkhah *et al.* (1992) and Chew *et al.* (1994), supported this interpretation and many subsequent experiments have shown little radial variation of effectiveness over substantial areas of the stator surface. However, this is not exact and depends on the configuration. For example, Orsini *et al.* (2025) show differences in ε on the stator of up to ~0.1 between measurements at $r/b = 0.943$ and 0.967, and corresponding differences in the shape of the ε versus $U_C/U_{C,min}$ curves. Measurements presented by Palermo *et al.* (2021) for the chute seal of Figure 9.9 show significant variation of ε with radius on the stator at higher flow coefficient and with strong ingestion. Horwood *et al.*'s (2020) chute seal measurements, shown in Figure 9.10, also show significant variation of ε on the stator.

Effectiveness measurements on the rotor surface are given by Cho *et al.* (2016) and Orsini *et al.* (2024). Cho *et al.* deduced effectiveness from transient measurements of rotor surface temperature using infrared sensors. Orsini *et al.* used a nonintrusive gas concentration method with a pressure-sensitive paint responsive to the partial pressure of oxygen. The surface distribution of effectiveness was obtained from analysis of the light emitted when the surface was subjected to bursts of laser light. Correlations of Cho *et al.*'s measurements were presented by Mear *et al.* (2016).

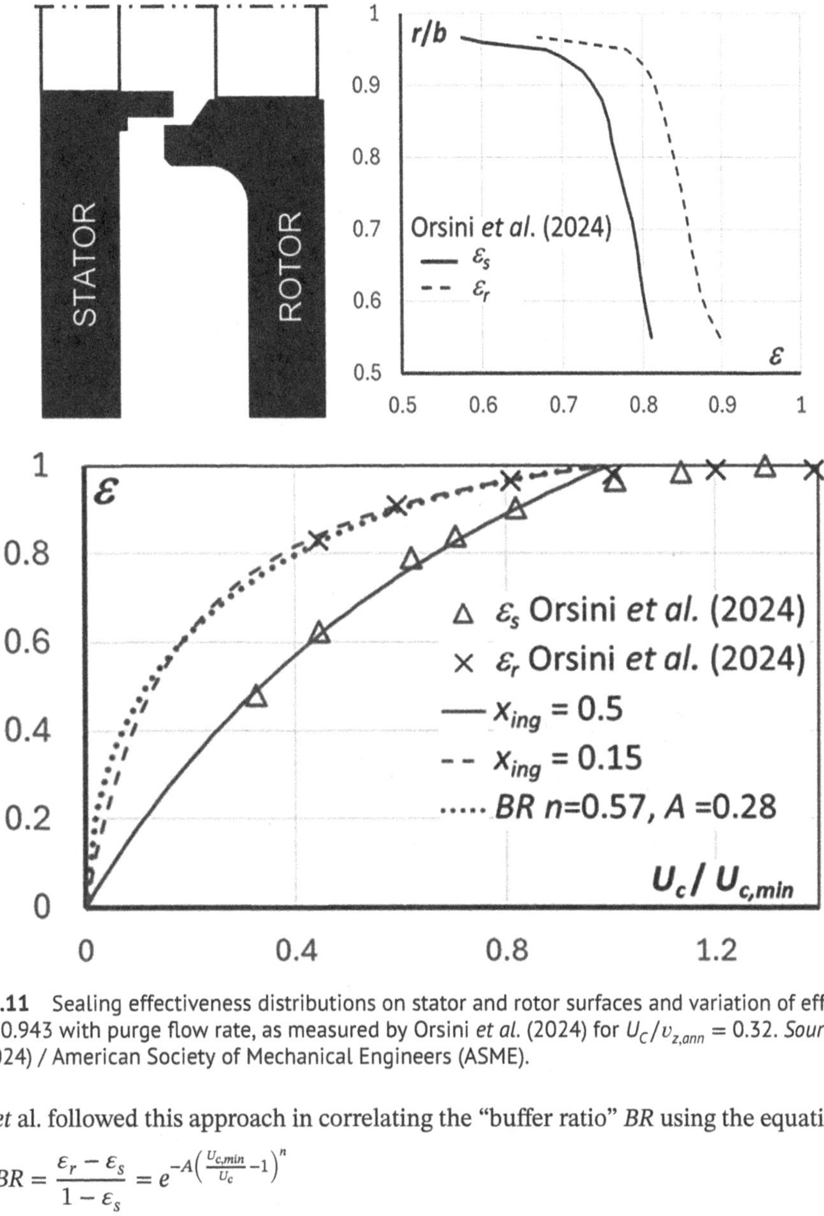

Figure 9.11 Sealing effectiveness distributions on stator and rotor surfaces and variation of effectiveness at $r/b = 0.943$ with purge flow rate, as measured by Orsini *et al.* (2024) for $U_C/v_{z,ann} = 0.32$. *Source:* Orsini *et al.* (2024) / American Society of Mechanical Engineers (ASME).

Orsini *et* al. followed this approach in correlating the "buffer ratio" *BR* using the equation

$$BR = \frac{\varepsilon_r - \varepsilon_s}{1 - \varepsilon_s} = e^{-A\left(\frac{U_{c,min}}{U_c} - 1\right)^n} \tag{9.3.1}$$

where ε_r and ε_s are effectiveness values on the rotor and stator, and A and n are empirical constants used to correlate each configuration. Equation 9.3.1 fits the data well, but there is considerable variation in A and n for different seals.

Examples from Orsini *et al.*'s results are given in Figure 9.11. The rotor blades in these experiments are symmetrical airfoils that minimise power absorption, and the seal has a radial minimum clearance. The radial distributions of ε_r and ε_s show significant shielding of the rotor by the disc pumping flow. The same minimum sealing flow rate is assumed for ε_r and ε_s in the bottom graph of the figure. If, for example, an effectiveness ε_r of 0.8 could be tolerated, this graph gives a 40% lower estimate of the required flow rate using the curve for ε_r rather than that for ε_s (neglecting the shielding effect). Reasonable fits to the measurements using equation 9.2.3 are obtained with the ingress fraction x_{ing} equal to 50% and 15% for the stator and rotor. Using equation 9.3.1 with Orsini *et al.*'s values for A and n for the operating point given shows a similar extrapolation to low flow rates as to the adjustment in x_{ing}. However, any extrapolation or extension of results to other seals and operating conditions should be treated with caution. Cho *et al.*'s measurements show different trends to those in Figure 9.11 with much reduced shielding at low flow rates and increased shielding at flows close to the sealing minimum.

9.3.4 Density Ratio Effects

The analysis above, and much of the experimental research, considers purge and annulus flows to have the same density. Annulus flow to purge flow temperature ratios may approach two in modern gas turbines giving similar purge flow to annulus flow density ratios. This introduces the density ratio $DR = \rho_{in}/\rho_{ann}$ as a further independent nondimensional parameter.

Orsini *et al.* (2024) noted the lack of previous systematic investigations of density ratio effects and presented measurements using nitrogen and carbon dioxide as the purge flow to show the effect of increasing DR from 1 to 1.52. Their data at the design flow coefficient is plotted against $\sqrt{DR}U_c/U_{c,min}$ and compared with a simple model in Figure 9.12. In plotting the data and in the model, it is assumed that $U_{C,min}$ is independent of DR. This assumption and the model have not been fully tested but are used to explore and illustrate the likely effects. The measurements for $DR = 1$ and 1.52 show a fair degree of convergence, suggesting that $\sqrt{DR}U_c/U_{c,min}$ may be a useful correlating parameter. The model indicates a slight increase in scatter would occur as DR is raised to 2.

The basis of the model is a fully mixed assumption for flow in the cavity with the ingress flow rate specified as a fraction of the shortfall between the supplied volume flow rate and the minimum sealing. Relating volume flow rates to mass flow rates, it is assumed that seal ingress has density ρ_{ann} and egress has density equal to the mixed out value for the cavity ρ_{cav}. The ingress flow rate is then given by

$$\dot{m}_{ing} = x_{ing}(\dot{m}_{min}/DR - \rho_{ann}\dot{m}/\rho_{cav}). \tag{9.3.2}$$

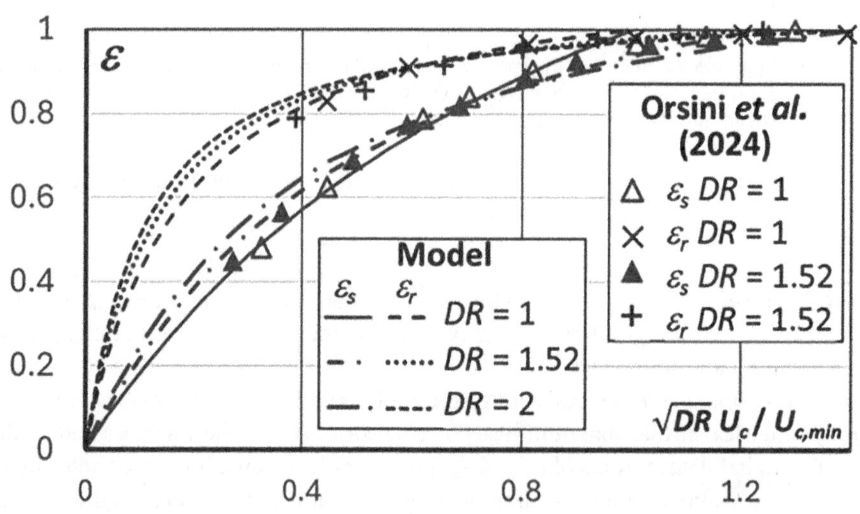

Figure 9.12 Variation of effectiveness at $r/b = 0.943$ with purge flow rate and density ratio, as measured by Orsini *et al.* (2024) for $U_C/\upsilon_{z,ann} = 0.32$ and as modelled. *Source:* Orsini *et al.* (2024) / American Society of Mechanical Engineers (ASME).

For isothermal and constant pressure mixing of the annulus and purge gases, the volume flow rate out of the cavity will equal the volume flow rate into the cavity. This gives the relation

$$\rho_{cav}/\rho_{ann} = (\dot{m}_{ing} + \dot{m})/(\dot{m}_{ing} + \dot{m}/DR). \tag{9.3.3}$$

Substituting from 9.3.3 into 9.3.2 gives the following quadratic equation for $\dot{m}_{ing}/\dot{m}_{min}$:

$$\left(\frac{\dot{m}_{ing}}{\dot{m}_{min}}\right)^2 + \left\{\left(\frac{\dot{m}}{\dot{m}_{min}}\right)(1 + x_{ing}) - \frac{x_{ing}}{DR}\right\}\left(\frac{\dot{m}_{ing}}{\dot{m}_{min}}\right) + \left\{\frac{x_{ing}}{DR}\left(\frac{\dot{m}}{\dot{m}_{min}}\right)\left[\left(\frac{\dot{m}}{\dot{m}_{min}}\right) - 1\right]\right\} = 0 \tag{9.3.4}$$

As U_C and $U_{C,min}$ are defined using the purge flow inlet density $U_c/U_{c,min} = \dot{m}/\dot{m}_{min}$. For $x_{ing} = 0.15$ and 0.5 (as in Figure 9.11) and given DR, and $U_c/U_{c,min}$ solutions of equation 9.3.4 and the relation $\varepsilon = \dot{m}/(\dot{m}_{ing} + \dot{m})$ were used to obtain the model results in Figure 9.12.

The treatment of density effects used here differs from the orifice model applied by Orsini *et al.* and the wave model used by Tang *et al.* (2025). In particular, $U_{C,min}$ is inversely proportional $\sqrt{DR}$ and assumptions in estimating $\dot{m}_{ing}$ differ in these alternative approaches. Nevertheless, reasonable matching with measurements is achieved in all cases. This again illustrates that the matching of simplified models with data does not signify that modelling assumptions are correct. All three modelling approaches are informed by the same set of measurements, and none have been tested in predictive mode. A general conclusion, consistent with the experiments and results from all three models, is that $\sqrt{DR}U_c/U_{c,min}$ may be a useful parameter in correlating results and extrapolating to engine conditions.

In engine applications, the density variation is principally due to temperature rather than gas composition. An energy balance for the cavity and the equation of state may be used to estimate ρ_{cav}, as discussed in Section 9.2. However, for a perfect gas, neglecting surface heat transfer, variations in gas specific heat and kinetic energy effects, equations 9.3.3 and 9.3.4 still apply.

9.3.5 Computational Fluid Dynamics (CFD)

CFD has been crucial in understanding rim seal flow mechanisms and is widely used in research and industry but has yet to be established as a fully reliable predictive tool for ingestion. Models are constrained by domain size, accuracy of boundary conditions and turbulence modelling limitations. Comparisons between CFD and measurements are also subject to experimental uncertainties including manufacturing tolerances and small amplitude, long wavelength circumferential flow asymmetries, which may significantly affect ingestion levels. Agreement between effectiveness measurements and unsteady RANS CFD solutions has been found in some cases, but other cases show discrepancies. LES has a stronger theoretical basis for these intrinsically unsteady flows and has been shown to improve on RANS models in several cases.

Most CFD models have assumed circumferential periodicity on annulus sectors chosen to resolve the circumferential asymmetry due to the vane or blades. However, Cao *et al.*'s (2004) unsteady RANS solutions showed that 90° and full 360° sector models neglecting vanes and blades gave much better predictions for the axial gap/seal "a" in Figure 9.6 (also shown in Figure 9.5). For the chute seal "b" in Figure 9.6, Boudet *et al.* (2005, 2006) found that a 120° sector RANS model without blades or vanes gave slightly lower effectiveness than small sector models, although this was still higher than the measured effectiveness. Extending this RANS modelling to a full 360° annulus with blades and vanes, O'Mahoney *et al.* (2011, 2012) found little change in flow structure or ingestion levels. More sensitivity was found to turbulence modelling with LES on smaller sector models showing considerable improvement in effectiveness when compared to measurements. Nketia *et al.* (2023) also show better prediction of effectiveness by LES than RANS CFD using a 22.5° sector model for an axial seal.

Detailed RANS solutions for 360° annulus models with blades and vanes are presented by Wang *et al.* (2013), Mirzamaghodam *et al.* (2014), Horwood *et al.* (2020) and Dawson *et al.* (2022). Wang *et al.*'s model for seal "g" in Figure 9.6 overestimated ingestion compared to measurements. This was considered a significant improvement on Zhou *et al.*'s (2011) small sector models, which underestimated ingestion and were unable to capture the large-scale flow structures given by the full model. Mirzamoghadam *et al.* reported 360° URANS solutions for the same Arizona State turbine rig, but with modified seal geometry. The clearance and overlap of the outer radial seal differed in the two cases, and the inner axial clearance seal for Wang *et al.*'s case was replaced by a much tighter radial seal. Again, the simulations tended to overestimate ingestion. Horwood *et al.* compared results from sector sizes of 30°, 60°, 90° and 360° for their chute seal (shown in Figure 9.10) with $U_C/\Omega b = 0.05$. Results show only a weak effect of sector size on sealing effectiveness despite changes in the unsteady flow frequency content. Their 30° model shows some agreement with the results in Figure 9.10 for $U_C/\Omega b = 0.05$ and 0.075. The results at $U_C/\Omega b = 0.1$, showing ε close to 1, do not capture the flattening of the experimental curve at $\varepsilon \approx 0.8$. At this $U_C/\Omega b$, the 30° model shows a dominant unsteady mode with ND12, suggesting that the sector size may be constraining the solution at this condition. Dawson *et al.* show agreement of a 360° RANS model with their effectiveness measurements capturing the effects of superimposed long wavelength pressure asymmetries. The authors state that for their double-lip radial seal, the upper of two seals denoted "h" in Figure 9.6, the unsteady fluctuations present in experiments and CFD are insufficient to cause ingestion in isolation.

LES on full 360°, bladed models has been conducted by Pogorelov *et al.* (2018, 2019) and Hösgen *et al.* (2020, 2021, 2023, 2024). These simulations are computationally expensive, using ~1 *billion* mesh points, but avoid known limitations of small sector and RANS models. For a turbine with 30 vanes and 62 blades, encouraging agreement with measured effectiveness is shown for the operating points selected for both an axial seal and a radial clearance seal with an extended overlap. The intrinsic unsteadiness is quite different in the two seals with a distinct inertial mode in the axial seal but not in the radial seal. The axial seal shows a 27ND mode in the seal and a 3ND acoustic mode in the cavity. Hösgen *et al.*'s (2020, 2021) examination of acoustic cavity modes observed for the radial seal suggests that they have little effect on ingestion, with lower frequency unsteadiness or turbulence dominating. For a turbine with 16 vanes and 32 blades and axial seals upstream and downstream of the rotor, Hösgen *et al.* (2023, 2024) show a strong 2ND structure in the upstream cavity with higher harmonics. The main driving mechanism for ingestion in the downstream cavity is a second set of vanes just downstream of the seal. However, ingestion in the outer section of the cavity is significantly enhanced by interaction with the 4ND unsteadiness from the upstream cavity. This decays significantly as it travels downstream through the rotor but may have a disproportionate effect when compared to other pressure asymmetries due to its high circumferential wavelength.

Comparisons of WMLES and RANS solutions for the axial seal of Figure 9.3 with three different annulus flow conditions are given in Figure 9.13. These simulations, from Burden *et al.* (2022), assume periodicity on a 10° sector and a cavity radius ratio $a/b = 0.9$. The top row shows solutions for a weak uniform axial flow imposed in the outer annulus at two purge flow rates. At the higher flow rate (on the left), the cavity is effectively sealed with ε adjusting from 1 in the cavity to 0 in the annulus across a thin layer. Some turbulence is evident in the WMLES. At the lower flow rate (on the right), the RANS solution shows large-scale coherent structures penetrating into the cavity and giving more ingestion than the WMLES. Fourier analysis of the WMLES shows evidence

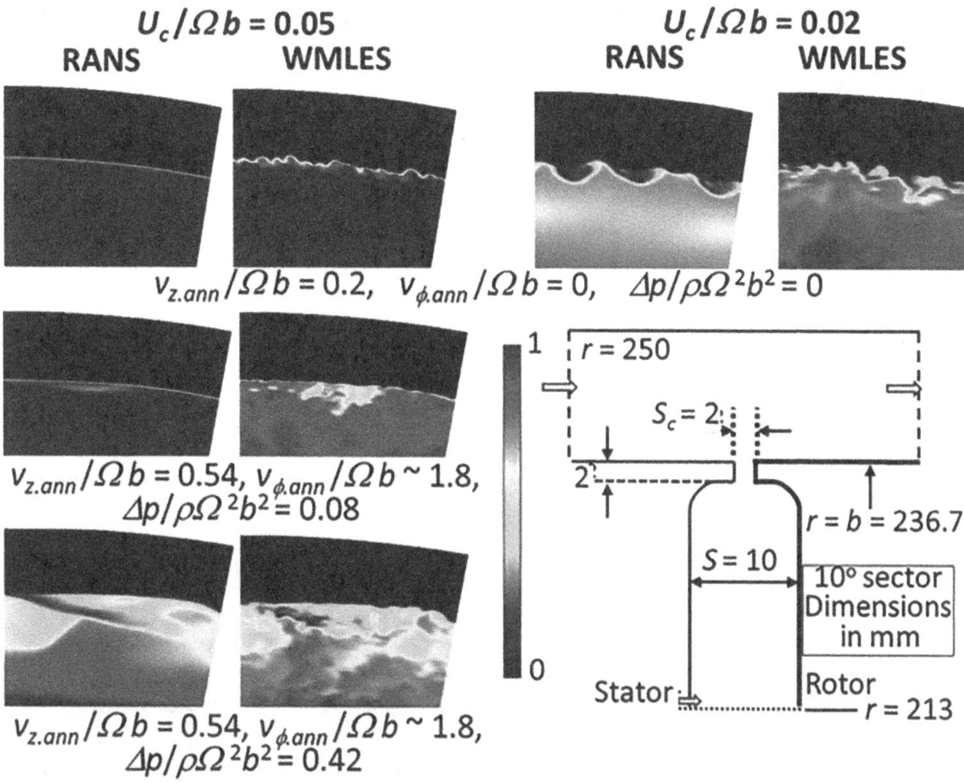

Figure 9.13 Instantaneous sealing effectiveness on mid-axial plane from RANS and WMLES for different flow conditions, $Re_\phi = 2.6 \times 10^6$, $s_c/b = 0.0084$, $a/b = 0.9$. *Source:* Burden *et al.* (2022) / American Society of Mechanical Engineers (ASME).

of large-scale structures, but these break down to smaller scale turbulence, as suggested by the contour plot. The results in middle and bottom rows have a higher flow coefficient and introduce inlet annulus swirl representative of Sangan *et al.*'s (2013a) turbine rig vane exit conditions. Circumferential flow asymmetry is also introduced at the annulus inlet with a wavelength corresponding to the sector size. The inlet condition was adjusted to set the pressure amplitude Δp just upstream of the seal. The value of Δp in the bottom row is representative of Sangan *et al.*'s rig. The levels of ingestion and the mean flows given by the RANS and WMLES are similar in this case. For the weaker pressure asymmetry in the middle row, the WMLES gives more ingestion than the RANS solution. In this case, the RANS solution is essentially steady and does not generate sufficient turbulence to drive ingestion into the cavity.

A further example, comparing RANS and WMLES with effectiveness measurements, is given in Figure 9.14. This test case corresponds to Horwood *et al.*'s result in Figure 9.10 with $U_C/\Omega b = 0.05$. Burden's (2024) RANS and WMLES include four vanes in a 30° sector, as shown, and employ the Spalart–Allmaras model and the Smagorinsky subgrid scale model, respectively. Horwood *et al.*'s RANS model used the $k-\omega$ SST model and included five rotating blades in the 30° sector, modelling their turbine stage with 48 vanes and 60 blades. Burden's purge flow inlet boundary conditions fixed the mass flow with zero swirl, while Horwood *et al.* specified the total pressure and flow angle at this inlet. Burden's solutions are for a slightly lower Reynolds number than those from Horwood *et al.* who found experimentally that there was little difference between results at these values of Re_ϕ.

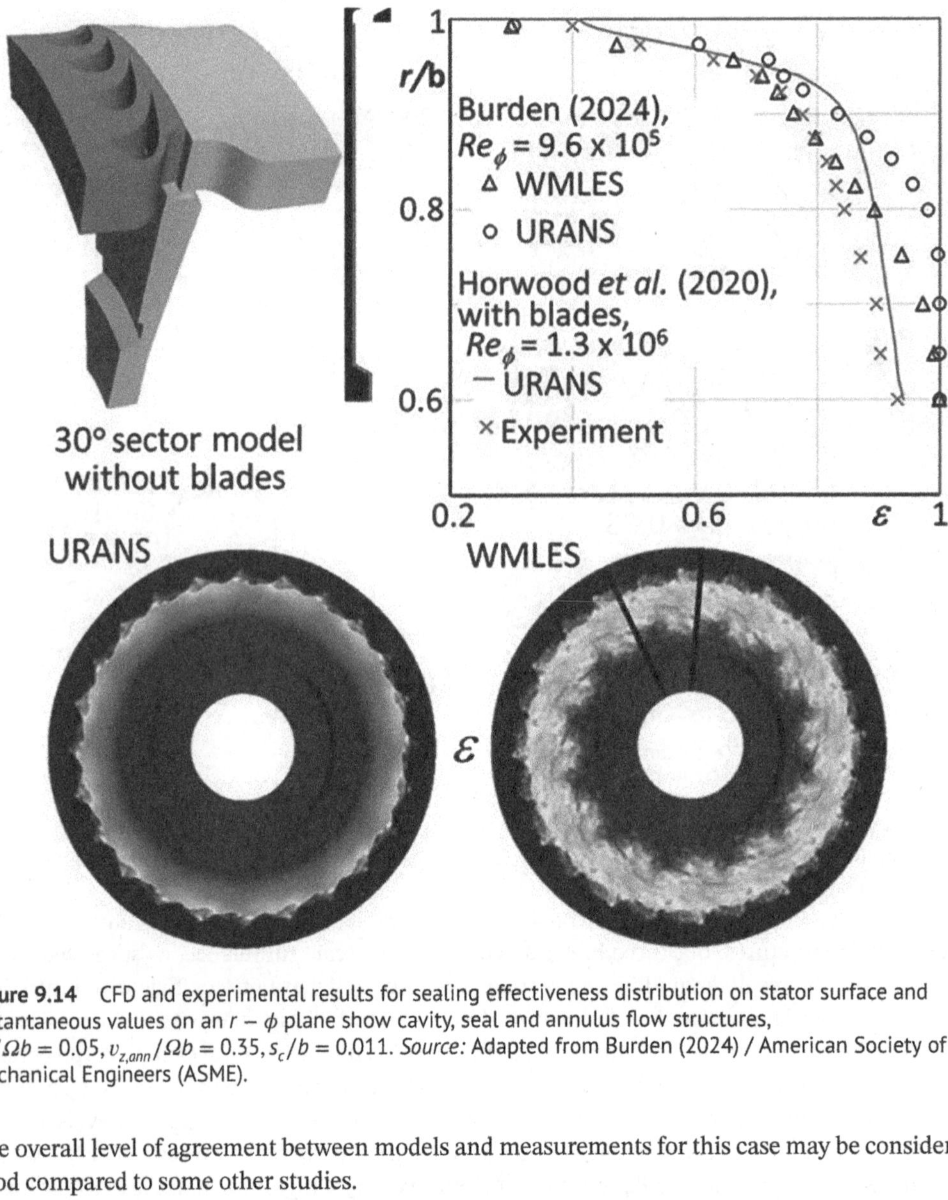

Figure 9.14 CFD and experimental results for sealing effectiveness distribution on stator surface and instantaneous values on an $r - \phi$ plane show cavity, seal and annulus flow structures, $U_c/\Omega b = 0.05, v_{z,ann}/\Omega b = 0.35, s_c/b = 0.011$. *Source:* Adapted from Burden (2024) / American Society of Mechanical Engineers (ASME).

The overall level of agreement between models and measurements for this case may be considered good compared to some other studies.

Considering Figure 9.14 in detail, the neglect of rotor blades in Burden's solution may account for the difference with Horwood *et al.*'s results for effectiveness on the stator near $r/b = 1$. All the CFD results converge around $r/b = 0.95$. Inboard of this point, the two RANS solutions diverge from the measurements, while the WMLES follows the measurement until $r/b \approx 0.85$. The two RANS solutions also diverge at this radius. A plausible interpretation of the results is that the rotating blades have little effect on ingestion, WMLES improves on RANS in calculating the turbulent mixing within the cavity, and that at lower radii, mixing is sensitive to purge inflow conditions. This is consistent with other rim sealing studies in which variable effects of rotor blades have been found and with observations on CFD in rotor–stator cavities, as discussed in Chapter 7. Burden's RANS and WMLES show 24ND modes in the seal. This compares to 24ND, 29ND and 21ND modes in Horwood *et al.*'s 30° sector CFD, 360° CFD model and experiments, respectively. The WMLES also show a 12ND mode around the radius where the cavity width reduces, and the effectiveness goes to unity in Burden's models. This feature is evident in the effectiveness contour plot and may be constrained by the sector size.

9.4 Seal Design

As shown in Figure 9.6, a wide variety of turbine rim seal geometries have been investigated. This reflects differences in requirements and design approaches. Aeroengines must operate through rapid accelerations and decelerations and may require relatively large clearances. Estimation of relative movements of the rotor and stator is a primary step in seal design and is itself challenging and subject to significant uncertainty and variation. Startup and shutdown of power generation turbines may be managed to achieve smaller clearances, but aeroengines are expected to tolerate rapid accelerations. Weight, cost, stress and heat transfer constraints also differ. Hence, seal design involves complex trade-offs in a large design space. The discussion here is limited to aerodynamic seal performance with reference to published studies addressing sealing effectiveness and turbine efficiency.

9.4.1 Seal Geometry

A number of studies have ranked seal designs in terms of effectiveness as, for example, in Figures 9.2 and 9.8. Figure 9.15 summarises the rankings obtained from experimental studies by various research groups. For each study, the sealing effectiveness improves moving from left to right, assuming the same purge flow rate $U_C/\Omega b$. Further details can be found in the references given. Generally, the trends observed are as might be expected. A more tortuous path through the seal improves effectiveness, and the influence of disc pumping can be interpreted as discussed in Section 9.1.1. Many practical arrangements include an inner seal with an outer/rim/buffer cavity or trench. The pressure asymmetries from the blades and vanes are expected to decay across this region, allowing a more effective inner seal. Larger-scale, low ND cavity modes are also less likely to form in the high radius ratio outer cavity. Results from several studies support this approach, with Dawson *et al.* (2022) having investigated the pressure asymmetry damping for Savov *et al.*'s double lip seal in some detail. While the examples give some general guidance on sealing effectiveness, it is important to note that the rankings do not account for the axial and radial clearance requirements for particular applications.

The angel wing seal from Scobie *et al.* (2016b) in Figure 9.15 resulted from a design study using unsteady RANS CFD. Experiments confirmed gains in effectiveness attributed to the two outer/buffer cavities created by the angel wing structure. The concept behind the finned seal from Sangan *et al.* (2014) was also damping of pressure asymmetries associated with higher swirl induced by the fins. The number of fins was selected as a prime number (37) to avoid resonance issues with the 41 rotor blades. Turbine endwall contouring was considered by Rose (1994) and Lott *et al.* (2009) as a means of reducing pressure asymmetry. Schuepbach *et al.* (2010), Regina *et al.* (2014), Schreiner *et al.* (2019) and Hänni *et al.* (2019) and others have considered the combined effect of purge flow and endwall contouring on turbine efficiency and endwall heat transfer. Hänni *et al.*'s rotor endwall contouring extends onto the rotor seal arm, illustrating increasing interest in integrating rim seal and turbine aerodynamic design.

Detailed RANS CFD studies of chute style seals, with emphasis on high-pressure turbine efficiency, have been presented by Chilla *et al.* (2013, 2016). Interaction of purge and main annulus

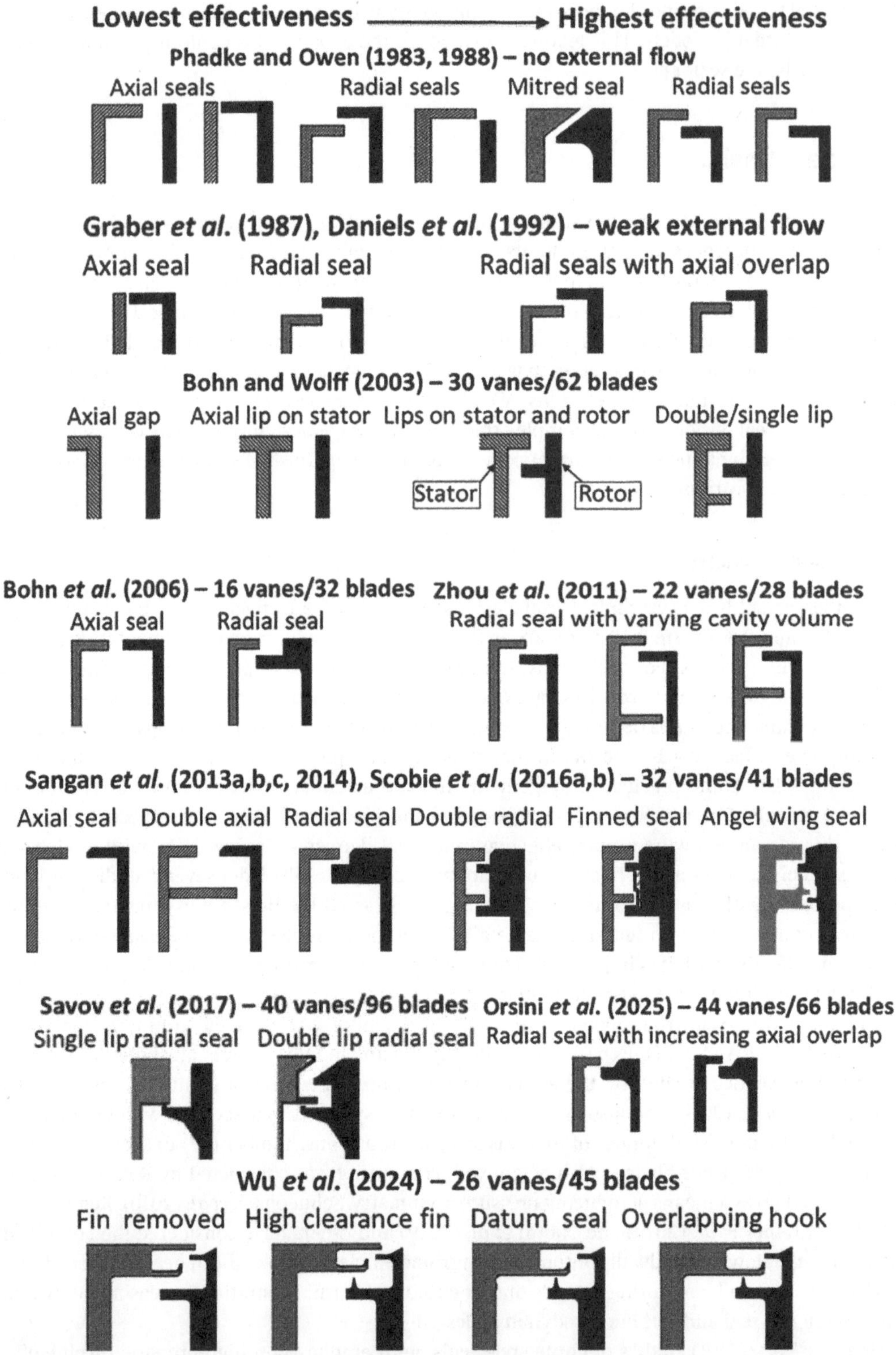

Figure 9.15 Ranking of rim seals in experiments on rim seal effectiveness. *Source:* Adapted from Burden *et al.* (2024) / American Society of Mechanical Engineers (ASME).

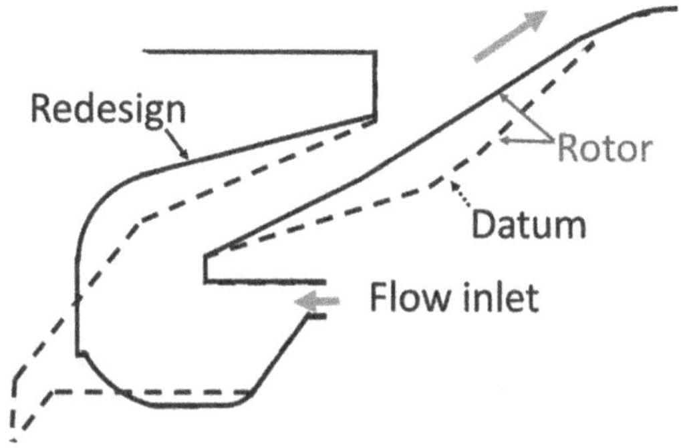

Figure 9.16 Chilla *et al.*'s (2016) seal redesign for improved aerodynamic and sealing performance. *Source:* Chilla *et al.*'s (2016) / American Society of Mechanical Engineers (ASME).

flow was found to be strongly influenced by the velocity deficit of the rim sealing flow relative to the annulus flow and by the pressure field due to the rotor blades. The flow interaction was stabilised by increasing either the sealing mass flow rate or the sealing tangential velocity. Turbine efficiency improved roughly 1% per 1% reduction in purge flow as a proportion of annulus flow. A 2D redesign of a datum seal and endwall contouring between the seal and the blade leading edge plane gave a predicted 0.2% improvement in stage efficiency at purge flow rates of 2% of annulus flow rate. At the nominal purge flow design value of 1% predicted benefits of the redesign remained but were reduced. As the redesign gave some improvement in predicted sealing performance, further benefits were expected from reduced purge flow requirements. The datum seal and its 2D redesign are shown in Figure 9.16. The redesign moves the minimum axial clearance radially outwards and streamlines the surfaces, aiming to present an accelerating purge flow as it approaches the annulus and encourage the purge flow to remain close to the rotor hub endwall.

WMLES and RANS CFD has been applied by Burden *et al.* (2024) to evaluate the sealing performance of different seals, with additional cases considered by Burden (2024). These simulations use a 10° sector model, as for the axial seal in Figure 9.13. In all cases, $Re_\phi = 2.6 \times 10^6$, $U_c/\Omega b = 0.02$, and the annulus flow was axial with $v_{z,ann}/\Omega b = 0.2$. The geometries considered, shown in Figure 9.17, include axial and radial seals (as in Figure 9.3). The aim in developing the combined chute/radial seals was to reduce ingestion by suppressing the rotating flow modes. As the rotating modes have been shown to be an important mechanism for ingestion under various conditions and will also induce aerodynamic losses, improvements obtained may apply to a wider range of conditions. The seals are ordered with effectiveness prediction from WMLES increasing from left to right and from top to bottom rows. RANS CFD results showed some difference in rankings to the WMLES, as might be expected from the discussion in Section 9.3.5, but agreed with WMLES in predicting the three seals in the bottom row to give less ingestion than those above.

Seals A and B in Figure 9.17 are based on the Horwood *et al.*'s and Bru Revert *et al.*'s 20° chute seals shown in Figure 9.10. Seals C, D and G modify the stator and rotor lips (from a base 20° chute) to give radial clearance sections at the seal inner radius. Seal C also has a higher seal clearance and the radial seal overlap for seal G is extended. Simulations for the extended overlap, seal G, indicate a benefit. Reducing the chute angle to 10° as in seals F and I also improves sealing. In seal I, a step was introduced on the rotor outer surface to avoid extending the seal beyond the blend point for the 20° chute. The benefit of the step may be due to the additional flow turbulence disrupting the formation of rotating modes. Seal J has an inner 20° chute blending with an outer radial clearance section. The improvement in effectiveness over the radial seal H may be due to the rotor lip deflecting disc

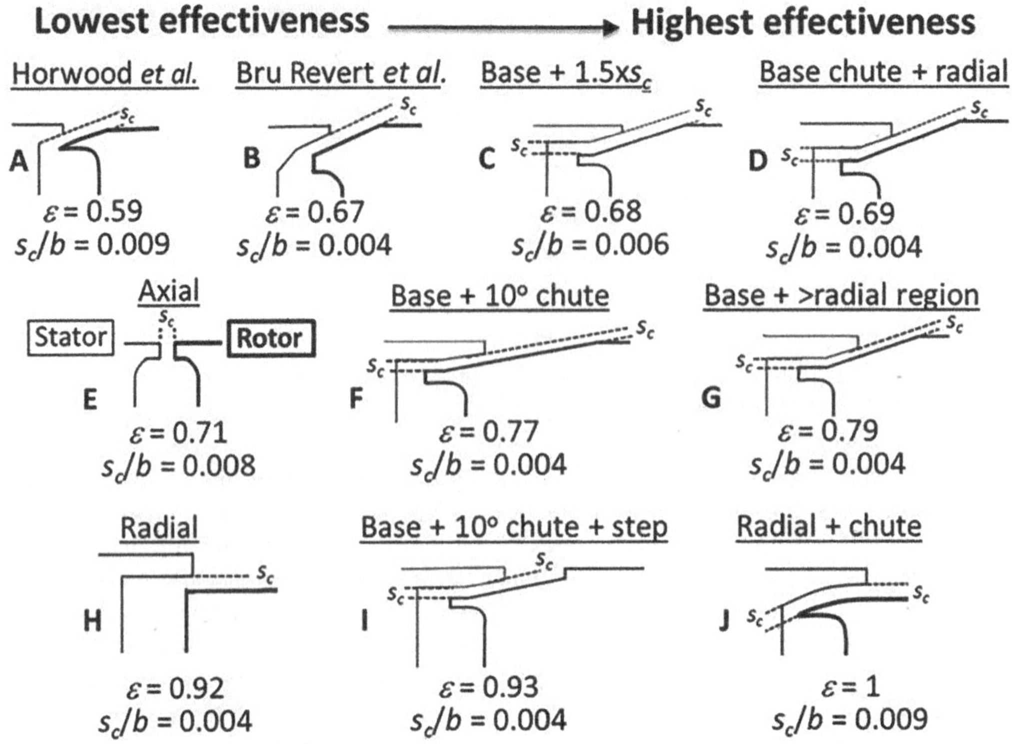

Figure 9.17 Sealing effectiveness at $r/b = 0.958$ for various seals from by Burden's (2024) WMLES, with $Re_\phi = 2.6 \times 10^6$, $U_c/\Omega b = 0.02$, $v_{z,ann}/\Omega b = 0.2$, $v_{\phi,ann}/\Omega b = 0$. *Source:* Burden's (2024) / American Society of Mechanical Engineers (ASME).

pumping flow away from the seal. Fourier analysis of pressure signals in the outer part of the seal showed the weakest distinct frequency signals for seals H and J. Seal I showed a lower amplitude dominant frequency compared to other chute seals with stronger higher harmonics.

9.4.2 Effect of Engine Configuration and Operating Conditions

A reasonable first design step is to use data from experiments and simulations, such as illustrated above, along with estimates for engine movements, stress and cooling requirements, to select a basic rim seal geometry. This may then be refined and tested using elementary modelling, simulations and experiment. For inner seals, the disc pumping correlations discussed in Sections 9.1.1 and 9.3.1 may be used. For outer seals, Scobie *et al.* (2016a) and others have suggested using correlations given by the orifice model to extrapolate from test rig to engine conditions with careful choice of the annulus location used to estimate the pressure asymmetry. With present understanding, use of the pressure asymmetry is highly questionable, but correlations may still be used by scaling the minimum sealing flow requirement using either of equation 9.2.1, or other combinations of the annulus flow angle and flow coefficient. The choice of scaling, and sensitivity to this choice should be considered in each case.

The correlations for sealing effectiveness compiled by Scobie *et al.* for configurations including annulus flow with vanes and blades are shown in Figure 9.18. The values of $U_{C,min}/\Omega b$ and Γ given by Scobie *et al.* were used in equation 9.2.2 to obtain these results apart from the radial seal curve

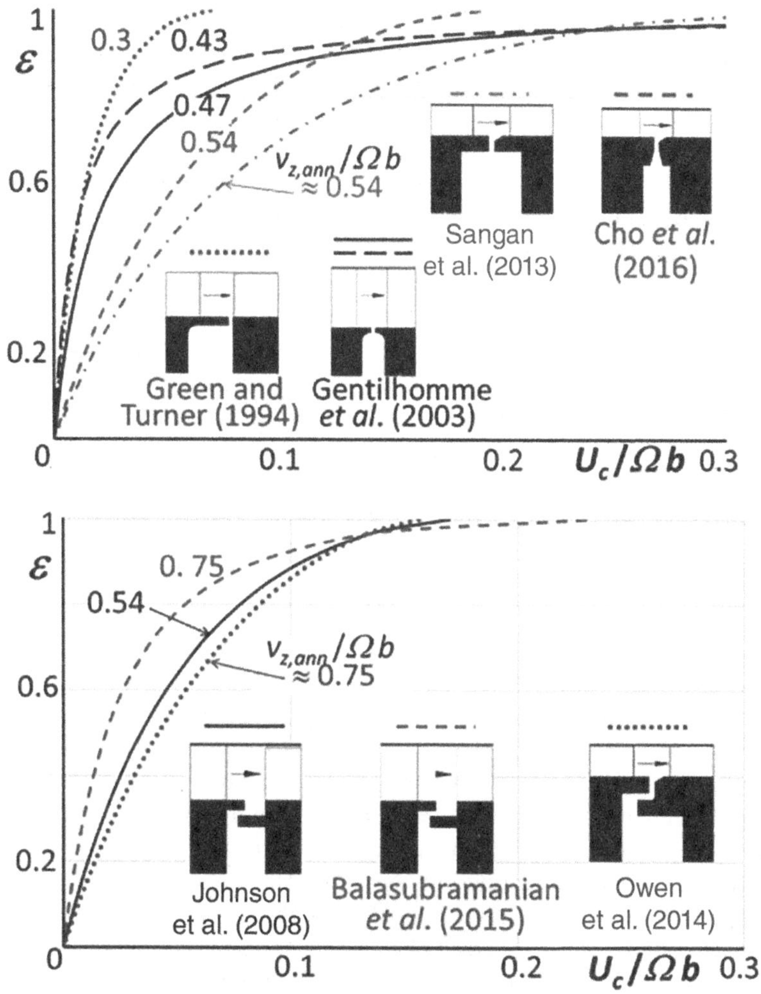

Figure 9.18 Scobie *et al.*'s 2016a correlations of sealing effective measurements for upstream axial (top) and radial (bottom) seals from various turbine test facilities.

from Owen *et al.* (2014). For this case $U_{C,min}/\Omega b$ was adjusted to compensate for the apparent use of axial clearance rather than minimum clearance in defining U_C. Approximate flow coefficients for each case are included in the figure. Approximate vane exit angles (to the axial) are 66° for the case with $v_{z,ann}/\Omega b \approx 0.3$, 70° for $v_{z,ann}/\Omega b \approx 0.43$ and 0.47, 73° for cases with $v_{z,ann}/\Omega b \approx 0.54$, and 68° for both cases with $v_{z,ann}/\Omega b \approx 0.75$. Other differences between some test cases include numbers and shape of blades and vanes, cavity and seal geometries, density ratio and Mach number. It should also be remembered that the correlations do not capture all experimental variations and may be sensitive to the selected measurement point on the stator.

The graphs in Figure 9.18 may be compared with Figure 9.9 from experiments on a chute seal having vanes but no rotating blades. The axial seal results in the top graph show considerable variation, but some consistency with Figure 9.9. For lower purge flows, effectiveness increases as flow coefficient reduces and approaches values expected from disc pumping with no annulus flow. This gives some support for the use of such charts for first order estimates of purge flow requirements.

Gentilhomme *et al.*'s (2003) tests show persistent ingestion at high purge flow rates, as also shown in Figure 9.9, and possibly due to low ND flow asymmetries. The difference between the two axial seal cases with $v_{z,ann}/\Omega b \approx 0.54$ in Figure 9.18 may be attributed to the different seal geometries.

The radial seal results in the bottom graph of Figure 9.18 show similar levels of ingestion from three studies. This is especially so when the sensitivity to the measurement position shown in Balasubramanian *et al.*'s (2015) paper is taken into account. The agreement may suggest a reduced dependency on flow coefficient for the radial seals, but differences in annulus flow angle are probably significant in this case. The flow angles for $v_{z,ann}/\Omega b \approx 0.54$ and 0.75 are ~68° and 73°, giving circumferential and resultant velocities ratios for $v_{z,ann}/\Omega b \approx 0.54$ about 5% and 8% higher, respectively, than those for $v_{z,ann}/\Omega b \approx 0.75$.

Estimates of sealing effectiveness based on the available published data will clearly be subject to significant uncertainty, so testing to verify acceptable performance or further optimisation may be required. CFD is currently used to complement experimental testing and can give valuable insight. The CFD design studies described in Section 9.1.1 illustrate the potential benefits of detailed seal design and refinement. These are significant in the context of highly optimised machines, but the exploitation of such methods is at an early stage. As capability continues to develop more detailed consideration of seal design for both sealing and turbine performance can be expected.

9.4.3 Flow Conditioning

A number of studies have looked at modifying the geometry or purge flow delivery to increase purge swirl, hence reducing the velocity difference and weakening interactions that promote ingestion and aerodynamic losses. Schädler *et al.* (2017) introduced 3D features on the stator and rotor in the rim cavity outboard of an axial seal. They measured improvements in stage efficiency of 0.35% and 0.39% for purge injection rates of 0.8% and 1.2% of annulus flow. For their unmodified configuration turbine efficiency improved roughly 0.5% per 1% reduction in purge flow as a proportion of annulus flow, and this sensitivity reduced slightly with the addition of the purge control features.

In their experiments simulating leakage flows through the stator at high r/b, Darby *et al.* (2021) showed that conditioning the leakage flow to have a positive swirl improved sealing effectiveness. It was suggested that this may be due to the reduced difference in velocity with the annulus flow. This is consistent with RANS CFD by Wang *et al.* (2019) and Qiang *et al.* (2023), who considered purge flow injection through the stator shroud directly into an axial rim seal. These and other studies, some of which have been discussed above, indicate that reducing the velocity deficit of the purge flow relative to the annulus flow can be beneficial for both sealing and overall aerodynamic efficiency for upstream turbine rim seals. This could vary with seal design and be constrained by system limits such as the available inlet pressure.

Iranidokht *et al.* (2020, 2021) specifically considered the suppression of rotating cavity modes for configurations with an axial gap between the stator and rotor discs. It was found that axial or radial flow deflectors suppressed the cavity modes suggesting that these could improve efficiency. Such approaches could perhaps complement swirl control in future seal designs.

9.5 Concluding Remark

The tendency of rotor–stator disc cavities to draw in fluid through gaps or seals between the rotor and stator is of practical importance and leads to complex flow interactions. This has attracted considerable, ongoing attention in axial turbine design and research. Even in the absence of imposed external flow and circumferential asymmetry, unsteady rotating flow modes develop in ingesting seals with alternating regions of ingress and egress around the seal circumference. These modes may propagate into the main cavity or couple with larger length scale (lower ND) cavity modes. For an axial turbine rim seal, rotating seal and cavity modes interact with the 3D, unsteady main annulus flow. This will include circumferential flow asymmetries from the turbine stage stationary vanes and rotating blades. Larger length scale pressure asymmetries originating elsewhere in the turbine may also propagate to the rim seal. Although these may be relatively weak, they may have disproportionate effects.

Several elementary models of rim seal ingestion have been proposed. These have had success in correlating measurements of sealing effectiveness but have limited predictive capability. For particular configurations and turbine operating conditions, sealing effectiveness is often reasonably correlated using the mean seal flow to disc velocity ratio, with Reynolds and Mach number effects being limited. Adjustment can be made for purge to annulus flow density ratio, based on the limited data available. The mean annulus flow has significant axial and swirl velocities. Characterisation of these introduces two more velocity ratios or, for a seal upstream of the rotor, the turbine flow coefficient and vane exit angle. Experiments, mostly for upstream seals, show variable effects of circumferential asymmetry due to vanes and blades. Some experiments show persistent ingestion at higher purge flow rates, which may be due to larger length scale asymmetries.

CFD modelling of rim sealing has proved challenging but has been crucial in revealing flow mechanisms. In many cases, variable performance in matching experimental data may be explained by annulus sector domains not capturing some flow mechanisms, turbulence modelling approximations and inaccuracies in boundary conditions or geometry. Further use of CFD in systematic research and seal design are anticipated, albeit constrained by the need to resolve unsteady flow features and the limitations of sector models.

Effective rim sealing is constrained by the extent of relative movements of the rotor and stator in engine operation, space available, component stresses, weight and costs. A corresponding wide variety of seal configurations are used in practice. Within this context, the broad understanding developed and data reviewed in this chapter provide some guidance in selecting seal types and purge flow rates. Possible effects of variations in disc cavity and turbine design have also been illustrated. Increasingly sophisticated CFD and experimental methods present future opportunities for refinement of sealing and aerodynamic performance.

References

Abe, T., Kikuchi, J., and Takeuchi, H., 1979. An investigation of turbine disc cooling. 3rd CIMAC Congress, Vienna, Paper GT30.

Balasubramanian, J., Junnarkar, N., Zhou, D. W., Roy, R. P., Kim, Y. W, and Moon, H. K., 2011. Experiments on aft-disk cavity ingestion in a model 1.5-stage axial-flow turbine. ASME Paper No. GT2011-45895. https://doi.org/10.1115/GT2011-45895.

Balasubramanian, J., Pathak, P. S., Thiagarajan, J. K., Singh, P., Roy, R. P., and Mirzamoghadam, A. V., 2015. Experimental study of ingestion in the rotor–stator disk cavity of a subscale axial turbine stage. ASME J. Turbomach., 137, 091010. https://doi.org/10.1115/1.4030099.

Bayley, F. J. and Owen, J. M., 1970. The fluid dynamics of a shrouded disk system with a radial outflow of coolant. J. Eng. Power, 92(3), 335–341. https://doi.org/10.1115/1.3445358.

Beard, P. F., Gao, F., Chana, K. S., and Chew, J. W., 2016. Unsteady flow phenomena in turbine rim seals. ASME J. Eng. Gas Turbines Power, 139(3), 032501–032501-10. https://doi.org/10.1115/1.4034452.

Bohn, D. and Wolff, M., 2003. Improved formulation to determine minimum sealing flow – $C_{W,min}$ for different sealing configurations. ASME Paper GT2003–38465, https://doi.org/10.1115/GT2003-38465.

Bohn, D., Rudzinski, B., Sürken, N., and Gärtner, W., 2000. Experimental and numerical investigation of the influence of rotor blades on hot gas ingestion Into the upstream cavity of an axial turbine stage. Proceedings of the ASME Turbo Expo 2000, Paper 2000-GT-0284. https://doi.org/10.1115/2000-GT-0284.

Bohn, D. E., Decker, A., Ohlendorf, N., and Jakoby, R., 2006. Influence of an axial and radial rim seal geometry on hot gas ingestion Into the upstream cavity of a 1.5-stage turbine. Turbo Expo: Power Land, Sea, Air, 4238, 1413–1422. https://doi.org/10.1115/GT2006-90453.

Boudet, J., Autef, V. N. D., Chew, J. W., Hills, N. J., and Gentilhomme, O., 2005. Numerical simulation of rim seal flows in axial turbines. Aeronaut. J., 109(1098), 373–383. https://doi.org/10.1017/S000192400070042X.

Boudet, J., Hills, N. J., and Chew, J. W., 2006. Numerical simulation of the flow interaction between turbine main annulus and disc cavities. ASME Turbo Expo 2006, paper GT2006-90307. https://doi.org/10.1115/GT2006-90307.

Boutet-Blais, G., Lefrancois, J., Dumas, G., Julien, S., Harvey, J.-F., Marini, R., and Caron, J.-F., 2011. Passive tracer validity for cooling effectiveness through flow computation in a turbine rim seal environment. ASME Turbo Expo, Paper GT2011-45654. https://doi.org/10.1115/GT2011-45654.

Bru Revert, A., Beard, P. F., Chew, J. W., and Bottenheim, S., 2021a. Performance of a turbine rim seal subject to rotationally-driven and pressure-driven ingestion. ASME J. Eng. Gas Turbines Power, 143, 081025. https://doi.org/10.1115/1.4049858.

Bru Revert, A., Beard, P. F., Chew, J. W., and Bottenheim, S., 2021b. Sealing performance of a turbine rim chute seal under rotationally-induced ingestion. J. Phys. Conf. Ser., 1909(1), 012035. https://iopscience.iop.org/article/10.1088/1742-6596/1909/1/012035.

Bru Revert, A., Beard, P. F., and Chew, H. W., 2023. Measurement of inertial and acoustic waves in a turbine chute rim seal cavity. ASME J. Eng. Gas Turbines Power, 145, 061021. https://doi.org/10.1115/1.4056552.

Burden, S., Chew, J. W., Gao, F., and Marxen, O., 2022. Effects of turbine conditions on rim seal performance and prediction. ASME Turbo Expo 2022, GT2022-83194. https://doi.org/10.1115/GT2022-83194.

Burden, S., Chew, J. W., Gao, F., and Marxen, O., 2024. Effect of rim seal geometry on rotationally-driven ingestion. ASME J. Eng. Gas Turbines Power, 146, 091013. https://doi.org/10.1115/1.4064730.

Burden, S. R., 2024. Evaluation and application of CFD for turbine rim sealing. PhD thesis. University of Surrey, https://doi.org/10.15126/thesis.901309.

Campbell, D. A., 1978. Gas turbine disc sealing system design. Proceedings AGARD Conference on Seal Technology in Gas Turbine Engines. AGARD-CP-237.

Cao, C., Chew, J. W., Millington, P. R., and Hogg, S., 2004. Interaction of rim seal and annulus flows in an axial flow turbine. ASME J. Eng. Gas Turbines Power, 126, 786–793. https://doi.org/10.1115/1.1772408.

Chew, J. W., 1991. A theoretical study of ingress for shrouded rotating disc systems with radial outflow. ASME J. Turbomach., 113, 91–97. https://doi.org/10.1115/1.2927742.

Chew, J. W., Dadkhah, S., and Turner, A. B., 1992. Rim sealing of rotor–stator wheelspaces in the absence of external flow. ASME J. Turbomach., 114, 433–438. https://doi.org/10.1115/1.2929162.

Chew, J. W., Gao, F., and Palermo, D. M., 2019. Flow mechanisms in axial turbine rim sealing. Proc. Inst. Mech. Eng., Part C: J. Mech. Eng. Sci., 233, 7637–7657. https://doi.org/10.1177/0954406218784612.

Chew, J. W., Green, T., and Turner, A. B., 1994. Rim sealing of rotor–stator wheelspaces in the presence of external flow. ASME Paper 94-GT-126. https://doi.org/10.1115/94-GT-126.

Chilla, M., Hodson, H., and Newman, D., 2013. Unsteady interaction between annulus and turbine rim seal flows. ASME J. Turbomach., 135, 051024. https://doi.org/10.1115/1.4023016.

Chilla, M., Hodson, H. P., Pullan, G., and Newman, D., 2016. High-pressure turbine rim seal design for increased efficiency. Proceedings of the ASME Turbo Expo. https://doi.org/10.1115/GT2016-57495.

Cho, G., Sangan, C. M., Michael Owen, J., and Lock, G. D., 2016. Effect of ingress on turbine disks. ASME J. Eng. Gas Turbines Power, 138, 042502. https://doi.org/10.1115/1.4031436.

Clark, K., Barringer, M., Johnson, D., Thole, K., Grover, E., and Robak, C., 2018. Effects of purge flow configuration on sealing effectiveness in a rotor–stator cavity. ASME. J. Eng. Gas Turbines Power, 140(11), 112502. https://doi.org/10.1115/1.4040308.

Clark, K., Barringer, M., Thole, K., Clum, C., Hiester, P., Memory, C., and Robak, C., 2017. Effects of purge jet momentum on sealing effectiveness. ASME J. Eng. Gas Turbines Power, 139, 031904. https://doi.org/10.1115/1.4034545.

Da Valle, L., Torre, A. F. M., Merli, F., Cernat, B. C., and Lavagnoli, S., 2025. Cavity instabilities in a high-speed low-pressure turbine stage. Int. J. Turbomach. Propul. Power, 10(1), 4. https://doi.org/10.3390/ijtpp10010004.

Dadkhah, S., Turner, A. B., and Chew, J. W., 1992. Performance of radial clearance rim seals in upstream and downstream wheelspaces. ASME J. Turbomach., 114, 439–445. https://doi.org/10.1115/1.2929163.

Daniels, W. A., Johnson, B. V., Graber, D. J., and Martin, R. J., 1992. Rim seal experiments and analysis for turbine applications. ASME J. Turbomach., 114, 426–432. https://doi.org/10.1115/1.2929161.

Darby, P. W., Mesny, A. W., De Cosmo, G., Carnevale, M., Lock, G. D., Scobie, J. A., and Sangan, C. M., 2021. Conditioning of leakage flows in gas turbine rotor–stator cavities. ASME J. Eng. Gas Turbines Power, 143, 021009. https://doi.org/10.1115/1.4049409.

Dawson, M. G., Otter, J. J., and Atkins, N. R., 2022. The effect of circumferential lengthscale on hot gas ingestion. Proceedings of the ASME Turbo Expo. https://doi.org/10.1115/GT2022-83345.

Gao, F., Chew, J. W., and Marxen, O., 2020. Inertial waves in turbine rim seal flows. Phys. Rev. Fluids, 5, 024802. https://doi.org/10.1103/PhysRevFluids.5.024802.

Gao, F., Chew, J. W., Beard, P. F., Amirante, D., and Hills, N. J., 2017. Numerical studies of turbine rim sealing flows on a chute seal configuration. Proceedings of 12th European Conference on Turbomachinery Fluid Dynamics & Thermodynamics. Stockholm, Sweden. Paper ETC2007-284. https://www.euroturbo.eu/publications/proceedings-papers/etc2017-284/.

Gao, F., Chew, J. W., Beard, P. F., Amirante, D., and Hills, N. J., 2018. Large-eddy simulation of unsteady turbine rim sealing flows. Int. J. Heat Fluid Flow, 30(70), 160–170. https://doi.org/10.1016/j.ijheatfluidflow.2018.02.002.

Gentilhomme, O., Hills, N. J., Chew, J. W., and Turner, A. B., 2003. Measurement and analysis of ingestion through a turbine rim seal. ASME J. Turbomach., 125, 505–512. https://doi.org/10.1115/1.1556411.

Graber, D. L., Daniels, W. A., and Johnson, B. V., 1987. Disk pumping test. Air Force Wright Aeronautical Laboratories, Technical Report AFWAL-TR-87-2050. https://apps.dtic.mil/sti/tr/pdf/ADA187199.pdf.

Graikos, D., Tang, H., Sangan, C. M., Lock, G. D., and Scobie, J. A., 2022. A new interpretation of hot gas ingress through turbine rim seals influenced by mainstream annulus swirl. ASME J. Eng. Gas Turbines Power, 144, 111005. https://doi.org/10.1115/1.4055457.

Green, T. and Turner, A. B., 1994. Ingestion Into the upstream wheelspace of an axial turbine stage. ASME J. Turbomach., 116, 327–332. https://doi.org/10.1115/1.2928368.

Grossmann, S., Lohse, D., and Sun, C., 2016. High–Reynolds number Taylor–Couette turbulence. Annu. Rev. Fluid Mech., 48, 53–80. https://doi.org/10.1146/annurev-fluid-122414-034353.

Hamabe, K. and Ishida, K., 1992. Rim seal experiments and analysis of a rotor-stator system with non-axisymmetric main flow. ASME Paper 92-GT-160. https://doi.org/10.1115/92-GT-160.

Hänni, D. D., Schädler, R., Abhari, R. S., Kalfas, A. I., Schmid, G., Lutum, E., and Lecoq, N., 2019. Purge flow effects on rotor hub endwall heat transfer with extended endwall contouring into the disk cavity. J. Global Power Propul. Soc., 3, 555–568. https://doi.org/10.33737/jgpps/109838.

Horwood, J. T. M., Hualca, F. P., Wilson, M., Scobie, J. A., Sangan, C. M., Lock, G. D., Dahlqvist, J., and Fridh, J., 2020. Flow instabilities in gas turbine chute seals. ASME J. Eng. Gas Turbines Power, 142, 021019. https://doi.org/10.1115/1.4045148.

Hösgen, T., Meinke, M., and Schröder, W., 2020. Large-eddy simulations of rim seal flow in a one-stage axial turbine. J. Global Power Propul. Soc., 4, 309–321. https://doi.org/10.33737/jgpps/129704.

Hösgen, T., Meinke, M., and Schröder, W., 2021. Analysis of wheel space ingress in a one-stage axial turbine. GPPS Xi'an21 Conference, Paper GPPS-TC-2021-0192. https://doi.org/10.33737/gpps21-tc-192.

Hösgen, T., Meinke, M., and Schröder, W., 2023. Analysis of single-blade passage and full circumference large-eddy simulations of turbine rim seal flows. ASME. J. Turbom., 146, 041002. https://doi.org/10.1115/1.4064086.

Hösgen, T., Meinke, M., and Schröder, W., 2024. Analysis of ingress Into the downstream wheel space of a 1.5-stage turbine by full circumference and single blade passage large-eddy simulations. ASME. J. Turbomach., 146, 091006. https://doi.org/10.1115/1.4065129.

Hualca, F. P., Horwood, J. T. M., Sangan, C. M., Lock, G. D., and Scobie, J. A., 2020. The effect of vanes and blades on ingress in gas turbines. ASME J. Eng. Gas Turbines Power, 142, 021020. https://doi.org/10.1115/1.4045149.

Iranidokht, V., Kalfas, A., Abhari, R., Senoo, S., and Momma, K., 2020. Sensitivity analysis on the impact of geometrical and operational variations on turbine hub cavity modes and practical methods to control them. J. Global Power Propul. Soc., 5, 66–78. https://doi.org/10.33737/jgpps/133736.

Iranidokht, V., Purwar, N., Kalfas, A. I., Abhari, R. S., and Senoo, S., 2021. Turbine hub cavity modes and their impact on efficiency. ASME J. Turbomach., 143(3), 031010. https://doi.org/10.1115/1.4050110.

Jakoby, R., Zierer, T., Lindblad, K., Larsson, J., deVito, L., Bohn, D. E., Funcke, J., and Decker, A., 2004. Numerical simulation of the unsteady flow field in an axial gas turbine rim seal configuration. Proceedings of the ASME Turbo Expo 2004, Paper GT2004-53829. https://doi.org/10.1115/GT2004-53829.

Johnson, B.V., Mack, G. J., Paolillo, R. E., and Daniels, W. A., 1994. Turbine rim seal gas path flow ingestion mechanisms. AIAA Paper 94, 1994. https://arc.aiaa.org/doi/abs/10.2514/6.1994-2703.

Johnson, B. V., Wang, C. Z., and Roy, R. P., 2008. A rim seal orifice model with two CDS and effects of swirl in seals. Proceedings of the ASME Turbo Expo, Paper GT2008-50650, https://doi.org/10.1115/GT2008-50650.

Julien, S., Lefrancois, J., Dumas, G., Boutet-Blais, G., Lapointe, S., Caron, J.-F., and Marini, R., 2010. Simulations of flow ingestion and related structures in a turbine disk cavity. ASME Turbo Expo, Paper GT2010-22729. https://doi.org/10.1115/GT2010-22729.

Lott, P. T., Hills, N. J., Chew, J. W., Scanlon, T., and Shahpar, S., 2009. High pressure turbine stage endwall profile optimisation for performance and rim seal effectiveness. Proceedings of the ASME Turbo Expo, 7, 1075-1087, ASME. https://doi.org/10.1115/GT2009-59923.

Mear, I. L., Owen, J. M., and Lock, G. D., 2016. Theoretical model to determine effect of ingress on turbine disks. ASME J. Eng. Gas Turbines Power, 138, 032502. https://doi.org/10.1115/1.4031315.

Merli, F., Krajnc, N., Hafizovic, A., Patinios, M., and Göttlich, E., 2023. Aerodynamics and sealing performance of the downstream hub rim seal in a high-pressure turbine stage. Int. J. Turbomach. Propul. Power, 8, 20. https://doi.org/10.3390/ijtpp8030020.

Mirzamaghodam, A. V., Kanjiyani, S., Riahi, A., Vishnumolakala, R., and Gundeli, L., 2014. Unsteady 360 computational fluid dynamics validation of a turbine stage mainstream/disk cavity interaction. ASME. J. Turbomach., 137(1), 011008–011008-9. https://doi.org/10.1115/1.4028248.

Nketia, S., Bryden, K. M., Dalton, R., and Shih, T. I.-P., 2023. Large eddy simulation of externally induced ingress about an axial seal by stator vanes. Energies, 16, 5985. https://doi.org/10.3390/en16165985.

O'Mahoney, T. S. D., Hills, N. J., and Chew, J. W., 2012. Sensitivity of LES results from turbine rim seals to changes in grid resolution and sector size. Prog. Aerosp. Sci., 52, 48–55. https://doi.org/10.1016/j.paerosci.2011.09.003.

O'Mahoney, T. S. D., Hills, N. J., Chew, J. W., and Scanlon, T., 2011. Large-eddy simulation of rim seal ingestion. Proc. IMechE, 225, 2881–2891. https://doi.org/10.1177/0954406211409285.

Orsini, L., Picchi, A., Facchini, B., Bonini, A., and Innocenti, L., 2024. Impact of the purge flow density ratio on the rim sealing effectiveness in hot gas ingestion measurements. J. Turbomach., 146, 021002. https://doi.org/10.1115/1.4063755.

Orsini, L., Picchi, A., Facchini, B., Bonini, A., and Innocenti, L., 2025. Comparison of rim sealing effectiveness in different geometrical configurations. ASME J. Eng. Gas Turbines Power, 147, 021022. https://doi.org/10.1115/1.4066536.

Owen, J. M., Wu, K., Scobie, J. A., Sangan, C. M., Cho, G., and Lock, G. D., 2014. Use of pressure measurements to determine effectiveness of turbine rim seals. ASME J. Eng. Gas Turbines Power, 137, 032510. https://doi.org/10.1115/1.4028395.

Palermo, D. M., Gao, F., Amirante, D., Chew, J. W., Bru Revert, A., and Beard, P. F., 2021. Wall-modeled large eddy simulations of axial turbine rim sealing. ASME J. Eng. Gas Turbines Power, 143, 061025. https://doi.org/10.1115/1.4049487.

Patinios, M., Ong, I. L., Scobie, J. A., Lock, G. D., and Sangan, C. M., 2018. Influence of leakage flows on hot gas ingress. ASME J. Eng. Gas Turbines Power, 141, 021010. https://doi.org/10.1115/1.4040846.

Phadke, U. P. and Owen, J. M., 1983. An investigation of ingress for an "air-cooled" shrouded rotating disk system with radial-clearance seals. ASME. J. Eng. Power, 1983(105), 178–182. https://doi.org/10.1115/1.3227382.

Phadke, U. P. and Owen, J. M., 1988. Aerodynamic aspects of the sealing of gas-turbine rotor-stator systems parts 1–3. Int. J. Heat Fluid Flow, 9, 98–117. https://doi.org/10.1016/0142-727X(88)90062-8.

Pogorelov, A, Meinke, M, and Schröder, W., 2018. Large-eddy simulation of turbine rim seal flow. Proceedings of the ASME Turbo Expo, paper GT2018-75022, https://doi.org/10.1115/GT2018-75022.

Pogorelov, A., Meinke, M., and Schröder, W., 2019. Large-eddy simulation of the unsteady full 3D rim seal flow in a one-stage axial-flow turbine. Flow Turbul. Combust., 102, 189–220. https://doi.org/10.1007/s10494-017-9827-9.

Qiang, D., Ruonan, W., Lei, X., Guang, L., and Zengyan, L., 2023. A novel design of rim sealing flow for improving sealing effectiveness. ASME J. Eng. Gas Turbines Power, 145, 071003. https://doi.org/10.1115/1.4056645.

Regina, K., Kalfas, A. I., Abhari, R. S., Lohaus, A., Voelker, S., and auf dem Kampe, T., 2014. Aerodynamic robustness of end wall contouring against rim seal purge flow. Proceedings of the ASME Turbo Expo 2014. https://doi.org/10.1115/GT2014-26007.

Rose, M. G., 1994. Non-axisymmetric endwall profiling in the HP NGV's of an axial flow gas turbine. Proceedings of the ASME 1994 International Gas Turbine and Aeroengine Congress and Exposition. https://doi.org/10.1115/94-GT-249.

Roy, A. S., Fridh, J., Scobie, J. A., Sangan, C. M., and Lock, G. D., 2021. Flow instability effects related to purge through a gas turbine chute seal. J. Global Power Propul. Soc., 5, 111–125. https://doi.org/10.33737/jgpps/140172.

Roy, R. P., Feng, J. J., Nazary, D. D., and Paolillo, R. E., 2005. Experiments on gas ingestion through axial-flow turbine rim seals. ASME J. Eng. Gas Turbines Power, 127(3), 573–582. https://doi.org/10.1115/1.1850499.

Sangan, C. M., Pountney, O. J., Zhou, K., Wilson, M., Owen, J. M., and Lock, G. D., 2013a. Experimental measurements of ingestion through turbine rim seals. Part I: Externally induced ingress. ASME. J. Turbomach., 135, 021012. https://doi.org/10.1115/1.4006609.

Sangan, C. M., Pountney, O. J., Zhou, K., Owen, J. M., Wilson, M., and Lock, G. D., 2013b. Experimental measurements of ingestion through turbine rim seals. Part II: Rotationally induced ingress. ASME J. Turbomach., 135, 021013. https://doi.org/10.1115/1.4006586.

Sangan, C. M., Pountney, O. J., Scobie, J. A., Wilson, M., Owen, J. M., and Lock, G. D., 2013c. Experimental measurements of ingestion through turbine rim seals—Part III: Single and double seals. ASME J. Turbomach., 135, 051011. https://doi.org/10.1115/1.4007504.

Sangan, C. M., Scobie, J. A., Owen, J. M., Lock, G. D., Tham, K. M., and Laurello, V. P., 2014. Performance of a finned turbine rim seal. ASME J. Turbomach., 136, 111008. https://doi.org/10.1115/1.4028116.

Savov, S. S. and Atkins, N. R., 2017. A rim seal ingress model based on turbulent transport. ASME Turbo Expo, Paper GT2017-63531. https://doi.org/10.1115/GT2017-63531.

Savov, S. S., Atkins, N. J., and Uchida, S. A., 2017. Comparison of single and double lip rim seal geometries. ASME J. Eng. Gas Turbines Power, 139(11), 112601–112601-13. https://doi.org/10.1115/1.4037027.

Scanlon, T., Wilkes, J., Bohn, D., and Gentilhomme, O., 2004. A simple method for estimating ingestion of annulus gas into a turbine rotor stator cavity in the presence of external pressure variations. ASME Paper GT2004-53097. https://doi.org/10.1115/GT2004-53097.

Schädler, R., Kalfas, A. I., Abhari, R. S., Schmid, G., auf dem Kampe, T., and Prabhu, S. B., 2017. Novel high-pressure turbine purge control features for increased stage efficiency. J. Global Power Propul. Soc., 1, 113–128. https://doi.org/10.22261/68MK5V.

Schädler, R., Kalfas, A. I., Abhari, R. S., Schmid, G., and Voelker, S., 2016. Modulation and radial migration of turbine hub cavity modes by the rim seal purge flow. ASME J. Turbomach., 139(1), 011011–011011-10. https://doi.org/10.1115/1.4034416.

Schreiner, B. D. J., Wilson, M., Li, Y. S., and Sangan, C. M., 2019. Design of contoured turbine endwalls in the presence of purge flow: A feature-based approach. Proceedings of the ASME Turbo Expo. https://doi.org/10.1115/GT2019-90443.

Schuepbach, P., Abhari, R. S., Rose, M. G., and Gier, J., 2010. Influence of rim seal purge flow on the performance of an endwall-profiled axial turbine. ASME J. Turbomach., 133, 021011. https://doi.org/10.1115/1.4000578.

Scobie, J. A., Sangan, C. M., Owen, J. M., and Lock, G. D., 2016a. Review of ingress in gas turbines. ASME J. En. Gas Turbines Power, 138(12), 120801. https://doi.org/10.1115/1.4033938.

Scobie, J. A., Teuber, R., Sheng Li, Y., Sangan, C. M., Wilson, M., and Lock, G. D., 2016b. Design of an improved turbine rim-seal. ASME J. Eng. Gas Turbines Power, 138, 022503. https://doi.org/10.1115/1.4031241.

Tang, E., Trébinjac, I., Ngo Boum, G., and Philit, M., 2017. Aerodynamic interactions between a high pressure turbine stage and a shroud cavity. Proceedings of 12th European Conference on Turbomachinery Fluid Dynamics & Thermodynamics. Stockholm, Sweden. Paper ETC2017-176. https://www.euroturbo.eu/publications/proceedings-papers/etc2017-176/.

Tang, H., Scobie, J. A., Wang, Z., and Lock, G. D., 2024. A theoretical model for ingress through turbine rim seals based on physically-observed unsteadiness. Int. J. Heat Fluid Flow, 106, 109300. ISSN:0142-727X. https://doi.org/10.1016/j.ijheatfluidflow.2024.109300.

Tang, H., Vella, S., Sangan, C. M., Scobie, J. A., and Lock, G. D., 2025. Ingress wave model with purge-mainstream density ratio. Int. J. Heat Mass Transfer, 237, 126372. https://doi.org/10.1016/j.ijheatmasstransfer.2024.126372.

Town, J., Averbach, M., and Camci, C., 2016. Experimental and numerical investigation of unsteady structures within the rim seal cavity in the presence of purge mass flow. ASME Turbo Expo, Paper GT2016-56500. https://doi.org/10.1115/GT2016-56500.

Wang, C.-Z., Mathiyalagan, S. P., Johnson, B. V., Glahn, J. A., and Cloud, D. F., 2013. Rim seal Ingestion in a turbine stage from 360 degree time-dependent numerical simulation. J. Turbomach., 136, 031007–031007-12. https://doi.org/10.1115/1.4024684.

Wang, R., Liu, G., Du, Q., Lian, Z., and Xie, L., 2019. An improved control method of rim seal based on auxiliary sealing holes. GPPS, Beijing, China, September 16–18, Paper No. BJ-2019-0096. https://gpps.global/wp-content/uploads/2021/02/GPPS-BJ-2019_paper_96.pdf.

Wu, Z., Liu, D., Wang, S., Zhang, Z., Bai, Y., and Luo, X., 2024. Sealing efficiency performance of four different rim seals in turbine cavity. Chin. J. Aeronaut., 37, 347–357. https://doi.org/10.1016/j.cja.2024.07.034.

Zhang, J. and Ma, H., 2013. Numerical investigation of improving turbine sealing effectiveness through slot width modification of the rim seal. ASME Turbo Expo, Paper GT2013-95570. https://doi.org/10.1115/GT2013-95570.

Zhou, D. W., Roy, R. P., Wang, C.-Z., and Glahn, J. A., 2011. Main gas ingestion in a turbine stage for three rim cavity configurations. J. Turbomach., 133, 031023. https://doi.org/10.1115/1.4002423.

Index